Molecular Biochemistry of Human Disease

Volume II

Author for Fundamentals of Clinical Diagnosis
George Feuer
Professor
Department of Clinical Biochemistry and Pharmacology
Banting Institute
University of Toronto
Toronto, Canada

Author for Pathological Supplements
Felix A. de la Iglesia
Director
Department of Pathology and Experimental Toxicology
Warner-Lambert/Parke-Davis Pharmaceutical Research
Ann Arbor, Michigan

CRC Press
Taylor & Francis Group
Boca Raton London New York

CRC Press is an imprint of the
Taylor & Francis Group, an **informa** business

CRC Press
Taylor & Francis Group
6000 Broken Sound Parkway NW, Suite 300
Boca Raton, FL 33487-2742

Reissued 2019 by CRC Press

A Library of Congress record exists under LC control number:

Publisher's Note
The publisher has gone to great lengths to ensure the quality of this reprint but points out that some imperfections in the original copies may be apparent.

Disclaimer
The publisher has made every effort to trace copyright holders and welcomes correspondence from those they have been unable to contact.

ISBN 13: 978-0-367-25216-8 (hbk)
ISBN 13: 978-0-367-25218-2 (pbk)
ISBN 13: 978-0-429-28659-9 (ebk)

Visit the Taylor & Francis Web site at http://www.taylorandfrancis.com and the
CRC Press Web site at http://www.crcpress.com

PREFACE

The primary aim of this series was to emphasize fundamental aspects of basic medical knowledge in view of the fact that recent trends in medical education have become more clinically oriented — thus, the need arose for a book that could provide an integrated approach to basic sciences with clinical applications. In fact, this second volume, while maintaining the objectives set out initially in the first volume, delves deeply into basic cellular dysfunction which will attract the medical student, basic scientist, and practicing physicians too. The themes covered include disease aspects of water and electrolyte metabolism, trace elements, and cellular organization abnormalities.

Alterations of water and electrolyte balance are key functional elements which can be superimposed upon other disease entities, resulting in complex diagnostic difficulties. Sometimes these subjects have been relegated to the interest of the specialist; thus the authors felt that a presentation of these problems should be made in a comprehensive fashion. Diagnostic elements of diseases due to these alterations are continually evolving, and it is anticipated that more knowledge will accumulate as analytical techniques are refined, improved, or some new ones created. Therefore, the technical aspects of diagnosis will have an important role in revealing unsuspected properties or mechanisms that will determine the clinical outcome and the relevance of treatment.

Significant detail has been added to the chapter on cellular alterations since the preparation of the first draft, although the initial outline is maintained. Modern trends in basic biologic disciplines have helped to clarify some intracellular processes and contributed to the elucidation of organelle function and dynamics. This approach led to understanding important mechanisms underlying various disease conditions which, although recognized for some time, have not been fully characterized until recently. According to this initial objective, these basic cellular alterations and the study of their mechanisms will contribute to understanding the pathogenesis of tissue damage, cell injury, and repair, thus relating to disease and therapeutic interventions.

Finally, a glossary has been included in this volume for practical purposes, with the intent to provide capsule summaries of disease entities and associated clinical or laboratory manifestations. There was no intention to prepare a comprehensive collection of terms, but rather to provide key diagnostic elements of biomedical significance.

In addition to those who provided support for the preparation of the first volume, we have been very fortunate to have disinterested colleagues and associates who reviewed the various chapters. Encouragement and support came from my close colleagues Dr. Emmanuel Farber and Dr. Peter A. Ward. A significant number of specimens and microphotographs were made available graciously by our Toronto colleagues Drs. Emmanuel Farber, John Kellen, and Norbert A. Kerenyi, and from the University of Michigan, by Drs. Kent Johnson, Paul Gykas, and Robert Gray. Solid support in the preparation of illustrations was provided by Drs. James E. Fitzgerald, Alec W. Gough, and Graham Smith. As the volume expanded, we relied on Jackie Ritchie and her Office Services staff who patiently attended to details of the manuscript. Barbara Becker, Kathy Owen, and the Research Literature staff provided invaluable support in checking every reference, a task dreaded by most scientists, and accomplished the task expeditiously. Roger Westland reviewed the chemical names and structures. As can be seen, the number of supporters grows significantly as the series grows, and we are very much indebted to them. We also attest to the high quality of the CRC staff who edited these series, and they deserve our recognition.

Finally, the completion of this volume marks our twentieth year of uninterrupted collaborative work. Working and writing with George Feuer has been an intense challenge, since behind his permanent grin hides an insatiable writer and the most agreeable, optimistic co-worker one ever wished to meet.

Ann Arbor, December 1985.   **Felix A. de la Iglesia**

THE AUTHORS

George Feuer, Ph.D., is a Professor of Clinical Biochemistry and Pharmacology (Toxicology) at the University of Toronto, Canada. Dr. Feuer received his B.Sc., M.Sc., and Ph.D. at the University of Szeged, Hungary from 1943 to 1944 and C. Med. Sc. at the Hungarian Academy of Sciences, Budapest, in 1952.

Dr. Feuer has worked in various fields of biochemistry. Originally, he was interested in the mechanism of muscular contraction and for eight years studied in this field. This area of study led to further investigations into the energy transfer process in relationship to acetylcholine synthesis in the brain. Interest in the metabolism of thyroid hormones commenced in 1955.

In 1960, Dr. Feuer was chairperson at a session of the Fourth International Goiter Conference in London, England. Between 1958 and 1963, he investigated the effects of endocrine glands on emotional behavior. From 1963 to 1968, he studied the effects of food additives on the liver. This area of study was extended to encompass the investigation into certain liver diseases such as neonatal hyperbilirubinemia and intrahepatic cholestasis.

Dr. Feuer's main interests, at present, are the investigations into the mechanisms of the development of intrahepatic cholestasis and hepatocarcinoma. He is also involved in the early detection of toxic side effects brought about by drugs and various other chemicals.

Dr. Feuer has written various books, reviews, and scientific papers including numerous publications on food and drug hepatic interactions and the biochemistry of liver diseases. Currently he is a member of the Editorial Boards of *Xenobiotica* and *Drug Metabolism and Drug Interactions.*

Felix A. de la Iglesia, M.D., is Vice President of Pathology and Experimental Toxicology at Warner-Lambert/Parke-Davis Pharmaceutical Research in Ann Arbor, Michigan, as well as Adjunct Professor of Toxicology at the School of Public Health and Research Scientist in the Department of Pathology, School of Medicine, at the University of Michigan, and Professor of Pathology, University of Toronto School of Medicine. He is a member of the Environmental Health Sciences Review Committee, External Consultant in Pathology and Toxicology to the National Institutes of Health, National Cancer Institute, and official expert in Pharmacology and Toxicology to the Ministries of Health of France and West Germany. In addition, he is Councilor, Society of Toxicologic Pathologists and Past President of the Michigan Chapter of the Society of Toxicology. Born in 1939 in Argentina, he received his M.D. degree in 1964 from the National University of Cordoba, Argentina, and completed postgraduate training in Experimental Pathology as Fellow of the Medical Research Council of Canada. In 1966, he joined the Warner-Lambert Research Institute of Canada in Ontario, Canada and developed various research and administrative positions until becoming Director of the Institute. After moving to the U.S. in 1977, he became Director of Pathology and Experimental Toxicology until assuming his present position. His research interests are in the area of toxicodynamics of subcellular organelle changes in drug-induced hepatic injury, toxicological aspects of novel anticancer chemotherapeutic agents, and safety assessment of novel chemical entities leading to the development of therapeutic agents. A Diplomate of the Academy of Toxicological Sciences, his publications include more than 250 articles, and currently he is the Editor of *Toxicologic Pathology,* and Editorial Board member of the journals *Toxicology and Applied Pharmacology, Drug Metabolism Reviews,* and *Toxicology.*

INTRODUCTION

In the course of recent decades a fundamentally new view has emerged in the analysis of the complexity of disease processes and their relationship to the regulation of cellular metabolism. It is now firmly established that many derangements of the normal structure and function of the organism stem from some impairment of the biochemical organization. Defects in normal biochemical processes proved to be the reason for the primary anomaly, even though they do not always result in immediate pathological conditions. However, when such biochemical changes persist they become irreversible and thus normal cells are progressively transformed into abnormal ones, resulting in pathological changes and clinical symptoms. Today, we understand a disease only if the impairment can be clearly identified with alterations of normal biochemical processes as recognized by various laboratory tests. Correlations between these biochemical tests and clinical manifestations provide key information on the basic etiology of human illness. The interpretation of clinical biochemical data is available for the diagnosis and at the same time this knowledge aids in management of the disease and in understanding the rationale behind drug therapy.

Although in many instances these interrelations have not yet been established, we now can consider that the task of correlating clinical signs and symptoms of diseases with laboratory data is unfinished, awaiting what further research will reveal about the connections between the manifestations of disease and impaired biochemical function. Clinical biochemistry has been growing steadily as new steps are discovered in the mechanisms of well-known diseases and new insights are being gained into the background and origin of abnormal biochemical processes. New methods are introduced in diagnosis, and new drugs are tested for therapy. Generally, there has been improved experience of disease mechanisms. Further progress in investigations of the biochemistry of human disease will disclose additional correlations, allowing us more knowledge of the primary biochemical lesions, which manifest in diseases of presently unknown origin or mechanism.

Although the title of this book might imply a wider academic scope, this book considers only basic problems regarding the correlation between altered biochemical processes and illness. Its purpose is to describe the development of abnormalities in biochemical reactions, as well as in the underlying mechanisms, and to illustrate how the action of drugs fits into reversing the changes of the progressing disorder caused by the derangement in normal reactions. Disease processes are usually complex; in these situations several interrelated systems function in an integrated manner. Although the main arguments of the book are based on biochemical information, an attempt is made to incorporate various other aspects originating from pathological and pharmacological studies into a uniform view. Considering the sick man as a unit, this work conveys an integrated picture of the biochemical changes associated with disease.

The primary aim of the book is to help medical, pharmacy, and advanced students in science to understand the growing importance of continuously advancing bhiochemical concepts in human disease. It may serve as a *vade mecum* for clinical biochemists to review the basis of their practical experience. At the same time it may also help physicians to brush up the clinical biochemistry learned during their years in medical school. Several excellent texts on general biochemistry are available; hence basic information will not be given in detail. However, there is reason to believe that many students and physicians as well would welcome a book in which the fundamental biochemistry underlying the course of disease is presented in an extended and readily understandable form. Thorough discussions on the interrelationships between organ, cell, or cellular organelle and disease, and more space than usual are devoted to the interpretation of the biochemical nature of human disease.

Essential knowledge of physiology, biochemistry, pathology, and pharmacology is assumed. The basic features of biochemistry, well described in standard textbooks, are omitted in order to focus the interest upon important issues of the relationships between impaired processes and disease. However, where necessary, limited background information is given to provide the reader with an introduction to the basis of a multitude of diseases with their various and often interrelated manifestations. At the same time more complex associations are also described and the defects in the molecular organization of the diseased cell or cellular organelle are discussed in depth. Relevant interactions with pharmacologically active substances, either produced by the body or applied by drug therapy, which may influence biochemical processes and the progress of disease are briefly mentioned. Essentials of diagnostic methods and interpretations, are also presented.

In general, the basic philosophy of the book could be summarized as follows:

1. Most (if not all) diseases originate from an impairment of biochemical molecular mechanism of the organism.
2. Biochemical processes affected by the disease and manifested through pathological lesions, may be revealed by clinical biochemistry tests.
3. The aim of therapy based on this knowledge is to repair the damage.
4. The specific purpose of drug administration is to restore normal conditions; the action of drugs lies in reversing the clearly discernible changes which manifest in the biochemical mechanism.
5. Prolonged injury of the biochemical mechanism leads to irreversible alterations.

The subject matter includes:

1. General changes characteristic of cellular components
2. The mechanism whereby these changes alter homeostasis
3. Specialized changes occurring in individual diseases, associated with particular organs
4. Changes peculiar to unique situations, inborn errors, diseases of the newborn, and aging

The course of diseases is described as a continuous process similar to a flow sheet:

$$\text{Normal cell} \nearrow \begin{array}{c} \text{composition} \\ \text{structure} \end{array} \searrow \text{biochemical process} \rightarrow$$

$$\text{impairment} \rightarrow \text{disease} \begin{array}{c} \nearrow \text{regeneration} \rightarrow \text{recovery} \\ \searrow \text{degeneration} \rightarrow \text{death} \end{array}$$

An understanding of the disease mechanism is essential for correct diagnosis and adequate therapy. This aim is presented in this book by summarizing our present knowledge on the molecular and cellular mechanisms of disease. This book is not comprehensive, because in several fields our knowledge is still fragmentary and no one could master all available information.

The various subjects have been arranged logically, starting from the participation of cellular elements in disease and continuing with disorders associated with a particular organ. The various topics may be read in sequence as they appear. Since, however, many biochemical findings accompanied by the progress of disease are not yet clearly understood, sometimes it may also be necessary to turn to earlier or later chapters. Hopefully, the background provided will be sufficiently clear to make it relatively easy to learn more about the various diseases from the general literature. A glossary of the essential terminology will be found in an Appendix. Each chapter is followed by references, plus suggestions for further reading. It should be understood that in order to avoid an encyclopedic aggregation it was necessary to limit the number of these references, including only basic illustrative examples, mainly from the latest available reports. Perhaps many important contributions have been omitted in the text. Individual references are mentioned only when they are fairly recent, and the references from our laboratory only indicate our interest in various areas. We hope that the references will provide the interested reader with a starting point for further enquiry.

BASIC REFERENCE BOOKS

BIOCHEMISTRY
Lehninger, A. L., *Biochemistry,* 2nd ed., Worth Publication, New York, 1975.
Montgomery, R., Dryer, R. L., Conway, T. W., and Spector, A. A., *Biochemistry: A Case-Oriented Approach,* Mosby Company, New York, 2nd ed., 1977.

CLINICAL BIOCHEMISTRY
Thompson, R. H. S. and Wootton, I. D. P., *Biochemical Disorders in Human Disease,* Academic Press, New York, 3rd ed., 1970.
Zilva, J. F. and Pannall, P. R., *Clinical Chemistry in Diagnosis and Treatment,* Year Book, Chicago, 2nd ed., 1975.
Cantarow, A. and Trumper, C., *Clinical Biochemistry,* W. B. Saunders, Philadelphia, 7th ed., 1975.
Gray, C. H. and Howarth, D., *Clinical Chemical Pathology,* Burroughs-Wellcome Foundation, 8th ed., 1977.
Gornall, A. G., *Applied Biochemistry of Clinical Disorders,* Harper and Row, New York, 1980.

CLINICAL CHEMISTRY
Tietz, N. W., *Fundamentals of Clinical Chemistry,* 2nd ed., W. B. Saunders, Philadelphia, 1976.
Varley, T. R., *Practical Clinical Biochemistry,* 5th ed., Heinemann, New York, 1976.
Henry, J. B., *Todd-Sanford-Davidsohn: Clinical Diagnosis and Management,* W. B. Saunders, Philadelphia, 16th ed., 1979.
Brown, S. S., Mitchell, F. L., and Young, D. S., *Chemical Diagnosis of Disease,* Elsevier/North Holland Biomedical Press, Amsterdam, 1979.

DISEASE
Walter, J. B., *An Introduction to the Principles of Disease,* W. B. Saunders Co., Philadelphia, 2nd ed., 1982.
Stanbury, F. B., Wyngaarden, F. B., and Fredrickson, D. S., *The Metabolic Basis of Inherited Disease,* McGraw-Hill, New York, 1978.

MOLECULAR BIOCHEMISTRY
of
HUMAN DISEASE

Volume I

Basis of Abnormal Biochemical Mechanisms
Abnormalities of Protein Synthesis and Metabolism
Abnormalities of Lipid Synthesis and Metabolism
Abnormalities of Carbohydrate Synthesis and Metabolism
Abnormalities of Nucleic Acid and Purine or Pyrimidine
Synthesis and Metabolism

Volume II

Abnormalities of Water and Electrolyte Metabolism
Abnormalities of Trace Element Metabolism
Abnormalities of Cellular Organization
Glossary
Appendix

TABLE OF CONTENTS

To my teachers who initiated my intellectual development,
and to my family who have given the most
because of my passion for science.

Chapter 1

ABNORMALITIES OF WATER AND ELECTROLYTE METABOLISM

I. INTRODUCTION

Disturbances affecting water and inorganic elements of our body occur in many diseases. The maintenance of constant composition of body fluids, the plasma, interstitial cells, and the fluid present within the cells (that is, the homeostasis) is of vital importance. It is essential to keep up the volume, strict chemical composition, osmotic pressure, and pH.[10,76,97,122,169] In disease conditions, the complete homeostasis is altered and a compensatory mechanism operates to achieve an optimal compromise under existing conditions.[20] Among these factors the osmotic pressure is the most essential, and it is regulated as long as possible.[64,68,137,147,167] The osmotic pressure of body fluids is mainly dependent on the number of ions in solution, especially the concentration of sodium, potassium, and chloride. The breakdown of homeostasis is therefore related to a defect in the metabolism of water, sodium, and potassium.[50,52,138] Among the other electrolytes, magnesium is an important intracellular cation; it participates in many cellular activities.[14,51] Calcium and phophorus are the principal elements of the skeleton, and disturbances of their metabolism are closely connected with disorders of bone calcification.[23,37,61,91,144]

II. WATER

A. Content

In the adult about 55 to 60% of the body weight is attributable to water. More than half of this amount is within cells and its constancy indicates a precise regulatory mechanism.[10,74,100,101,151]

The water content of the fetus is higher, and decreases in the full-term newborn. In early infancy there is a further rapid fall and at the age of 6 months water content reaches the adult level. There is a slow continuous decrease of total body water with age, particularly in the extracellular compartment; however, a difference exists here between men and women (Figure 1). The daily fluctuation of body water is normally very small, approximately 0.2% of body weight. The infant has the largest total body water and an excess of extracellular volume related to body weight; still, water deficit and dehydration have much greater detrimental action in the infant and young child than in the adult. In infants the concentrating capacity of the kidney is less efficient than in the adult. Therefore, the daily fluid requirement is relatively greater. The body reserve is, however, smaller in proportion to the daily requirement, and hence the infant is less able to withstand water deprivation and loss.[19,105]

Body water is not static; there is a continous interchange of individual molecules between the cellular fluid, interstitial fluid, and the plasma. There is a considerable secretion of fluid into the alimentary canal in the form of digestive juices amounting to approximately 8 ℓ daily. This fluid is usually reabsorbed, but if it is lost by vomiting or diarrhea it results in very serious disturbances of body hydration.[52] Although the kidney maintains a maximal conservation of water, the daily turnover is fairly large, due to continuous loss through the skin, lungs, and gastrointestinal tract. Some part of the body water is removed in the urine, feces, and expired air and through the skin as perspiration. The latter amount is extremely variable, being increased in fever and considerably elevated in normal circumstances in tropical climates and certain occupations.[49,86] The water lost is replaced by fluid drunk, by water in the food, and water produced in metabolic processes (Figure 2). The normal intake of water with food and drink is about 2.1 to 2.5 ℓ/day; the minimum requirement is 650

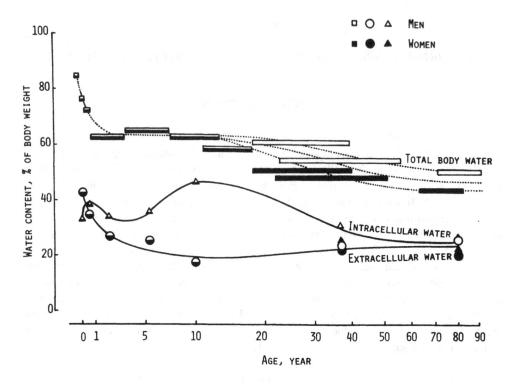

FIGURE 1.   Changes in water compartments during life span. The water content of the fetus is high; there is a rapid fall at birth and early infancy, followed by a slow drying of the tissues with age. The total body water of women is lower than that of men. The loss of extracellular water runs parallel with that of total body water, but the intracellular water shows some variations. Modified from References 173 through 176. Most loss occurs in muscle cells.[177]

mℓ. In addition to the exogenous water uptake the secretions of various digestive glands provide water in the alimentary canal. This fluid is rapidly and almost completely reabsorbed.

B. Metabolism and Regulation

The total body water can be divided into two parts existing in two compartments: extra- and intracellular fluids. A number of substances such as inulin, mannitol, and sodium thiocyanate, if injected into the body, occupy a volume which is much less than the total body water.[10,138] These substances do not cross the cell membrane or enter the cell except in very small amounts. The space these substances occupy is the extracellular fluid, which includes all body water external to the cells. It is about 20% of total body weight in adult man. The extracellular fluid is composed of several parts: the intravascular portion makes up the blood plasma and the extravascular part, which is the true interstitial fluid, and the lymph are actually in contact with the cells. These fluids are in rapid exchange with the plasma.[59] Some parts of the interstitial fluid are slowly exchanging, including dense connective tissue and cartilage water. Some water is practically inaccessible (such as water bound to bone structure). The average normal plasma volume in an adult is 3 ℓ and the interstitial fluid is 12 ℓ.

The division of total body water into intra- and extracellular water represents an oversimplification. In a healthy subject there is additional water in other compartments, such as the cerebrospinal fluid, occular fluid, and in joints and serous cavities.[143] This is produced from the extracellular fluid by passage, and this liquid is also called transcellular water. In several diseases large amounts of transcellular water accumulate in the peritoneal, pericardial, and pleural cavities. This water is effectively lost from the extracellular fluid and must be replenished.[83,84,111]

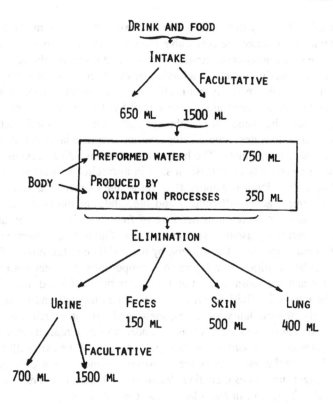

FIGURE 2. Daily water turnover in man.

Table 1
APPROXIMATE COMPOSITION OF
NORMAL PLASMA AND CELLULAR
FLUID

	Plasma (meq/ℓ)	Cellular fluid (meq/ℓ cell water)
Sodium	142	10
Potassium	5	148
Calcium	5	0
Magnesium	2	36
Chloride	102	0
Bicarbonate	27	10
Phosphate, inorganic	2	0
Phosphate, organic	0	84
Sulfate	1	20
Organic acids	4	0
Protein	18	80

Note: Concentrations in the cells are expressed as meq/ℓ cell water. Cell fluid is 70% water.

The electrolyte composition of the various body fluids shows important variations (Table 1). The interstitial fluid is similar to the plasma but contains only a trace of protein, slightly more chloride, and less calcium and magnesium. The composition of extracellular and intracellular fluids differs remarkably. The principal cation of the cells is potassium, whereas in the extracellular fluid it is sodium. However, these major electrolytes, sodium and potassium, are found in the body in a state of dynamic exchange.[122] The membranes separating the cell and extracellular space are permeable both to sodium and potassium ions and the maintenance of the differential concentration is dependent on metabolic activity of the cell. Other electrolytes are partly bound to each other or to proteins, or form complexes.[47,85,107]

The cellular membranes are freely permeable, and the net movement of water between cells and surrounding interstitial fluid is determined by the osmolalities of these compartments. In normal man the osmolality or the concentration of osmolally active solutes in body fluids and total body water is remarkably constant despite a large variation in the intake and excretion of water and solutes. The body water contains 285 to 290 mOsm solute per kilogram, consisting primarily of potassium salts in intracellular fluids and sodium salts in extracellular fluids. The identical osmolality of intra- and extracellular fluids is maintained by free water movement across all cellular and subcellular membranes controlled only by osmosis and diffusion. An increase or decrease of free water is shared by all major compartments, e.g., interstitial, vascular, and intracellular. The free movement of water is only controlled in the distal tubules of the nephron by the antidiuretic hormone.[45,90] The concentration of the circulating antidiuretic hormone is proportional to water permeability of the tubular cells. This and the sensation of thirst regulates the intake and loss of water. If the osmolality of the interstitial fluid decreases, water enters the cell and cellular volume increases. Conversely, if osmolality increases, associated with an increase of solutes, water leaves the cell and cellular volume decreases. Under normal circumstances the regulation of this volume and water metabolism maintain the osmolality of the extracellular fluid within narrow limits. The cellular membranes are, however, permeable not only to water but to low-molecular-weight substances including electrolytes. Therefore, the regulation of cellular ionic content is also important in considering water metabolism.

The regulation of body water is dependent on the function of the kidney and its relationship with a receptor system which responds to the need of water ingestion.[10,22,53,87,102,136,152,154,158] Thirst is the signal of this sytem, and the sensation of thirst is linked with the excitation of cortical centers which are further connected with various areas of the hypothalamus. A small decrease of total body water following mild cellular dehydration and simultaneous rise of osmolality stimulates these centers. The release of antidiuretic hormone from storage is integrated with the thirst mechanism. The antiduretic hormone or arginine vasopressin is synthesized in specialized hypothalamic nuclei and in the neurohypophyseal tract and plays a fundamental role in the homeostatic regulation of the volume and osmolality of body fluids.

The release of antidiuretic hormone is connected with physiological and other inhibitors.[160] Physiological inhibitors are related to the reduction of intracellular osmolality and expansion of the plasma or extracellular fluid. Some anticholinergic or α-adrenergic drugs such as norepinephrine also inhibit the release of antidiuretic hormone.[11,12] Ethyl alcohol is a potent inhibitor of release, probably through neurohypophyseal nuclei. In the circulation, arginine vasopressin has a half-life of about 20 min. About 10% of the vasopressin present in the blood is excreted by the kidney in active form. The other part is inactivated by the kidney and liver in about equal amounts. The renal effect on antidiuretic hormone will be discussed in a later chapter. Disturbances of water metabolism and fluid retention associated with renal[10,22] or hepatic disease[57] may be related to heart failure[33,48] and can be detected by measuring water balance.[123] Physiological fluctuations may occur in different phases of the menstrual cycle; in particular a premenstrual fluid retention and menstrual diuresis are apparent.[37,165]

The metabolism of water and electrolytes is also influenced by several hormones. These include mineralo- and glucocorticoids derived from the adrenal cortex, antidiuretic hormone, and other factors of the posterior pituitary gland.[164] Other hormones may affect electrolyte metabolism either indirectly as insulin deficiency or locally at the level of the target organ as the effect of estrogens on the water uptake of the uterus. Estrogen administration leads to sodium retention in tissues.[47,128] Hypothyroidism is connected with a defect in water excretion and may be associated with an impairment in the secretion of antidiuretic hormone.

III. SODIUM METABOLISM

Approximately one half of the total sodium content is in the extracellular water, making up 90% of the total solute in this body compartment. Most of the remaining sodium is bound to the crystalline bone structure and it is poorly exchangeable.[18,50] Due to active transport of sodium from the cells, the intracellular fluid contains only about 10 meq/ℓ. Potassium salts are the major solutes of this compartment. In contrast to sodium, close to 90% of potassium in the body is in the cell and exchangeable. Increasing age is linked with a reduction of the exchangeable potassium pool: the sodium pool is unaltered. In spite of the different electrolyte composition, the total solute and osmotic concentrations are always the same in the intra- and extracellular compartments throughout the body, due to the rapid equilibration of water between those spaces. The only exceptions are the fluid in the renal medullary interstitium, which is hypertonic; the sweat, which is hypotonic; and the final urine, which varies from hypo- to hypertonic.

Water balance is regulated primarily to maintain the osmolality to body fluids within a narrow range.[54,74,111,113,121] This is, however, a function of the solute concentration at a level of about 140 meq/ℓ. The constant volume is therefore related to the retention or excretion of sodium. Changes in sodium excretion are directly related to changes of the extracellular volume in order to maintain the nearly constant volume of this compartment.[18,44,79] The ingestion of sodium salts without water expands the extracellular compartment resulting from the osmotic movement of water from the cells. Sodium salts also stimulate thirst and the release of antidiuretic hormone which subsequently initiates water intake and retention in the body in sufficient amounts to maintain isotonicity. When the regulation of water balance is normal, during further processing the increased extracellular space provides the initiative to the system for the excretion of excess ingested sodium. Conversely, if plasma sodium concentration is elevated as a consequence of dehydration, the reduced volume of the extracellular fluid is associated with decreased sodium excretion. The response to change in dietary sodium is very slow, reflecting a gradual increase or decrease in the volume of extracellular fluid. The entire mechanism regulating this balance has not yet been clarified. It seems to be independent of cardiac output or arterial pressure, although sodium is retained in heart failure when arterial pressure is elevated.[154] Hypertension produced by vasoactive compounds may be associated with diminished sodium excretion.[24,27,163] The regulation of sodium balance is controlled by an intrarenal mechanism. This is under the influence of multiple factors which may include changes in blood composition, arterial pressure, renal vascular resistance, aldosterone, and other hormone actions (Figure 3). These affect the relationship between the intake, excretion, and resorption of sodium.[40,54,92,104]

The physical sign of excess extracellular fluid is the appearance of edema. Minor volume changes may occur and can be detected by a rapid gain in body weight or a disproportion between the intake and urinary output of fluid. However, edema becomes manifest when the extracellular water is increased by 10%. Whenever there is an excess of fluid in the body, there is an approximately proportionate excess of sodium and chloride. Fluid accumulations in the pleural or peritoneal cavity have an electrolyte composition similar to the extracellular fluid. In some instances the primary disturbance of edema production is responsible for the impairment of sodium balance, and in others a defect in extracellular water metabolism.[58,115,133,171] Excess sodium input is rarely the only factor of edema, but it aggravates edema in many diseases. Probably an interference with normal sodium excretion is the direct cause. Hydrocortisone derivatives applied in the treatment of Addison's disease

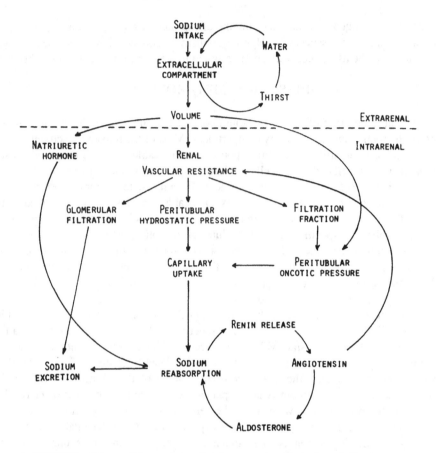

FIGURE 3.   Scheme of the regulation of sodium excretion. The pathway indicates many variables which probably play a role in the expansion of the extracellular compartment and the decrease of sodium resorption and increase of sodium excretion.

are often accompanied by edema. In cardiac edema the primary defect is probably associated with renal excretion of sodium.[127,154] Widespread edema in acute nephritis is linked with subnormal sodium levels. In malignant peritonitis or lymphatic edema, water retention is the primary factor in edema formation and the disturbance of electrolytes is secondary. In the nephrotic syndrome, malnutrition, and liver disease, the edema is usually correlated with low plasma albumin levels and the parallel sodium retention is regarded as a secondary phenomenon.[70,71,102]

Depletion of water and sodium occurs occasionally.[106,125,158,161] Water depletion usually results from water deprivation. In this condition the reduction of the sodium content of the body is minimal. It occurs in the comatose patient or in patients with lesions of the esophagus or pharynx that prevent swallowing. Excess renal loss of water occurs in diabetes insipidus due to failure in the production of antidiuretic hormone.[123] The patients usually compensate for polyuria by an enormous intake of water.[45,63] Water depletion develops in shipwrecked sailors and in exposure to high temperatures when water supplies are cut off.[49,86,168]

IV. POTASSIUM METABOLISM

The distribution of potassium in the body is uneven; however, at least 95% of body potassium is within the cells.[97] Most is contained in the muscle cells, some in the liver and erythrocytes, and a small amount is present in the extracellular fluid (Figure 4). The accumulation of potassium in the muscle can be explained by a possible mechanism whereby

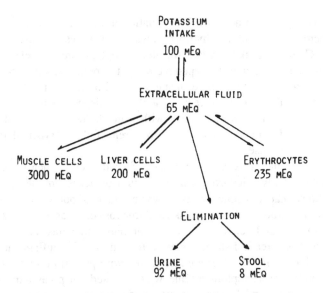

POTASSIUM
INTAKE
100 MEQ

EXTRACELLULAR FLUID
65 MEQ

MUSCLE CELLS
3000 MEQ

LIVER CELLS
200 MEQ

ERYTHROCYTES
235 MEQ

ELIMINATION

URINE
92 MEQ

STOOL
8 MEQ

FIGURE 4.   Distribution and metabolism of potassium in the body of a normal man.

the relative affinity of radicals for potassium is greater than for sodium. The maintenance of the ionic gradient in the muscle requires the presence of adequate amounts of energy provided by adenosine triphosphate and creatine phosphate. The influx of potassium is linked with the active ion transport requiring adenosine triphosphatase which functions properly in the presence of sodium and potassium ions. This enzyme acts probably by alternating the affinity for potassium and sodium in different phases of the transport cycle. This is responsible for the uneven distribution of potassium and sodium in the extra- and intracellular space. Another substance, adenosine 3′,5′-cyclic monophosphate (cyclic AMP), is also linked with electrolyte transport. The antidiuretic hormone activates adenyl cyclase which converts adenosine triphosphate to cyclic AMP. Cyclic AMP is especially involved in potassium transport.

Under normal conditions potassium balance in the body is maintained by the urinary excretion of 90% or more of the potassium intake.[3,46,149,150,159,170] Some is eliminated in the sweat and feces. The latter contain only a negligible amount of sodium. Potassium is handled by the kidney. The plasma potassium level reflects the concentration in the extracellular fluid. Any major deviation from the normal range indicates a disease condition, particularly when associated with neuromuscular or cardiac symptoms. Alkalosis causes a shift of potassium from the extracellular space into the cells; conversely, acidosis increases plasma potassium content. In both metabolic and respiratory alkalosis, hypokalemia is aggravated by increased renal excretion of potassium. Increased storage of glycogen is associated with lower plasma potassium level. Insulin, aldosterone, and epinephrine also cause hypokalemia. Hypokalemia is connected with irritability of the contractile tissue. Muscular weakness and occasional attacks of paralysis are associated with a fall of plasma potassium. The disturbance of potassium metabolism affects the function of the heart. Hypokalemia as well as hyperkalemia causes arrhythmias manifested in changes of the electrocardiogram. Large decreases in intracellular potassium of the heart bring about cardiac irregularity and failure, especially in severe depletion. Structural damage to the heart muscle also occurs in these circumstances. Potassium intoxication exercises an effect on heart function. This is not a common disturbance, but is nevertheless important. As plasma potassium level rises, progressive electrocardiographic changes occur even causing sudden death from cardiac arrest at about 10 meq/ℓ.

Hyperkalemia is apparent in acidosis. The situation in this case is more complex than in alkalosis. The increased excretion of hydrogen ions may suppress potassium elimination in the distal tubule. Cellular breakdown is associated with the increased release of potassium into the extracellular fluid leading to hyperkalemia if the renal excretion cannot cope with the overload. Vasopressin, acetylcholine, histamine, and thyroxine produce hyperkalemia. Patients with acute renal failure have reduced ability to dispose of potassium. Anuria and extreme oliguria are found in most patients with acute kidney failure and the output of potassium is negligible. It is not common with chronic renal failure. Hyperkalemia is regularly linked with Addison's disease.

Potassium depletion may arise in some diseases.[21,99] It occurs when there is wasting of the lean tissue of the body. However, true potassium depletion implies only a deficit of potassium in the body unaccompanied by tissue waste. Impaired potassium intake or vomiting can be regarded as a mixture of abortive intake and true loss of potassium; they are associated with anorexia, lethargy, and coma. Gastrointestinal causes of potassium depletion include diarrhea, biliary and pancreatic fistula, secretion from tumors, or inflammatory exudate in ulcerative colitis. Some renal diseases are also linked with potassium loss. Short periods of potassium deprivation are asymptomatic and well tolerated. In potassium depletion there may be an increased thirst and impaired secretion of antidiuretic hormone.

V. ACID-BASE METABOLISM

The regulation of hydrogen ion activity is dependent on the function of the lungs and kidneys.[13] Although the hydrogen ion concentration in body fluids is negligible as compared to that of sodium, its activity must be regulated within very narrow limits. This is essential for the various enzyme processes which maintain the intact interrelationship between many structures and organs. Slight changes in hydrogen ion activity may dramatically alter enzyme action, and subsequent changes in substrate molecules may upset homeostasis of the body. The changes in hydrogen ion concentration are regulated by buffer systems which maintain the body fluids at a relatively constant pH. Two buffering mechanisms operate within the body. One is the renal system which removes acid or base load directly. This response is slow and cannot provide the rapid protection needed in many acute acid-base disturbances.[16] The second system is the respiratory buffering which gives a quick adjustment to pH changes (Figure 5).

The normal pH range of the blood may vary between 6.8 and 7.7, but the more precise maintenance of a narrow pH range is characteristic of various cells. Deviation of the arterial pH below 7.0 and above 7.45 is considered as acidosis or alkalosis, respectively. The pH change in the blood in acid-base disturbances is the outcome of a deviation in the interaction between the metabolic (bicarbonate content) and respiratory (carbon dioxide tension, pCO_2) components. There are different conditions related to the acid-base status of a patient. He may have a primary chronic respiratory acidosis with decreased pCO_2. These deviations may occur as compensatory phenomena due to a primary metabolic anomaly. The patient may have a compensatory respiratory alkalosis connected with a primary metabolic acidosis with a compensatory hyperventilation and a decreased pCO_2. Conversely, metabolic alkalosis shows a trend toward mild hypoventilation and increased pCO_2 which may represent a compensatory respiratory acidosis.

A. Intracellular Hydrogen Ion Concentration

The hydrogen ion concentration shows variations in the cell resulting from the heterogeneity of metabolic processes producing these ions. Temporarily accumulated metabolites and compartmentalization of subcellular organelles may also cause a change of the accumulation of hydrogen ions so that the interior milieu of these particles may have a different

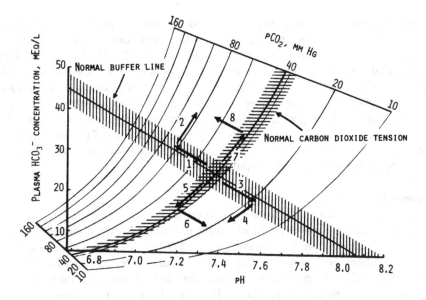

FIGURE 5. Graphic representation of disturbances in hydrogen ion equilibrium. Alveolar carbon dioxide tension, pCO_2, has been calculated; the horizontal shaded area along the line $pCO_2 = 40$ mmHg gives the normal range. The vertical shaded area indicates the normal buffer line representing changes and normal range of plasma bicarbonate and hydrogen ion concentration that occur if pCO_2 is altered by varying alveolar ventilation without any compensatory adjustments. The central double-shaded rectangle represents the normal resting conditions. Abnormal changes result in four types of disease states. Respiratory acidosis is associated with reduced pulmonary ventilation. Alveolar and arterial pCO_2 are increased and hydrogen ion concentration is also increased; pH falls (1). The kidney retains bicarbonate ions and excretes hydrogen and, therefore, tends to restore pH towards normal (2). In respiratory alkalosis pulmonary ventilation is increased (3), alveolar and arterial pCO_2 are decreased, and hydrogen ion concentration is also decreased. The kidney excretes more bicarbonate ions and retains hydrogen, and therefore tends to restore normal pH (4). In metabolic acidosis excess production of hydrogen ions causes a fall in pH. Bicarbonate concentration is reduced due to the combination of increased hydrogen ions with bicarbonate to form carbonic acid (5). The fall of pH stimulates breathing so the excess carbon dioxide is excreted by the lung and the pH returns to normal (6). In metabolic alkalosis loss of hydrogen ions causes an increase of bicarbonate ions (7). The increased pH depresses breathing, reduced amounts of carbon dioxide are eliminated, and pCO_2 rises, which restores the pH to normal (8). (Modified from Reference 172.)

pH from that of the cytoplasm. Even the cytoplasm may show variations due to high concentration of polyelectrolytes. Different tissues may also vary in their average pH values.

The cellular membranes are freely permeable to weak organic acids and bases; however, when they form undissociated compounds, their ionic forms usually do not pass through. Conditions which alter the pH of the extracellular fluid also affect the pH of the intracellular space and shift the distribution of partially ionized compounds between the intra- and extracellular water, resulting in an equilibrium of pH changes between these spaces. The small dimensions of the cell and the mechanism of flow and mixing of intracellular water render the pH changes minimal within the cell not separated by membrane barriers. However, the various processes occurring in subcellular particles may lead to nonuniform local pH.

Various disturbances of the extracellular acid-base balance affect the pH of the intracellular space, spinal fluid, and brain tissue, these changes influence the clinical status of a patient.[94] These changes are associated with carbon dioxide tension rather than extracellular bicarbonate concentration. Carbon dioxide readily diffuses throughout all tissues within a relatively short time. Extracellular bicarbonate penetrates slowly into intracellular space. The intracellular pH is profoundly altered by changes in carbon dioxide tension. The intracellular pH therefore depends on a complex association between pCO_2, extracellular concentration of bicarbonate, cellular storage of potassium, and degree of hydration of the cells. The response of the cell to alteration of pH shows variations. The intracellular pH is more protected against acidosis than against alkaline changes. While a fall to pH 6.9 to 7.0 linked with a decrease of extracellular bicarbonate is rapidly compensated, a rise to pH 7.55, whether it is associated with decreased pCO_2 or the presence of elevated extracellular bicarbonate, results in an increase of the intracellular pH. When the hydrogen ion equilibrium of the body is affected, homeostatic mechanisms limit pH changes. These include buffer systems and compensatory adjustments of pulmonary and renal excretion.

B. Hydrogen Ion Balance

The metabolic production of hydrogen ions is connected with several processes.[60] Most fats and carbohydrates are converted finally to the end products carbon dioxide and water. These metabolic pathways produce an average of 20,000 mmol carbon dioxide daily in a normal adult. Carbonic acid is volatile and its concentration is regulated by ventilation and adjustment of the pCO_2. If the generated carbon dioxide is eliminated completely through the lung, no excess hydrogen is produced. However, intermediary metabolism of these substrates leads to various organic acids which generate hydrogen ions. Furthermore, hydrogen ions are also formed from the oxidation of sulfur-containing amino acids — primarily cysteine and methionine — to sulfuric acid, and from the oxidation and hydrolysis of phosphoproteins and phosphate esters to phosphoric acid. These nonvolatile acids must be eliminated by the kidney to maintain normal buffer balance.

The excess acid produced by these metabolic processes is secreted into the urine. This is filtered through the glomerulus by a sodium-hydrogen exchange process. The intracellular hydrogen ions derived from the dissociation of carbonic acid (Figure 6) within the tubular cells of the kidney are dependent on pCO_2. Hydration of carbon dioxide produces carbonic acid. If pCO_2 is high and the enzyme activity of carbonic anhydrase is inhibited, the dissociation of carbonic acid to hydrogen and bicarbonate ions is limited and the hydrogen ions produced by metabolism cannot be neutralized by the bicarbonate to carbonic acid. Other substances such as phosphate and ammonia also participate in the renal buffering in the urine. Ammonia originates from synthesis in the kidney and from the breakdown of glutamine to glutamate.

C. Respiratory Acidosis and Alkalosis

Abnormalities of hydrogen ion balance cause various disease conditions such as respiratory acidosis or alkalosis and metabolic acidosis or alkalosis (Figure 5). Some impairment is linked with the production of carbon dioxide in the tissues. If the rate of synthesis exceeds the rate of removal, acute respiratory acidosis or acute hypercapnia occurs.[42] Many conditions are associated with respiratory acidosis, such as asphyxia, generalized pulmonary disease (emphysema, asthma, fibrosis), disorders of the respiratory muscles (muscular dystrophy, poliomyelitis, excessive potassium loss), central nervous system lesions involving the medullary respiratory center, obstruction of airways by extreme obesity, and, occasionally, congestive heart failure. Some drugs (alcohol, morphine, anesthetics, sedatives) may cause respiratory acidosis. The symptoms of this disease condition are drowsiness, confusion, tremor, and coma. When the removal of carbon dioxide by the lungs exceeds the production, pCO_2 is lowered and results in respiratory alkalosis or hypocapnia. Clinical states of respiratory alkalosis occur in hypoxia due to high altitude residence,[49] alveolar capillary block, congenital heart disease,[48] anemia, disturbance of the medullary respiratory center, hypermetabolic states (thyrotoxicosis, fever, delirum tremens), drug intoxication (salicylate, 2,4-dinitrophenol), cirrhosis,[57] reflex hyperventilation (pulmonary hypertension, pneumothorax),

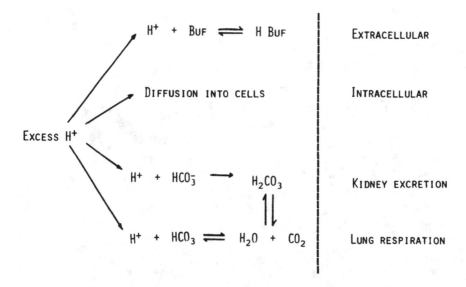

FIGURE 6.   Disposition of excess acid load in the body. All routes buffering acidosis function simultaneously. Extracellular buffering of excess hydrogen ions is instantaneous. Most of this buffering is done by plasma phosphate and proteins, particularly by hemoglobin. Hemoglobin accounts for about 30% of the immediate extrarenal defense against pH changes. Diffusion into cells and intracellular buffering take 2—4 hr. The renal buffering system is slow; acid excretion takes hours and days. Respiratory buffering by the lung occurs within minutes. In alkalosis, the reactions in the kidney and lung proceed in the reverse direction.

or Gram-negative sepsis. In the pathogenesis of this condition, important factors include decreased cerebral flow, pH changes in most body fluids, and altered ionization of calcium ions in the plasma. Plate 1 illustrates pathological changes due to anemic kidney infarction.

D. Metabolic Acidosis

Metabolic acidosis is due to a disturbance of metabolism and is characterized by increased numbers of hydrogen ions in the extracellular space originating from acids other than carbonic acid.[60,94,139] Plasma bicarbonate content is low and the degree of lowering the pH of the blood depends on the severity of the metabolic impairment, the buffering capacity of the blood and tissue, and the efficiency of the compensating respiratory mechanism. The clinical signs of metabolic acidosis are associated with the appearance of marked hyperpnea. The measurement of anion concentration or anion gap provides information on the possible origin of metabolic acidosis. If plasma chloride concentration is normal, other unmeasured anions are increased, and this gives rise to normochloremic acidosis. The causes of this condition are diabetic ketoacidosis, lactic acidosis, ingestion of drugs (salicylate, ethylene glycol, methyl alcohol, paraldehyde), and azotemic renal failure. In diabetic ketoacidosis the major organic acids are β-hydroxybutyric and acetoacetic acid accumulating from the incomplete metabolism of fat. The elevation of these keto acids in the circulation accounts for the widening of the anion gap. The diagnosis of this symptom is based on the increased ketone bodies in the plasma associated with hyperglycemia. Potassium depletion occurs frequently in diabetic ketoacidosis due to enhanced renal elimination and anorexia. Lactic acidosis develops from a disturbance in the balance between tissue synthesis and hepatic degradation of lactic acid resulting in an increased lactate level in the extracellular fluid. It is related to hypoxia which results in an increase in the reduced nicotinamide adenine dinucleotide and a consequent accumulation of lactic acid (Figure 7). Increased blood lactate has been found in a number of disease conditions such as acute hypoxemia, severe anemia, leukemia, diabetes mellitus, or circulatory insufficiency and in other circumstances such as following the admin-

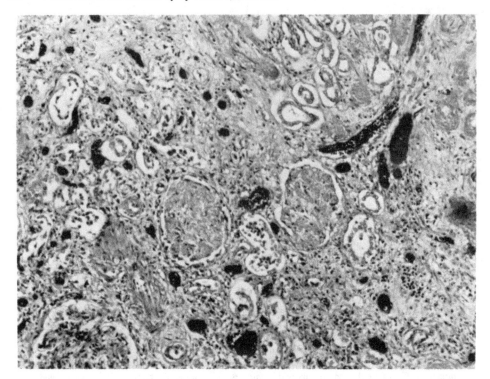

PLATE 1.   Anemic infarction of kidney. Left lower corner shows surviving glomeruli; 75% of the figure, however, reveals necrotic kidney tissue with shadows of glomeruli and tubuli.

istration of epinephrine or phenformin, alcohol ingestion, or exercise.[62] A small increase can occur in situations when pyruvate concentration is raised since the lactate/pyruvate ratio is maintained at a relatively constant level. These situations include acute hyperventilation, type I glycogen storage disease, and infusions of glucose, pyruvate, and sodium bicarbonate. Ischemic hypoxia of tissue during circulatory insufficiency may elevate blood lactate concentration. This is a secondary lactic acidosis. The normal acid-base equilibrium can be restored by the improvement of circulatory dynamics.

Hyperchloremic metabolic acidosis is due to diarrhea, loss of bicarbonate in the stool, or loss of alkaline gastrointestinal juices by the tube drainage, ileostomy or colostomy, draining fistulas, proximal or distal renal tubular acidosis, chronic pyelonephritis, administration of carbonic anhydrase inhibitors, or ammonium chloride ingestion by hepatic metabolism which leads to the release of hydrogen ions into the extracellular fluid. Renal tubular acidosis results from the inability of the distal nephron to maintain a steep hydrogen ion gradient in the lumen of the tubules. Normally following a small plasma bicarbonate reduction the urine pH is lowered to pH 5.3 or below and the titratable acid and ammonium is increased. Patients with distal renal tubular acidosis cannot acidify their urine to a pH less than 6.0; their acid and ammonia excretion is also diminished. Patients with proximal renal disorder may respond with normally capability to reduce the urine pH, but they still have a low bicarbonate threshold.

Chronic pyelonephritis occasionally causes hyperchloremic metabolic acidosis associated with greater dysfunction of tubular reabsorption than dysfunction of the glomerular filtration.[141] In severe tubular disease, due to the relatively intact glomerular function, the ammonia-producing capacity and anion retention are reduced. Linked with enhanced tubular chloride resorption, hyperchloremia develops in association with decreased plasma bicarbonate concentration. The reduction of plasma bicarbonate is partially compensated for by

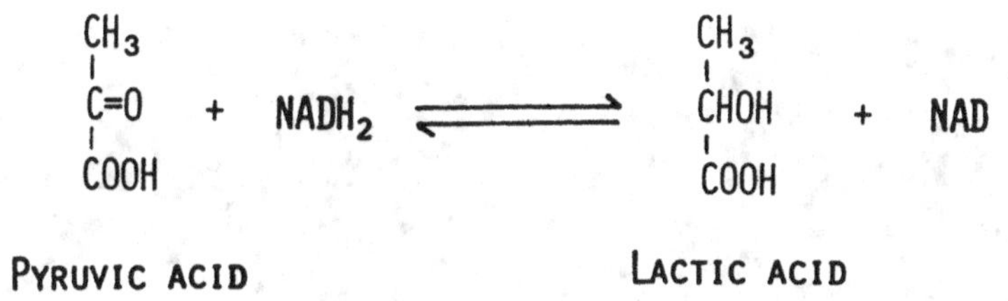

$$\begin{array}{ccc} \underset{\overset{|}{\underset{COOH}{|}}}{\overset{CH_3}{\underset{|}{C=O}}} + NADH_2 & \rightleftharpoons & \underset{\overset{|}{\underset{COOH}{|}}}{\overset{CH_3}{\underset{|}{CHOH}}} + NAD \end{array}$$

PYRUVIC ACID   LACTIC ACID

FIGURE 7. Lactic acid production by pyruvate oxidation. The reversible process is catalyzed by lactic dehydrogenase.

prompt decrease of pCO_2; however, the efficiency of the respiratory defense response lessens progressively with the increasing severity of plasma bicarbonate diminution. An example of morphological changes associated with pyelonephritis is given (Plate 2).

In contrast to metabolic acidosis resulting from renal disease of tubular origin, in azotemic kidney failure the glomerular filtration rate is reduced. This reduction is associated with anion retention; however, a normal plasma chloride concentration is maintained on account of reduced bicarbonate concentration. Metabolic acidosis occurs, therefore, in the azotemic state with a widened anion gap. In this chronic kidney disease there is a reduction of renal tubular mass which limits the amount of ammonia that is generated by the tubular cells. Reduced ammonia is connected with the inadequate neutralization of hydrogen ions in the urine. The metabolic acidosis in azotemic renal disease results, therefore, from the chronic positive acid balance. In some patients there is a bicarbonate-wasting defect. This bicarbonate leak may be related to excess parathyroid hormone or vitamin D resistance. In some cases patients with azotemia can maintain a stable low plasma bicarbonate level. In these instances bones provide large amounts of alkaline salts, such as calcium carbonate and calcium phosphate, for buffering of hydrogen ions continually retained in the tissues. Bone salts generally contribute to the buffering capacity of the extracellular fluid. They are dissolved in response to acid overload in normal people, and in acidosis induced by starvation or kidney dysfunction. Chronic conditions, such as chronic glomerulonephritis, cause several structural changes (Plate 3).

E. Metabolic Alkalosis

The clinical condition of metabolic alkalosis is characterized by an elevated plasma bicarbonate concentration, alkaline arterial pH, and normal or slightly increased pCO_2.[140,142] This condition differs from normal physiological elevation of plasma bicarbonate resulting from chronic hypercapnia. Primary metabolic alkalosis is an abnormality as a consequence of disturbed plasma bicarbonate concentration occasionally leading to a secondary hypoventilation. The secondary action is a compensatory mechanism to minimize pH changes brought about by the rise of plasma bicarbonate. Causes of metabolic alkalosis are many, including loss of acid gastric contents by vomiting or gastric drainage, depletion of chloride, severe depletion of potassium, administration of excessive alkali, or administration of excess exogeneous mineralocorticoids. Loss of gastric acid causes a loss of hydrogen and bicarbonate ions generated by the stomach secretory cells. Usually metabolic alkalosis does not occur in these conditions if the kidney sustains an adequate urinary excretion of bicarbonate. However, chloride and potassium depletion and loss of extracellular volume may interfere with the ability of the kidney to eliminate excess alkali from body fluid.

Administration of exogenous mineralocorticoids, Cushing's syndrome, primary hyperaldosteronism, ACTH-secreting tumors, and hyperplasia of the juxtaglomerular apparatus can be associated with metabolic alkalosis and potassium depletion. In alkalosis induced by mineralocorticoid excess, an enhanced potassium and hydrogen ion secretion occurs in exchange for sodium. This results in a rejection of chloride by the renal tubular cells and

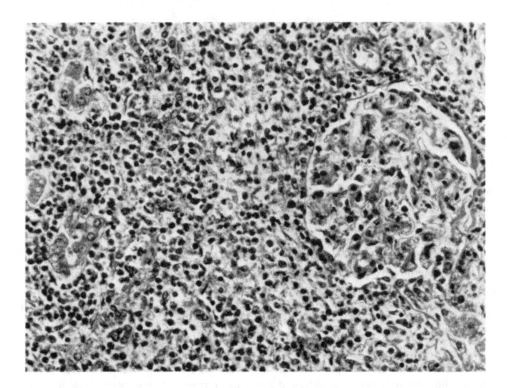

PLATE 2.   Pyelonephritis. Parenchyma is densely infiltrated with inflammatory cells. A glomerulus and tubuli are still recognizable.

its appearance in the urine. Consequently, this condition is connected with an expanded plasma volume rather than a reduction characteristic of metabolic alkalosis associated with chloride depletion.

VI. CALCIUM METABOLISM

The calcium content of the body is about 1.9% of total weight because of its abundance in bone.[29,30,112,155] It also takes part in muscular excitation. The important functions of calcium ions in the cell are related to the control of the activity of cell membranes and coupling of the chemical or electrical excitatory processes to the responses of contraction, electrical stimulation, and secretory function. The cell membrane requires constant amounts of calcium ions for normal activity. Calcium ions also control the permeability of the cell surface.[132] The cell surface is, however, relatively impermeable to calcium due to an active extrusion process from the intracellular fluid into extracellular space against an electrochemical gradient. The amount of calcium in the cell is bound in many different forms and the intracellular ionic calcium concentration is therefore low.[80,156] Part of the bound calcium is located on the surface of the cells and can be removed only with membrane proteins. There are further binding sites, including sites of active calcium transport out of the cell and sites of storage and removal, which limit the adverse effect of excess free calcium ions. The accumulation of calcium is probably an energy-dependent process and associated with stimultaneous phosphate movement.[2,4] The release is influenced by parathyroid hormone[72,108,124] (Figure 8) and vitamin D.[73,148] From the intestinal cell a vitamin-D-dependent, relatively specific calcium-binding protein has been isolated.

Calcium is needed for many cellular activities.[15] It is necessary for the promotion of cellular adhesion between cells and to maintain intracellular connections or cytoplasmic bridges on the membranes of excitable tissues. The existence of specific binding sites has

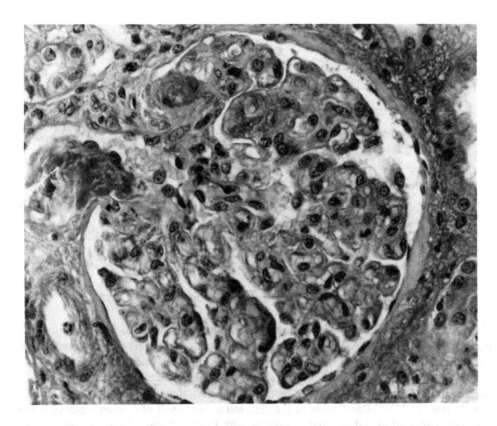

PLATE 3. Chronic glomerulonephritis. Thickened Bowman capsule with cellular lobulated glomerulus.

been postulated. These are the places where calcium enters the cell, regulates sodium permeability and transport, and controls the initiation and propagation of an action potential. The presence of calcium ions in cholinergic nerve endings is an absolute requirement for the release of acetylcholine in response to action potential changes. If calcium ions are removed, acetylcholine remains bound in the nerve tissue. Calcium also affects muscular function. A calcium-activated adenosine triphosphatase is present in muscle which may play a role in the adenosine-triphosphate-dependent movement of calcium from binding sites in the sarcoplasmic reticulum. Actomyosin fibers are probably the calcium-binding sites. The process of binding may be responsible for the effect of excitation and for the initiation of muscular contraction. The removal of calcium leads to relaxation.

A calcium-inhibited adenosine triphosphatase has been identified in erythrocytes. The production of characteristic secretions by the anterior and posterior pituitary, or the islet cells of the pancreas, adrenal cortex and medulla, or salivary glands in response to stimulation all depends on the presence of calcium ions. The various actions of calcium probably involve its association with specific cellular components. Calcium can be bound to simple organic acids and to a variety of acidic groups on proteins or carbohydrate derivatives by covalent bond or chelation. It also binds to nucleic acids, mucopolysaccharides, and other complex macromolecules, particularly if they contain large amounts of phosphate or sulfate moieties. The binding of calcium to phospholipids alters the properties of lipid-water interfaces, which may represent an important step in calcium transport across membranes. This transport is probably influenced by calcitonin largely at specific sites present in only one system. Besides the hormonal control, serum phosphate concentration and tissue phosphoproteins can also affect the equilibrium between ionized and nonionized calcium in the blood, and the calcium bound to the surface of cellular and subcellular membranes.

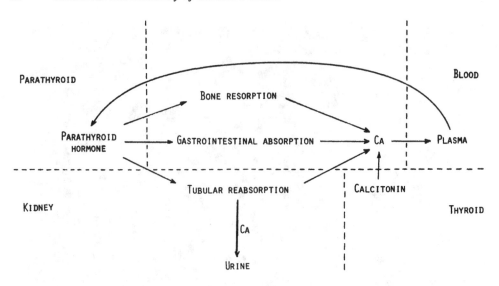

FIGURE 8.   Regulation of calcium metabolism.

Raised plasma calcium concentration may be linked with increased absorption from the gut,[98,108,110,129] abnormal mobilization from the bones, and occasionally with abnormal plasma globulins.[25,38,41,69,77,78,82,131] Increased absorption leading to elevated plasma level is commonly produced by administration of vitamin D in great excess.[55,73] Parathyroid gland overactivity usually connected with adenoma often causes abnormal mobilization of calcium from bones.[5,26,75,93,162] Rarefaction of the skeleton as the result of calcium immobilization has been described in children and in patients with Paget's disease. In these cases, plasma calcium is rarely raised. However, patients with multiple myeloma develop elevated plasma calcium levels. Similarly, widespread secondary carcinomatosis is connected with hypercalcemia due to bone destruction.[61] Increased serum calcium levels are associated with various symptoms such as abdominal pain, nausea, vomiting, anorexia, and coronary heart disease.[39,122] Renal calcium excretion is enchanced, leading primarily to diuresis rather than to calcification of the tubules, and finally to kidney failure.[114] Metastatic calcifications often manifest around joints, and other secondary biochemical changes occur in final stages of these diseases.[31,135,153]

Low plasma calcium concentrations may be linked with defects of intestinal absorption, excess urinary loss, low plasma protein level, or increased deposition of calcium in the skeleton.[29,30,108] Abnormal absorption is due to lack of vitamin D or resistance to this factor, or to an excess of anions in the gut precipitating calcium and preventing the mobilization, such as increased amounts of fatty acids due to pancreatic dysfunction, steatorrhea, celiac disease, or long-standing obstructive jaundice. Hypoparathyroidism causes increased deposition of skeletal calcium and hypocalcemia.[88] This condition also occurs from excess renal excretion associated with the prolonged use of mercurial diuretics.[63] Prolonged acidosis of kidney origin and slowly progressing renal failure lead to calcium depletion.[2,69,70] In many cases of hypocalcemia, secondary hyperplasia of the parathyroid gland causes an increased urinary excretion of phosphate and mobilization of phosphate and calcium from the bone.[7]

VII. MAGNESIUM METABOLISM

Among the common electrolytes, the body contains the smallest amount of magnesium; it comprises only 0.05% of body weight. Its major function is the activation of many enzyme systems. Magnesium is necessary for the synthesis of proteins and nucleic acids. Several enzymes do not function without magnesium.[56,109] Moreover, there are various adenosine triphosphatases that are activated by sodium, potassium, or calcium; yet, they are active without these electrolytes. However, the presence of magnesium is an absolute requirement for their action. There is an important relationship between metabolic activity of the cell and electrolyte distribution. Increased activity shows a parallel with elevated magnesium to calcium and potassium to sodium ratios and total phosphate content.

The cellular transport of magnesium is an energy-related process which can be greatly reduced from the serum and extracellular fluids without any considerable clinical consequences. The depletion of serum magnesium usually represents a relatively small loss of total body magnesium. More excessive losses are related to various disease symptoms. Magnesium loss may occur in postoperative patients receiving parenteral feeding for a prolonged period. It has been reported in hyperparathyroidism and primary aldosteronism, and also found in chronic alcoholics.[8,28,116,134]

VIII. PHOSPHATE METABOLISM

The phosphate content of the body is about 1% of total weight. Phosphate is an absolutely essential element in organic compounds containing high energy bonds and in phosphorylated nucleotides.[73] These compounds provide the possibility of replication and transfer of chemical energy which are essential features of the cell. The process of phosphate uptake into the cell is very rapid.[23] It is quickly incorporated into nucleotides which are key components in the intermediary metabolism of carbohydrates, phospholipids, and phosphoproteins, and in the synthesis of various nucleic acids. If phosphate intake is not satisfactory, nucleotide synthesis becomes deficient and, in particular, inadequate amounts of adenosine triphosphate are produced. In the presence of low tissue adenosine triphosphate and nucleotide levels, growth rate is retarded and energy metabolism and energy utilization for muscular function are restricted. There is a positive correlation between cellular phosphate concentration and growth rate. Inorganic phosphate is very important in bone formation.[23] A small amount of inorganic pyrophosphate is present in the body, which influences the crystallization of hydroxyapatite, one of the major components of the bone. Pyrophosphate, therefore, may play an important role in the regulation of bone metabolism.

IX. REGULATION OF CALCIUM, MAGNESIUM, AND PHOSPHATE METABOLISM

There is some interrelationship in the action and metabolism between these constituents, and their metabolism is linked mainly to the function of the parathyroid gland. Other hormones are also involved, such as thyroid hormones, sex and adrenal steroids, growth hormone, insulin, and glucagon.[1,29,96,103,119,120,145,146,166]

The regulation of calcium concentration in extracellular fluids is linked with the uptake and continuous release of calcium from intestinal absorption, renal tubular resorption, and bone resorption, and with the removal of calcium by intestinal secretion, renal glomerular filtration, and utilization for bone formation.[41,117,156] In addition, in metabolic alkalosis the removal of calcium is associated with chloride depletion. The amount of ionized calcium is very small compared to the total body calcium, but a constant amount is necessary for the normal performance of many important functions as discussed before. The bidirectional movement of calcium in each of these systems is partly controlled by parathyroid hormone[43,130,148] (Figure 8). The secretion of parathyroid hormone influences magnesium ion concentration but there is no efficient feedback control. Magnesium is actively taken up by the cells. In conditions of reduced magnesium intake or deficiency the intracellular concentration is maintained by active transport at the expense of the extracellular concentration. The mechanism of this transport is probably associated with intracellular factors; however, the exact mechanism has not yet been established.

The phosphate concentration of the serum is not regulated as well as that of calcium or magnesium, and many factors influence its level, e.g., growth rate and development.[157] There is no feedback regulation of phosphate metabolism. Acute changes alter the amount of phosphate in the blood, such as the release of insulin or glucose which causes a rapid fall, probably linked with increased intracellular synthesis of sugar-phosphate derivatives. Some adverse relationship exists between phosphate and calcium, and thus the parathyroid regulation of calcium metabolism influences serum phosphate concentration.[6]

In addition to direct regulation by hormones and cellular transport, some interactions occur with other ions. Hydrogen ion concentration is especially essential in the indirect regulation of calcium, magnesium, and phosphate metabolism. When hydrogen ion concentration is decreased in tissue due to alkalosis, there is an increase of neuromuscular excitability resembling that which is found in hypocalcemia. There is a reciprocal movement between calcium and hydrogen ions in membrane transport and between the effect of calcium and that of hydrogen ions on excitability. This relationship may have an effect on the action of hydrogen ions by reducing calcium binding to nerve membranes.

The synthesis and secretion of parathyroid hormone play the major role in regulating the extracellular concentration of divalent cations. Conversely, changes in calcium and magnesium concentrations exercise a specific control over the activities of the parathyroid gland. Both calcium and magnesium ions alter parathyroid function. The effective feedback regulation is, however, connected with changes manifest in the serum calcium concentration present in ionic form.[30] When the amount of calcium is lowered, both hormone synthesis and secretion are increased very rapidly. This effect is continuous, responding to calcium and magnesium changes in the blood with appropriate hormone synthesis and release. Morphological observations have shown that parathyroid hormone is produced and stored in secretory granules and that low calcium concentration stimulates the release of these granules from the cells. Acute changes in magnesium ion concentration exercise a similar effect on parathyroid hormone secretion, but changing the serum phosphate level elicits no direct action on the synthesis or blood level of parathyroid hormone.[8,9] Phosphate ions exert an indirect effect on the gland through their action on ionic calcium. There are, however, differences in action between calcium and magnesium. When serum calcium concentration is decreased, the resulting discharge of parathyroid hormone from the gland restores the normal serum calcium level by various actions; directly by increase of calcium transport from the cells into the blood, elevated intestinal absorption, and increased renal tubular and bone resorption, and indirectly by lowering the concentration and elevating renal excretion of serum phosphate. Low serum magnesium concentration also stimulates the secretion of parathyroid hormone, but it is followed by only slight change in magnesium transport from the cells, intestinal tract, kidney, or bones. Due to this unresponsiveness in the case of hypomagnesemia, hypocalcemia may also occur.

The interfollicular or C cells of the thyroid gland produce a hormone, calcitonin, which opposes the metabolic changes associated with hypocalcemia.[15] The effect of this hormone thus differs from the action of the parathyroid hormone. Thyroid C cells are rich in secretory granules and show similarities to other polypeptide-secreting cells such as those of the pituitary and pancreas. The parathyroid gland and thymus may also contain calcitonin-secreting cells, but the relative importance of these sites has not been established.

The effect of calcium ions on calcitonin secretion is the opposite of their action on parathyroid hormone. There is a direct linear relationship between calcium concentration and calcitonin secretion. It seems, however, that in the regulation of serum calcium level the role of parathyroid hormone is more important than that of calcitonin. Calcitonin secretion is increased in hypercalcemic stress; the action is probably connected with inhibition of bone resorption. Calcitonin apparently does not play any role in changes of serum magnesium concentration.

In addition to the regulatory action of parathyroid hormone and calcitonin on serum bivalent cation concentration, many other factors influence the blood level of ionic calcium, magnesium, and phosphate.[61,116] These include several hormones and local factors which cause changes in ion transport in different organs. Fatty acids and bile acids alter intestinal cation absorption; organic acids and sugar derivatives affect the renal excretion of calcium, magnesium, and phosphate; organic acids, heparin, and oxygen influence the release of calcium from the bone.

Deficiency of thyroid hormones causes growth retardation and decreased responsiveness of bones to local stimulation.[1,17,66] In the absence of thyroid and parathyroid hormones, localized osteoporosis develops. Thyroxine and triiodothyronine elicit an acute increasing action on urinary phosphate excretion, and serum phosphate concentration is also raised. The latter effect can be related to the release of phosphate from soft tissues, but indirect action by enhancing bone resorption and releasing calcium also leads to an increase of serum phosphate. Thyroid hyperactivity causes a negative magnesium balance and loss of serum magnesium concentration.

Growth hormone increases serum phosphate concentration, probably through decreased urinary excretion of phosphate related to increased tubular reabsorption. The retention of magnesium is elevated both in soft tissue and bone by growth hormone. This hormone thus increases bone formation. Sex steroids also influence bone metabolism by accelerating the closure of the epiphyses resulting in a linear growth at puberty. Under certain conditions, estrogens and androgens bring about calcium and phosphate rentention.[35-37,80,81] Adrenal steroids play a regulatory role in bone metabolism. Glucocorticoids can decrease intestinal absorption and increase urinary excretion of calcium. Insulin and glucagon can alter serum phosphate concentration. Insulin causes a reduction, probably by the entry of phosphorylated sugars into the cell.[67] Glucagon elicits transient hypocalcemia, perhaps indirectly by stimulating calcitonin secretion.

In the regulation of calcium, magnesium, and phosphate metabolism the bone plays a considerable part as the main reservoir of these minerals.[65,89] The occurrence of hyper- and hypocalcemia due to disorders of the skeletal system indicates that its solid constituents maintain an active equilibrium with their counterparts in the extracellular fluid. The relationship is intrinsic; bone diseases cause abnormalities in electrolyte composition of the plasma and of urine, and conversely, disturbances in the composition of the extracellular fluid may cause secondary disorders in the bone.[34] This association and further details of anomalies in bone formation will be discussed in Volume 3.

X. ASSESSMENT OF DISTURBED METABOLISM

In the study of disturbances of the body fluids it is necessary to assess whether water or any of the principal electrolytes are present in excess or are deficient.[32,76,95,118] The total body water can be measured by a dilution technique using substances which diffuse evenly throughout all water in the tissues. These substances, such as heavy water, urea, antipyrine, or aminoantipyrine, are given orally or by intravenous injection. After complete diffusion throughout the body the concentration of the test substance is measured in the serum and the space occupied by them in the body can be calculated. If the substance is evenly distributed in all tissues, this space is equivalent to the total body water. A similar technique can be employed for measuring the volume of the various compartments of the total body water. Inulin or thiosulfate can be used for the extracellular fluid, Evans blue for the plasma, and radioactive chromium label for the measurement of the red cell mass.

Comparable methods are available for the measurement of body electrolytes. In the case of sodium and potassium radioactive isotopes can be used, and, after a period of equilibrium, the ratio of radioactive to stable isotope can be measured in the plasma or urine.[126] From the original amount given, corrected for excretion, it is possible to calculate the total amount of sodium or potassium. Total body chloride can also be measured by a dilution technique using isotope chloride or bromide. In certain situations, especially in the assessment of preoperative patients and those with chronic disease, measurements of total body sodium, potassium, and red cell mass are of value in diagnosis and preparation for surgery. The concentrations of electrolytes in the plasma, however, provide no information about the plasma volume and what is happening in the cells.

Biochemical examination of the urine may give indirect evidence about fluid and electrolyte equilibrium, especially the 24-hr collection. There are, however, a number of conditions in which the results of urine analysis may be misleading.[38] In patients with untreated Addison's disease there is a sodium and chloride deficiency accompanied by low plasma concentrations. On the other hand, severe diabetic ketosis leads to similar tissue depletion, but plasma levels of these ions are above normal range. Patients with edema due to acute nephritis or cardiac failure have a great excess of body sodium, yet plasma sodium level is often decreased. The urinary output can also be very low in congestive heart failure in spite of an edematous condition. Addison's disease causes severe sodium depletion and considerable urinary excretion, whereas head exposure and exhaustion are associated with tissue losses of sodium but urinary sodium concentration is low. Thus the biochemical finding should be interpreted in conjunction with the clinical findings.

REFERENCES

1. **Adams, P. H., Jowsey, J., Kelley, P. J., Riggs, L., Kenney, V. R., and Jones, D. J.,** Effects of hyperthyroidism on bone and mineral metabolism in man, *Q. J. Med.,* 36, 1, 1967.
2. **Agus, Z. S.,** Renal tubular transport of calcium: update, in *Homeostasis of Phosphate and Other Minerals,* Massry, S., Ritz, F., and Rapado, A., Eds., Plenum Press, New York, 1978, 37.
3. **Aizman, R. I.,** Effect of a water and potassium loading test on kidney function, *Hum. Physiol.,* 7, 295, 1981.
4. **Alevizaki, C. C., Ikkos, D. G., and Singhelakis, P.,** Progressive decrease of true intestinal calcium absorption with age in normal man, *Nucl. Med.,* 14, 760, 1973.
5. **Aloia, J. F., Zani, I., Ellis, K., Jowsey, J., Roginsky, M., Wallach, S., and Cohn, S. H.,** Effects of growth hormone in osteoporosis, *Clin. Endocrin. Metab.,* 43, 992, 1976.
6. **Alston, W. C., Allen, K. R., and Tovey, J. E.,** A comparison of nephrogenous cyclic AMP, total urinary cyclic AMP and the renal tubular maximum reabsorptive capacity for phosphate in the diagnosis of primary hyperparathyroidism, *Clin. Endocrinol.,* 13, 17, 1980.
7. **Alvarez-Ude, F., Feest, R. G., Ward, M. K., Pierides, A. M., Ellise, H. A., Peart, K. M., Simpson, W., Weightman, D., and Kerr, D. S.,** Hemodialysis bone disease: correlation between clinical, histologic and other findings, *Kidney Int.,* 14, 68, 1978.
8. **Anast, C. S., Mohs, J. M., Burns, T. W., and Kaplan, S. L.,** Evidence for parathyroid failure in magnesium deficiency, *Science,* 177, 606, 1972.
9. **Anast, C. S., Winnacker, J. L., Forte, L. R., and Burns, T. W.,** Impaired release of parathyroid hormone in magnesium deficiency, *J. Clin. Endocrinol. Metab.,* 42, 707, 1976.
10. **Anderson, B.,** Regulation of water intake, *Physiol. Rev.,* 58, 582, 1978.
11. **Arieff, A. I. and Guisado, R.,** Effects on the central nervous system of hypernatremic and hyponatremic states, *Kidney Int.,* 10, 104, 1976.
12. **Arieff, A. I., Llach, F., and Massry, S. G.,** Neurological manifestations and morbidity of hyponatremia: correlation with brain water and electrolytes, *Medicine,* 55, 121, 1976.
13. **Astrup, P., Siggaard-Anderson, O., Jørgensen, K., and Engel, K.,** The acid-base metabolism — a new approach, *Lancet,* 1, 1035, 1960.
14. **Austic, R. E. and Calvert, C. C.,** Nutritional interrelationships of electrolytes and amino acids, *Fed. Proc.,* 40, 63, 1981.
15. **Austin, I. A. and Oldham, S. B.,** Calmodulin: its role in calcium-mediated cellular regulation, *Miner. Electrolyte Metab.,* 8, 1, 1982.

16. **Battle, D. and Arruda, J. A. L.**, The renal tubular acidosis syndromes, *Miner. Electrolyte Metab.*, 5, 83, 1981.

17. **Bayley, T. A., Harrison, J. E., McNeill, K. G., and Mernagh, J. R.**, Effect of thyrotoxicosis and its treatment on bone mineral and muscle mass, *J. Clin. Endocrinol. Metab.*, 50, 916, 1980.

18. **Bay, W. H. and Ferris, T. F.**, Hypernatremia and hyponatremia: disorders of tonicity, *Geriatrics*, 31, 53, 1976.

19. **Bell, E. F. and Oh, W.**, Fluid and electrolyte balance in very low birth weight infants, *Clin. Perinatol.*, 6, 139, 1979.

20. **Berl, T., Anderson, R., McDonald, K., and Schrier, R.**, Clinical disorders of water metabolism, *Kidney Int.*, 10, 117, 1976.

21. **Berl, T., Linas, S. O., and Aisenbrey, G. A.**, On the mechanism of polyuria in potassium depletion: the role of polydipsia, *J. Clin. Invest.*, 60, 620, 1977.

22. **Bernard, D. B. and Alexander, F. A.**, Edema formation in the nephrotic syndrome: pathophysiologic mechanisms, *Cardiovasc. Med.*, 4, 605, 1979.

23. **Bijvoet, O. L. M., Morgan, D. B., and Fourman, P.**, The assessment of phosphate reabsorption, *Clin. Chim. Acta*, 26, 15, 1969.

24. **Birkenhager, W. H. and DeLeeuw, P. W.**, Pathophysiological mechanisms in essential hypertension, *Pharmacol. Ther.*, 8, 297, 1980.

25. **Bordier, P., Ryckewart, A., Gueris, J., and Rasmussen, H.**, On the pathogenesis of so-called idiopathic hypercalciuria, *Am. J. Med.*, 63, 398, 1977.

26. **Broadus, A. E., Horst, R. L., Littledike, E. T., Mahaffey, J. E., and Rasmussen, H.**, Primary hyperparathyroidism with intermittent hypercalcaemia: serial observations and simple diagnosis by means of oral calcium tolerance test, *Clin. Endocrinol.*, 12, 225, 1980.

27. **Brody, M. J., Haywood, J. R., and Touw, K. B.**, Neural mechanisms in hypertension, *Annu. Rev. Physiol.*, 42, 441, 1980.

28. **Buckle, R. M., Care, A. D., Cooper, C. W., and Gitelman, H. J.**, The influence of plasma magnesium concentration on parathyroid hormone secretion, *Endocrinology*, 42, 529, 1968.

29. **Bullamore, J. R., Gallagher, J. C., Walkinson, R., Nordin, B. E. C., and Marshall, D. H.**, Effect of age on calcium absorption, *Lancet*, 2, 535, 1970.

30. **Burritt, M. F., Pierides, A. M., and Offord, K. P.**, Comparative studies of total and ionized serum calcium values in normal subjects and patients with renal disorders, *Mayo Clin. Proc.*, 55, 606, 1980.

31. **Burt, M. E. and Brennan, M. F.**, Incidence of hypercalcemia and malignant neoplasm, *Arch. Surg.*, 115, 704, 1980.

32. **Caniggia, A. and Vittimo, A.**, Kinetics of ^{99m}technetium-tin-methylene-diphosphonate in normal subjects and pathological conditions: a single index of bone metabolism, *Calc. Tissue Int.*, 30, 5, 1980.

33. **Cannon, P. J.**, The kidney in heart failure, *N. Engl. J. Med.*, 296, 26, 1977.

34. **Carr, D., Davidson, J. E., McMillan, M., and Davidson, J.**, Renal osteodystrophy: an underdiagnosed condition, *Clin. Radiol.*, 31, 55, 1980.

35. **Christiansen, C., Brandt, M. J., Ebbesen, F., Sardemann, H., and Trolle, D.**, Bone mineral content during pregnancy in epileptics on anticonvulsant drugs and in their newborns, *Acta Obstet. Gynecol. Scand.*, 60, 501, 1981.

36. **Christiansen, C., Christensen, M. S., and Transbøl, L.**, Bone mass in postmenopausal women after withdrawal of oestrogen/gestogen replacement therapy, *Lancet*, 1, 459, 1981.

37. **Christiansen, C., Christensen, M. S., McNair, P., Gagen, C., Stocklund, K., and Transbol, L.**, Prevention of early postmenopausal bone loss: controlled 2-year study in 315 normal females, *Eur. J. Clin. Endocrinol. Metab.*, 10, 273, 1980.

38. **Christensson, T., Hellstrom, K., and Wengle, B.**, Clinical and laboratory findigns in subjects with hypercalcaemia, *Acta Med. Scand.*, 200, 355, 1976.

39. **Chipperfield, B.**, Water hardness, myocardial metal concentrations, and sudden coronary death, *Am. Heart J.*, 103, 1085, 1982.

40. **Chouko, A. M., Bay, W. H., Stein, J. H., and Ferris, T. F.**, The role of renin and aldosterone in the salt retention of edema, *Am. J. Med.*, 63, 881, 1977.

41. **Coburn, J. W., Kopple, M. H., Brickman, A. S., and Massry, S. G.**, Study of intestinal absorption of calcium in patients with renal failure, *Kidney Int.*, 3, 264, 1973.

42. **Cohen, J. J., Madias, N. E., Wolf, C. J., and Schwartz, W. B.**, Regulation of acid-base equilibrium in chronic hypocapnia: evidence that the response of the kidney is not geared to the defense of extracellular $|H^+|$, *J. Clin. Invest.*, 57, 1483, 1976.

43. **Cohn, S. H., Roginsky, M. S., Aloia, J. E., Ellis, K. J., and Skukla, K. K.**, Alternations in skeletal calcium and phosphorus in dysfunction of the parathyroids, *J. Clin. Endocrinol. Metab.*, 36, 750, 1973.

44. **Conley, S. B., Brocklebank, J. T., Taylor, I. T., and Robson, A. M.**, Recurrent hypernatremia: a proposed mechanism in a patient with absence of thirst and abnormal excretion of water, *J. Pediatr.*, 89, 898, 1976.

45. **Cooke, C. R., Turin, M. D., and Walker, W. G.**, The syndrome of inappropriate antidiuretic hormone secretion (SIADH): pathophysiologic mechanisms in solute and volume regulation, *Medicine*, 58, 240, 1979.

46. **Costill, D. L., Cote, R., and Fink, W. J.**, Dietary potassium and heavy exercise: effects on muscle water and electrolytes, *Am. J. Clin. Nutr.*, 36, 266, 1982.

47. **Davies, F. B.**, Water metabolism in diabetes mellitus, *Am. J. Med.*, 70, 210, 1981.

48. **Davies, J. O,**, The pathogenesis of peripheral cardiac edema, *Contrib. Nephrol.*, 21, 68, 1980.

49. **De, A. K.**, Serum electrolyte changes in mountaineers at various altitudes of expedition, *J. Sports Med. Phys. Fitness*, 20, 23, 1980.

50. **DeWardener, H. E.**, The control of sodium excretion, *Am. J. Physiol.*, 4, F163, 1978.

51. **di Sant' Agnese, P. A. and Lalamo, R. C.**, Pathogenesis and pathophysiology of cystic fibrosis of the pancreas, *N. Engl. J. Med.*, 227, 1287, 1967.

52. **Dobbins, J. W. and Binder, H. J.**, Pathophysiology of diarrhoea: alternations in fluid and electrolyte transport, *Clin. Gastroenterol.*, 10, 605, 1981.

53. **Dorhout Mees, E. J., Roos, J. C., Boer, P., Yoe, O. H., and Simatupang, T. A.**, Observations on edema formation in the nephrotic syndrome in adults with minimal lesions, *Am. J. Med.*, 67, 378, 1979.

54. **Discala, V. A. and Kinney, M. J.**, Effects of myxedema on the renal diluting and concentrating mechanism, *Am. J. Med.*, 50, 325, 1971.

55. **Eastwood, J. B., Harris, E., Stamp, T. C. B., and De Wardener, H. E.**, Vitamin D-deficiency in the osteomalacia of chronic renal failure, *Lancet*, 2, 1209, 1976.

56. Editorial, Magnesium deficiency, *Lancet*, 1, 523, 1976.

57. **Epstein, M.**, Deranged sodium homeostasis in cirrhosis, *Gastroenterology*, 76, 622, 1979.

58. **Eriksson, F. and Robson, M. C.**, New pathophysiologic mechanism explaining the generalized edema after a major burn, *Surg. Forum*, 28, 540, 1977.

59. **Fatthi, A., El-Fouli, E. M., Soliman, T., El-Khodary, M., and El-Roubi, O.**, The electrolyte pattern of human saliva in health and disease, *J. Egypt Med. Assoc.*, 61, 61, 1978.

60. **Felig, P.**, Diabetic ketoacidosis, *N. Engl. J. Med.*, 290, 1360, 1974.

61. **Felsenfeld, A. J., Harrelson, J. M., Wells, S. A., and Gutman, R. A.**, Severe osteomalacia with hypercalcemia following subtotal parathyroidectomy, *Kidney Int.*, 16, 952, 1979.

62. **Ferguson, E. R., Blachley, J. D., and Knochel, J. P.**, Skeletal muscle ionic composition and sodium transport activity in chronic alcoholism, *Trans. Assoc. Am. Physicians*, 94, 61, 1981.

63. **Fichman, M. P., Vorherr, H., Kleeman, C. R., and Telfer, N.**, Diuretic-induced hyponatremia, *Ann. Intern. Med.*, 75, 853, 1971.

64. **Fiorotto, M. and Coward, W. A.**, Pathogenesis of oedema in protein-energy malnutrition: the significance of plasma colloid osmotic pressure, *Br. J. Nutr.*, 42, 21, 1979.

65. **Fogelman, I., Bessent, R. G., Cohen, H. M., Hart, D. M., and Lindsay, R.**, Skeletal uptake of diphosphonate. Method for prediction of post-menopausal osteoporosis, *Lancet*, 2, 667, 1980.

66. **Fraser, S. A., Anderson, J. B., Smith, D. A., and Wilson, G. M.**, Osteoporosis and fractures following thyrotoxicosis, *Lancet*, 1, 981, 1971.

67. **Frier, B. M.**, Misleading plasma electrolytes in diabetic children with severe hyperlipidaemia, *Arch. Dis. Child.*, 55, 771, 1980.

68. **Gennari, F. J. and Kassirer, J. P.**, Osmotic diuresis, *N. Engl. J. Med.*, 291, 714, 1974.

69. **Gipstein, R. H., Coburn, J. W., Adams, D. A., Lee, D. B. N., Parsa, K. P., Sellers, A., Suki, W. N., and Massry, S. G.**, Calciphylaxis in man: a syndrome of tissue necrosis and vascular calcification in 11 patients with chronic renal disease, *Arch. Intern. Med.*, 136, 1273, 1976.

70. **Glassock, R. J.**, The nephrotic syndrome, *Hosp. Pract.*, 14, 105, 1979.

71. **Golden, M. H. N., Golden, B. F., and Jackson, A. A.**, Albumin and nutritional oedema, *Lancet*, 1, 114, 1980.

72. **Goldsmith, R. S., Jowsey, J., Dube, W. J., Riggs, B. L., Arnaud, C. B., and Kelly, P. J.**, Effects of phosphorus supplementation on serum parathyroid hormone and bone morphology in osteoporosis, *J. Clin. Endocrinol. Metab.*, 43, 523, 1979.

73. **Greer, F. R., Searcy, J. E., Levin, R. S., Steichen, J. J., and Steichen-Asche, P. S.**, Effect of increased calcium, phosphorus and vitamin D intake on bone mineralization, *J. Pediatr.*, 100, 951, 1982.

74. **Guyton, A. C., Young, D. B., Manning, R. D., Lohmeier, T. E., and DeClue, J. W.**, An overview of water and electrolyte distribution in the body, *Contrib. Nephrol.*, 21, 6, 1980.

75. **Habener, J. F. and Potts, J. T., Jr.**, Relative effectiveness of magnesium and calcium in the control of parathyroid hormone secretion and biosynthesis, *Endocrinology*, 98, 209, 1976.

76. **Harrington, J. T.**, Evaluation of serum and urinary electrolytes, *Hosp. Pract.*, 17, 28, 1982.

77. **Heaney, R. P., Recker, R. R., and Saville, P. D.**, Calcium balance and calcium requirements in middle-aged women, *Am. J. Clin. Nutr.*, 30, 1603, 1977.

78. **Hesp, R., Hulme, P., Williams, D., and Reeve, J.,** The relationship between changes in femoral bone density and calcium balance in patients with involutional osteoporosis treated with human parathyroid hormone fragment (hPTH 1-34), *Metab. Bone Dis. Related Res.,* 2, 331, 1981.
79. **Hogan, G. R.,** Hypernatremia, problems in management, *Pediatr. Clin. N. Am.,* 23, 569, 1976.
80. **Horsman, A., Gallagher, J. C., Simpson, M., and Nordin, B. E. C.,** Prospective trial of oestrogen and calcium in post-menopausal women, *Brit. Med. J.,* 2, 789, 1977.
81. **Horsman, A., Nordin, B. E. C., and Crilly, R. G.,** Effect on bone of withdrawal of oestrogen therapy, *Lancet,* 2, 33, 1979.
82. **Hosking, D. J., Cowley, A., and Bucknall, C. A.,** Rehydration in the treatment of severe hypercalcaemia, *Q. J. Med.,* 50, 473, 1981.
83. **Hürter, T., Bröcker, W., and Bosma, H. J.,** Investigations on vasogenic and cytotoxic brain edema, *Microsc. Acta,* 85, 285, 1982.
84. **Illner, H. and Shires, G. T.,** Changes in sodium, potassium and adenosine triphosphate contents of red blood cells in sepsis and septic shock, *Circ. Shock,* 9, 259, 1982.
85. **Kaissling, B.,** Structural aspects of adaptive changes in renal electrolyte excretion, *Am. J. Physiol.,* 243, 211, 1982.
86. **Kirsch, K. A., von Ameln, H., and Wicke, H. J.,** Fluid control mechanisms after exercise dehydration, *Eur. J. Appl. Physiol.,* 47, 191, 1981.
87. **Klahr, S. and Alleyne, G. A. O.,** Effects of chronic protein-calorie malnutrition on the kidney, *Kidney Int.,* 3, 129, 1973.
88. **Kleerekoper, M., Rao, D. S., and Frame, B.,** Hypercalcemia, hyperparathyroidism, and hypertension, *Cardiovasc. Med.,* 3, 1283, 1978.
89. **Knochel, J. P.,** The pathophysiology and clinical characteristics of severe hypophosphatemia, *Arch. Intern. Med.,* 137, 203, 1977.
90. **Kokko, J.,** The role of the renal concentrating mechanisms in the regulation of serum sodium concentration, *Am. J. Med.,* 62, 165, 1977.
91. **Krabbe, S., Transbol, I., and Christiansen, C.,** Bone mineral homeostasis, bone growth, and mineralization during years of pubertal growth, *Arch. Dis. Child.,* 57, 359, 1982.
92. **Kramer, H. J.,** Natriuretic hormone — a circulating inhibitor of sodium — and potassium-activated adenosine triphosphatase, *Klin. Wochenschr.,* 59, 1225, 1981.
93. **Kraut, J. A., Shinaberger, J., Singer, F. R., Sherrard, D. J., Saxton, J. N., Hodsman, A. B., Miller, J. H., Kurokawa, K., and Coburn, J. W.,** Reduced parathyroid response to acute hypocalcemia in dialysis osteomalacia, *Clin. Res.,* 29, 102A, 1981.
94. **Kreisberg, R. A.,** Lactate homeostasis and lactic acidosis, *Ann. Intern. Med.,* 92, 227, 1980.
95. **Krensky, A. M., Harmon, W. E., Ingelfinger, J. R., Kirkpatrick, J. A., and Grupe, W. E.,** Elevated nephrogenous cyclic adenosine monophosphate to monitor early renal osteodystrophy, *Clin. Nephrol.,* 16, 245, 1981.
96. **Krolner, B. and Nielsen, P. S.,** Bone mineral content of the lumbar spine in normal and osteoporotic women, *Clin. Sci.,* 62, 329, 1982.
97. **Kunau, T. R. and Stein, J. H.,** Disorders of hypo- and hyperkalemia, *Clin. Nephrol.,* 7, 173, 1977.
98. **Ladenson, J. H., Lewis, J. W., McDonald, J. M., Slatopolsky, E., and Boyd, J. C.,** Relationship of free and total caclium in hypercalcemic conditions, *J. Clin. Endocrinol. Metab.,* 48, 393, 1979.
99. **Lantz, B., Carlmark, B., and Reizenstein, P.,** Electrolytes and whole body potassium in acute leukemia, *Acta Med. Scand.,* 206, 45, 1979.
100. **Lee, M. R.,** Effects of drugs on water metabolism, *Br. J. Clin. Pharmacol.,* 12, 289, 1981.
101. **Lespier-Dexter, L. E.,** Electrolyte abnormalities as complication of drug therapy, *Contrib. Nephrol.,* 27, 45, 1981.
102. **Levinsky, N. G.,** Pathophysiology of acute renal failure, *N. Engl. J. Med.,* 296, 1453, 1977.
103. **Lindergard, B.,** Changes in bone mineral content before the start of active uremia treatment, *Clin. Nephrol.,* 16, 126, 1981.
104. **Little, R. A., Savic, J., and Stoner, H. B.,** H_2 receptors and traumatic edema, *J. Pathol.,* 125, 201, 1978.
105. **Lorenz, J. M., Kleinman, L. T., Kotagal, U. R., and Reller, M. D.,** Water balance in very low-birth weight infants, *J. Pediatr.,* 101, 423, 1982.
106. **Macaron, C. and Famuyiwa, O.,** Hyponatremia of hypothyroidism, *Arch. Intern. Med.,* 138, 820, 1978.
107. **MacGregor, G. A., Markandu, N. D., Roulston, J. E., Jones, J. C., and DeWardener, H. E.,** Is "idiopathic" oedema idiopathic?, *Lancet,* 1, 397, 1979.
108. **Malluche, H. H., Werner, E., and Ritz, E.,** Intestinal absorption of calcium and whole body calcium retention in incipient and advanced renal failure, *Miner. Electrol. Metab.,* 1, 263, 1978.
109. **Manthey, J., Stoeppler, M., Morgenstern, W., and Nussel, E.,** Magnesium and trace metals: risk factors for coronary heart disease?, *Circulation,* 64, 722, 1981.

110. **Marx, S. J., Stock, J. L., Attie, M. F., Downs, R. W., Jr., Gardner, D. G., Borwn, E. M., Spiegel, A. M., Doppman, J. L., and Brennan, M. F.,** Familial hypocalciuric hypercalcemia: recognition among patients referred after unsuccessful parathyroid exploration, *Ann. Intern. Med.*, 92, 351, 1980.

111. **Mason, J., Beck, F., Dorge, A., Rick, R., and Thurau, K.,** Intracellular electrolyte composition following renal ischemia, *Kidney Int.*, 20, 16, 1981.

112. **Meunier, P. J., Courpron, P., Edouard, C., Bernard, J., Bringuier, J., and Vignon, G.,** Physiological senile involution and pathological rarefaction of bone. Quantitative and comparative histological data, *Clin. Endocrinol. Metab.*, 2, 239, 1973.

113. **Milla, P. J.,** Disorders of electrolyte absorption, *Clin. Gastroenterol.*, 11, 31, 1982.

114. **Mioni, G., D'Angelo, A., Ossi, E., Bertaglia, E., Marcon, G., and Maschio, G.,** The renal handling of calcium in normal subjects and in renal disease, *Eur. J. Clin. Biol. Res.*, 16, 881, 1971.

115. **Morgan, D. B. and Thomas, T. H.,** Water balance and hyponatremia, *Clin. Sci.*, 56, 517, 1979.

116. **Nilsson, B. E. and Westlin, N. E.,** Changes in bone mass in alcoholics, *Clin. Orthop.*, 90, 229, 1973.

117. **Nordin, B. E. C.,** Calcium balance and calcium requirement in spinal osteoporosis, *Am. J. Clin. Nutri.*, 10, 384, 1962.

118. **Nordin, B. E. C. and Fraser, R.,** Assessment of urinary phosphate excretion, *Lancet*, 1, 947, 1960.

119. **Nordin, B. E. C., Horsman, A., Crilly, R. G., Marshall, D. H., and Simpson, M.,** Treatment of spinal osteoporosis in postmenopausal women, *Br. Med. J.*, 280, 451, 1980.

120. **O'Brien, P. M., Selby, C., and Synonds, E. M.,** Progesterone, fluid, and electrolytes in premenstrual syndrome, *Br. Med. J.*, 280, 1161, 1980.

121. **Oh, M. S. and Carroll, H. J.,** The anion gap, *N. Engl. J. Med.*, 297, 814, 1977.

122. **Olesen, K. H.,** Exchangeable electrolytes in heart disease, *Acta Med. Scand. Suppl.*, 647, 47, 1981.

123. **Oliver, R. A. and Jamison, R. L.,** Diabetes insipidus: a physiological approach to diagnosis, *Postgrad. Med.*, 68, 120, 1980.

124. **Pak, C. Y. C.,** The hypercalciurias: causes, parathyroid functions, and diagnostic criteria, *J. Clin. Invest.*, 54, 387, 1974.

125. **Penney, M. D., Murphy, D., and Walters, G.,** Resetting osmoreceptor response as cause of hyponatraemia in acute idiopathic polyneuritis, *Br. Med. J.*, 279, 1474, 1979.

126. **Price, P. A., Parthemore, J. G., and Deftos, L. J.,** New biochemical markers for bone metabolism. Measurement by radioimmunoassay of bone GLA protein the plasma of normal subjects and patients with bone disease, *J. Clin. Invest.*, 66, 878, 1980.

127. Proceedings Stockholm Symposia, Electrolytes and cardiac arrhythmias, *Acta Med. Scand. Suppl.*, 647, 1, 1981.

128. **Rabast, U., Vornberger, K. H., and Ehl, M.,** Loss of weight, sodium and water in obese persons consuming a high- or low-carbohydrate diet, *Ann. Nutr. Metab.*, 25, 341, 1981.

129. **Recker, R. R., Saville, P. D., and Heaney, R. P.,** Effect of estrogens and calcium carbonate on bone loss in postmenopausal women, *Ann. Intern. Med.*, 87, 649, 1977.

130. **Reeve, J., Meunier, P. J., Parsons, J. A., Bernat, M., Bijvoet, O. L. M., and Courpron, P.,** Anabolic effect of human parathyroid hormone fragment on trabecular bone in involutional osteoporosis: a multicentre trail, *Br. Med. J.*, 280, 1340, 1980.

131. **Riggs, B. L., Randall, R. V., Wahner, H. W., Jowsey, J., Kelly, P. J., and Singh, M.,** The nature of the metabolic bone disorder in acromegaly, *J. Clin. Endocrinol. Metab.*, 34, 911, 1972.

132. **Robertson, W. G. and Marshall, R. W.,** Ionized calcium in body fluids, *Crit. Rev. Clin. Lab. Sci.*, 15, 85, 1981.

133. **Ross, F. J. and Christie, S. B. M.,** Hypernatremia, *Medicine*, 48, 441, 1969.

134. **Rude, R. K., Oldham, S. B., and Singer, F. R.,** Functional hypoparathyroidism and parathyroid hormone end-organ resistance in human magnesium deficiency, *Clin. Endocrinol.*, 5, 209, 1976.

135. **Rude, R. K., Sharp, C. F., Fredericks, R. S., Oldham, S. B., Elbaum, N., Link, J., Irwin, L., and Singer, F. R.,** Urinary and nephrogenous adenosine 3', 5' monophosphate in the hypercalcaemia of malignancy, *J. Clin. Endocrinol. Metab.*, 52, 765, 1981.

136. **Sachs, G.,** Ion pumps in the renal tubule, *Am. J. Physiol.*, 2, 359, 1977.

137. **Schettini, A., Stahurski, B., and Young, H. F.,** Osmotic and osmotic-loop diuresis in brain surgery. Effects on plasma and CSF electrolytes and ion excretion, *J. Neurosurg.*, 56, 679, 1982.

138. **Schrier, R. W. and Szatalowicz, V. I.,** Disorders of water metabolism, *Contrib. Nephrol.*, 21, 48, 1980.

139. **Sejersted, O. M., Medbo, J. I., and Hermansen, L.,** Metabolic acidosis and changes in water and electrolyte balance after maximal exercise, *Ciba Found. Symp.*, 87, 153, 1982.

140. **Seldin, D. W. and Rector, F. C., Jr.,** The generation and maintenance of metabolic alkalosis, *Kidney Int.*, 1, 306, 1972.

141. **Sebastin, A., McSherry, E., and Morris, R. C., Jr.,** Impaired renal conservation of sodium and chloride during sustained correction of systemic adidosis in patients with Type I, classic renal tubular acidosis, *J. Clin. Invest.*, 48, 454, 1976.

142. **Shear, I. and Brandman, I. S.**, Hypoxia and hypercapnia caused by respiratory compensation for metabolic alkalosis, *Annu. Rev. Respir. Dis.*, 107, 836, 1973.

143. **Simon, R. P. and Freedman, D. D.**, Neurologic manifestations of osmolar disorders, *Geriatics*, 35, 71, 1980.

144. **Simpson, W., Ellis, H. A., Kerr, D. N. S., McElroy, M., McNary, R. A., and Peart, K. N.**, Bone disease in long-term haemodialysis: the association of radiological with histologic abnormalities, *Br. J. Radiol.*, 49, 105, 1976.

145. **Slatopolsky, E., Rutherford, W. E., Hruska, K., Martin, K., and Klahr, S.**, How important is phosphate in the pathogenesis of renal osteodystrophy?, *Arch. Intern. Med.*, 138, 848, 1978.

146. **Smith, D. A., Fraser, S. A., and Wolson, G. M.**, Hyperthyroidism and calcium metabolism, *Clin. Endocrinol. Metab.*, 2, 333, 1973.

147. **Smithline, N. and Gardner, K. D.**, Gaps-anionic and osmolal, *J. Am. Med. Assoc.*, 236, 1594, 1976.

148. **Somerville, P. J. and Kaye, M.**, Resistance to parathyroid hormone in renal failure: role of vitamin D metabolites, *Kidney Int.*, 14, 245, 1978.

149. **Sopko, J. A. and Freeman, R. M.**, Salt substitutes as a source of potassium, *J. Am. Med. Assoc.*, 238, 608, 1977.

150. **Southon, S. and Heaton, F. W.**, Changes in cellular and subcellular composition during potassium deficiency, *Comp. Biochem. Physiol.*, 72, 415, 1982.

151. **Stein, G., Marsh, A., and Morton, J.**, Mental symptoms, weight changes and electrolyte excretion in the first post post partum week, *J. Psychosom. Res.*, 25, 394, 1981.

152. **Stein, J. H., Lipschitz, M. D., and Barnes, L. D.**, Current concepts on the pathophysiology of acute renal failure, *Am. J. Physiol.*, 243, F171, 1978.

153. **Stewart, A. P., Horst, R., Deftos, L. J., Cadman, E. C., Lang, R., and Broadus, A. E.**, Biochemical evaluation of patients with cancer-associated hypercalcaemia: evidence for humoral and non-humoral groups, *New Engl. J. Med.*, 303, 1377, 1980.

154. **Stone, J. G., Wicks, A. E., Hennessey, J. W., and Bendixen, H. H.**, Water clearance patterns after open heart surgery, *Bull. N.Y. Acad. Med.*, 56, 610, 1980.

155. **Suki, W. N.**, Calcium transport in the nephron, *Am. J. Physiol.*, 237, Fl, 1979.

156. **Sutton, R. A. L. and Dirks, J. H.**, Renal handling of calcium: overview, *Adv. Exp. Med. Biol.*, 81, 15, 1977.

157. **Swenson, R. S., Weisinger, J. R., and Ruggeri, J. L.**, Evidence that parathyroid hormone is not required for phosphate homeostasis in renal failure, *Metabolism*, 24, 199, 1975.

158. **Symposium**, Disturbance of water and electrolyte metabolism, *Contrib. Nephrol.*, 21, 1, 1980.

159. **Tanne, R. L.**, Symposium on potassium homeostasis, *Kidney Int.*, 11, 389, 1977.

160. **Thibonnier, M., Marchetti, J., Corvol, P., Menard, J., and Milliez, P.**, Abnormal regulation of antidiuretic hormone in idiopathic edema, *Am. J. Med.*, 67, 67, 1979.

161. **Thomas, T. H., Morgan, D. B., and Swaminathan, R.**, Severe hyponatraemia, *Lancet*, 1, 621, 1978.

162. **Thorgersson, U., Costa, J., and Marx, S. J.**, The parathyroid glands in familial hypocalciuric hypercalcemia, *Human Pathol.*, 12, 229, 1981.

163. **Tobian, L.**, Salt and hypertension, *Ann. N.Y. Acad. Sci.*, 189, 178, 1978.

164. **Trump, D. L., Abeloff, M. D., and Hsu, T. H.**, Frequency of abnormalities of cortisol secretion and water metabolism in patients with small cell carcinoma of the lung and other malignancies, *Chest*, 81, 576, 1982.

165. **Ullrich, I. and Lizarralde, G.**, Amenorrhea and edema, *Am. J. Med.*, 64, 1080, 1978.

166. **Walker, D. A., Davies, S. J., Siddle, K., and Woodhead, J. S.**, Control of renal tubular phosphate reabsorption by parathyroid hormone in man, *Clin. Sci. Molec. Med.*, 53, 431, 1977.

167. **Wesson, L. G.**, Diurnal circadian rhythms of electrolyte excretion and filtration rate in end-stage renal disease, *Nephron*, 26, 211, 1980.

168. **Williams, E. S., Ward, M. P., Milledge, J. S., Withey, W. R., and Older, M. W.**, Effect of the exercise of seven consecutive days hill-walking on fluid homeostasis, *Clin. Sci.*, 56, 305, 1979.

169. **Widmer, B., Berhardt, R. E., Harrington, J. T., and Cohen, J. J.**, Serum electrolyte and acid base composition. The influence of graded degrees of chronic renal failure, *Arch. Intern. Med.*, 139, 1099, 1979.

170. **Wright, F. S. and Giebisch, G.**, Renal potassium transport: contributions of individual nephron segments and populations, *Am. J. Physiol.*, 235, F515, 1978.

171. **Yanagihara, T. and McCall, J. T.**, Ionic shift in cerebral ischemia, *Life Sci.*, 30, 1921, 1982.

172. **Campbell, E. J. M.**, Hydrogen ion regulation, in *Clinical Physiology*, Campbell, E. J. M. and Dickinson, C. J., Eds., Blackwell Scientific, Oxford, 1960, 187.

173. **McCance, R. A. and Widdowson, E. M.**, A method of breaking down the body weights of living persons into terms of extracellular fluid, cell mass and fat, and some applications of it to physiology and medicine, *Proc. R. Soc. Ser. B.*, 138, 115, 1951.

174. **Novak, L. P.,** Total body water in man, in *Compartments, Pools and Spaces in Medical Physiology,* Bergner, P. E. E. and Lushbaugh, C. C., Eds., U. S. Atomic Energy Commission Symposium, Washington, 11, 197, 1967.
175. **Parker, H. V., Olesen, K. H., McMurrey, J. and Priis-Hansen, B.,** Body water compartments throughout the life-span, in *Ciba Foundation Colloquia on Ageing,* Vol. 4., Wolstenholme, G. E. W. and O'Connor, M. Eds., Ciba Foundation, Churchill, London, 1958, 102.
176. **Widdowson, E. M. and Dickerson, J. T. W.,** Chemical composition of the body, *Mineral Metabolism,* Comar, C. L. and Bronner, F., Eds., Vol. 2, Part A, Academic Press, London, 1964, 1.

FURTHER READINGS

Avioli, L. P., Bordier, P., Fleisch, H., Massry, S., and Slatopolsku, E., Eds., *Phosphate Metabolism, Kidney and Bone,* Armour Montagu, Paris, 1975.

Andreoli, T. E., Grantham, J. J., and Rector, S. C., *Disturbance in Body Fluid Osmolality,* American Physiological Society, Bethesda, Md., 1977.

Brenner, B. M. and Rector, F. C., Eds., *The Kidney,* W. B. Saunders, Philadelphia, 1981.

Brenner, B. M. and Stein, J. H., *Sodium and Water Homeostasis,* Churchill-Livingstone, London, 1978.

Coe, F. L., Ed., *Nephrolithiasis,* Year Book Medical Publishers, Chicago, 1978.

Earley, I. E. and Gottschalk, C. W., Eds., *Strauss and Welt's Diseases of the Kidney,* Little Brown, Boston, 1979.

Fleisch, H., Robertson, W. G., Smith, L. H., and Vahlensieck, W., Eds., *Urolithiasis Research,* Plenum Press, New York, 1976.

Klahr, S., Ed., *The Kidney and Body Fluids in Health and Disease,* Plenum Press, New York, 1981.

Klahr, S. and Massry, S., Eds., *Contemporary Nephrology,* Plenum Press, New York, 1981.

Maxwell, M. H. and Kleeman, C. R., *Clinical Disorders of Fluid and Electrolyte Metabolism,* 3rd ed., McGraw-Hill, New York, 1980.

Nordin, B. E. C., Ed., *Calcium, Phosphate and Magnesium Metabolism,* Churchill Livingstone, Edinburgh, 1977.

Rose, B. D., Ed., *Clinical Physiology of Acid-Base and Electrolyte Disorders,* McGraw-Hill, New York, 1977.

Chapter 2

ABNORMALITIES OF TRACE ELEMENT METABOLISM

I. INTRODUCTION

Abnormalities of trace element metabolism are relatively rare.[91,101] This may be due to the presently inadequate knowledge of the importance of these constituents in experimental animals as well as in man. Health may be affected by either an excess or a deficiency of an essential trace element. (See References 2, 11, 18, 26, 28, 31, 37, 39, 49, 57, 60, 64, 92, 97, 122, 125, 142, 150, and 151.) Certain clinical conditions such as malabsorption, severe burns, or marked proteinuria are frequently associated with deficiency of many trace elements, and these conditions influence the degree of deficiency in various tissues.[23,34,35,61,95,123] Health may also be impaired, chiefly through environmental contamination by some nonessential elements at trace levels.[70,82,90,91,127]

The trace elements present in tissues in small amounts include zinc, copper, manganese, cobalt, selenium, molybdenum, chromium, tin, iodine, fluorine, and silicone.[36,63,80,86] They are required for the function of many biochemical processes. Some of these are an integral part of body constituents such as cobalt in vitamin B_{12}, selenium as the active site of glutathione peroxidase, and the relationship between zinc and the action of insulin. Several other elements have been described as beneficial but they are not essential and their function can be taken over by another element. These include nickel, vanadium, and bromine. There are some other trace metals which are neither essential nor beneficial; they are, rather, linked with the chronic development of several diseases. These elements include lead, mercury, cadmium, arsenic, and beryllium. Exposure to any trace element in excessive amounts causes disorders. It is, however, difficult to establish these abnormalities due to complications caused by biochemical interactions between various elements or between the trace metal and many compounds present in tissues.[27,33,40,67,68,96,102,128,154] The abnormalities occur rarely; however, there is evidence for the physiological role and relevance to disease of many essential trace elements[34,37,38] (Table 1). Some have been well studied, such as iron and iodine, and will not be considered here in detail. A number of these trace elements are metallic ions and function as enzyme activators. This role is not exclusive evidence of essentiality, since many metals can substitute for another in enzyme activation. Many elements are found in human and animal tissue and, have no known role as yet. Several elements elicit toxic effects rather than performing any physiologic function. Even essential elements in excessive amounts are likely to exert toxicity.[45]

II. EFFECT OF ENVIRONMENT AND HOMEOSTASIS

The presence of trace elements in our body depends on the geochemical environment.[9,47,100,146,153] The food of animals and man is usually derived from a limited geographical area and therefore it reflects the trace element composition of the local soil and water.[19,43,87,125] Regional occurrences of trace-element-related disorders in grazing animals, the consequences of iodine deficiency, and mercury contamination of fish eaten by man have provided evidence for such a relationship.[123,137] Moreover, throughout centuries man has altered the natural environment, resulting in a new set of trace elements. A number of them, mainly heavy metals, have not been abundant during the early development of life and thus we have not developed an adequate protective mechanism against them. Consequently, excessive intake of these trace elements may cause toxic symptoms and disease.[31] The spleen represents one of the sites where foreign chemicals impair normal cellular metabolism (Plate 1).

Table 1
TRACE ELEMENT CONTENT
OF HUMAN BODY

Element	Av amount[a]
Iron	5000
Zinc	2300
Zirconium	230
Silicon	200
Copper	150
Iodine	30
Bromine	30
Vanadium	30
Cadmium	30
Manganese	20
Molybdenum	20
Arsenic	18
Nickel	10
Chromium	6
Titanium	4
Lead	$0 \rightarrow 4$[b]
Mercury	$0 \rightarrow 2$[b]
Cobalt	2
Tin	$0 \rightarrow 1$[b]
Fluorine	Variable
Aluminum	0.1

[a]   Each element comprising less than 0.01% of total body weight. Amounts are given in milligrams per body weight. Some values are variable according to geographical region.
[b]   Some elements are not detectable at birth and are accumulated during childhood.

Calculated and modified from several publications, and from a major source.[161]

The metabolism of the essential trace metals is under homeostatic control.[16,62,121] This control regulates absorption, tissue retention, and excretion in spite of the wide variations in our dietary supply and it exists to avoid both deficiency and the toxicity resulting from excess digestion.[13,30,54,88,111] Due to this extremely effective regulatory mechanism, disorders linked with essential trace elements are rare. However, if they occur, diseases associated with essential metals are characterized by a failure of homeostasis, due mainly to hereditary defects.[7,59] In contrast, in the case of nonessential metals such a homeostatic mechanism does not develop, although many similarities exist between the essential and nonessential trace elements in their toxic potential (Table 2). Essential metal ions are needed in health and life for a number of biochemical purposes, their absorption and metabolism are actively facilitated, and they constitute integral parts of specific protein molecules. Nonessential trace metals are not linked to any mechanism which stimulates or prevents accumulation and no specific proteins are synthesized that bind or detoxify these ions. All nonessential heavy metals are toxic, whereas essential heavy metals elicit toxic action only in unusual circumstances.[71]

III. TERATOLOGIC EFFECTS

Deficiency of trace metals has been shown to cause deformation of the fetus associated with abnormalities in the homeostatic mechanism which regulates the formation of bio-

Table 2
COMPARISON OF METABOLISM AND TOXICITY
OF TRACE COPPER AND MERCURY

Parameter	Copper	Mercury
Essentiality	Yes	No
Toxic form	Cu^{2+}	Hg^{2+}, organic Hg^{2+}, metallic Hg
Basis of toxicity	SH-binding	SH-binding
Active absorption sites	Yes	No
	GI tract, lung, skin, uterine mucosa	GI tract, lung, skin
Active excretion sites	Yes	No
	GI tract, kidney	GI tract, kidney, lung (metallic Hg)
Metalloprotein formation	Yes	No
Industrial hazard	Negligible	Appreciable
Toxic serum level	10 μg/dℓ	20 μg/dℓ
Affected organs	Liver, kidney, brain, eye	GI tract, kidney, brain, eye, skin
Therapeutic agents	Penicillamine, BAL	Penicillamine, BAL, N-acetylpenicillamine

membranes.[55,58,74,84,135,157] Some trace elements form an integral part of tissues and cells and participate in their reactivity. They influence the endocrine function of the organism at each major level of the system. Depression of several metal ions in trace amounts accompanies biochemical manifestation of most infectious illnesses of a disseminated nature. Many enzyme reactions involved in the metabolism of foreign compounds are influenced by the dietary levels of trace minerals. Deficiencies in some trace elements are linked with congenital malformations.[51] Lack of manganese in the diet during prenatal development causes ataxia in the offspring characterized by incoordination and lack of equilibrium. This defect is related to abnormal development of otoliths. The major defect in skeletal abnormalities may be in cartilage production associated with the biosynthesis of mucopolysaccharides, in particular with sulfomucopolysaccharide production. Animal experiments with a manganese-deficient diet have shown various abnormalities in cellular structures, such as dilatation and disorganization of the rough endoplasmic reticulum and hypertrophy of the Golgi apparatus. These changes are especially pertinent since these organelles constitute the sites of mucopolysaccharide synthesis.

Lack of zinc produces a wide variety of gross congenital defects.[51,53,157] It causes a drastic reduction of the newborn's body weight. Zinc deficiency affects primarily the fast-proliferating tissues such as gonads and skin. This effect is probably associated with a depressed synthesis of DNA. Some experiments have revealed that in zinc-deficient animals thymidine kinase activity is decreased leading to depressed nucleic acid synthesis which in turn diminished protein synthesis. Chromosomes obtained from the liver of zinc-deficient fetuses show aberrations, thus providing further evidence for the interaction between zinc and nucleic acids.

Newborns of copper-deficient animals show anemia and abnormal development of the brain (mainly cerebellum and cortical areas) and skeletal muscle. The skin shows lack of hair follicles and the texture of the hair is abnormal.

IV. EFFECTS ON ENDOCRINE ORGANS

Several physiological and biochemical interrelationships exist between trace metals and

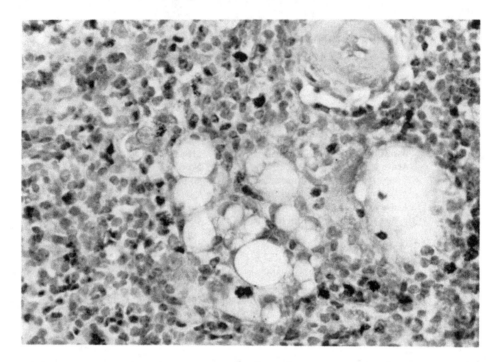

PLATE 1   Mineral oil deposit in spleen. Groups of mineral oil droplets are at the periphery of Malpighian corpuscles.

hormones. Many trace metals have been shown to influence hormones at several sites of their action including their secretory target tissue binding and activity. Conversely, several hormones have been found to alter the metabolism of trace elements at several levels including absorption, transport, and excretion (Figure 1).

Trace metals influence the endocrine system in different ways, depending upon whether the metal is in excess or depleted.[118,133] Excessive amounts of heavy metals such as mercury, lead, cadmium, copper, nickel, and iron cause an adverse effect.[73,132]

Excess intake occurs by gastrointestinal ingestion of contaminated food such as fish contaminated with a high level of mercury, paints containing lead, or following accidental or purposeful poisoning.[69,70] Some occupations or environmental exposure to fuels containing metal complexes may be associated with the hazard of inhaling excessive quantities. Insufficient intake of some metallic ions, such as manganese, zinc, and copper, elicits an impairment of the endocrine system. Deficiency may be linked generally with malabsorption, but increased loss of these metals can occur with intrinsic kidney disease or changes in the function of the glomerules or tubules, with defects in metal binding and transport by plasma proteins, on through inborn errors of amino-acidurias which may cause excessive loss of metals.

Lead poisoning adversely influences the pituitary-thyroid axis in patients. Iodine uptake and the conversion to the protein-bound form are retarded. Exogenous administration of thyrotropic hormone usually reverses this effect, suggesting that the excessive lead may act at both the pituitary and thyroid gland and impair thyroid function. Lead contamination also produces decreased secretion of other pituitary hormones and inhibits adrenal function. The responsiveness to exogenous adrenocorticotropic hormone is depressed. Excessive quantities of lead are injurious to germ cells in both sexes and may cause decreased fertility, spontaneous abortion, and stillbirths.

Incubation of the pituitary gland with copper has been shown to increase the release of thyrotropic, luteotropic, adrenocorticotropic, and growth hormones.[79] This experiment in-

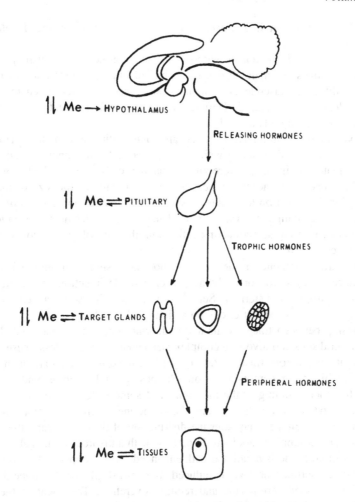

FIGURE 1. Interaction between trace metals and the endocrine system. Increasing or decreasing concentration of trace elements affects the secretion of releasing hormones in the hypothalamic-pituitary axis, the discharge of tropic hormones in the anterior pituitary-peripheral target gland axis, uptake of hormones into the target organs, the synthesis and secretion of hormones at several peripheral organs, and the action at target tissues including hormone binding and utilization. The interaction between the trace metals and various organs exists both ways, perhaps with the exception of the hypothalamus. No direct effect on metals has yet been shown by releasing factors produced by this organ.

dicates a direct action. Intravenous or intracerebral administration of copper into rabbits produces ovulation. The application of an intrauterine device containing copper has consistently caused infertility in women.[132] From the intrauterine device, the copper ions are released and local tissue absorption has been suggested as the contraceptive agent; its action may be due to a local failure in the implantation of the ovum into the uterus wall.[22] However, the action may be indirect since copper also affects the function of progesterone-binding protein.[14] Copper ions are found in the blood, liver, and kidney and may suggest a possibility for indirect action. Mercury, silver, and zinc also produce decreased progesterone binding.

Excessive quantities of nickel increase the release of the prolactin-inhibiting factor which in turn decreases the release of prolactin from the pituitary gland.[78,79] The hypothalamus contains relatively large amounts of nickel. This trace element has also been suggested to play a role in reproduction. High levels of nickel in the diet cause a decrease in the litter size in rats. In the male rat it produces seminal tubular shrinkage and disintegration of spermatocytes.

Manganese deficiency is associated with sterility and absence or irregularity of the estrus cycle in female animals. In males it causes decreased libido, degeneration of the seminal tubules, and sterility. Decreased levels of manganese adversely influence pancreatic function; oral glucose tolerance is impaired. Deficiency brings about pancreatic hypoplasia and a decreased number of beta cells and islets.

Reduced levels of zinc in animals and in man are linked with hypogonadism. This condition is associated with decreased testicular weight, decreased testicular function, and a decreased number of spermatozoa. In rats, severe zinc deficiency causes diminished growth and development of the testes and accessory sex organs and depresses the size of the pituitary. The effect of zinc depletion on testicular atrophy is irreversible; however, supplementation with zinc reverses the changes in other organs. The deficiency has no effect on testosterone and dihydroepiandrosterone secretion, indicating that the lack of zinc is connected with an extratesticular mechanism.

Several other trace elements influence the endocrine system. Chromium is essential in glucose tolerance. Excess iron or molybdenum or a single injection of cadmium produce marked testicular atrophy and sterility. Selenium in the drinking water adversely affects the reproduction of the female rat but elicits no action on the male.

The relationship between trace metals and endocrine functions is manifold. Interactions take place at several sites and involve a complex mechanism. Nevertheless, these interactions may throw light on the role trace metals may play in hormone action, in the structure-function relationship, the receptor-inhibition relationship, and receptor binding. They may contribute to the understanding of the mechanism of some endocrine diseases.

The interaction between trace metals and the endocrine function is reciprocal. Changing levels of circulating hormones may alter the distribution of these elements, their binding to ligands, and hepatic or renal metabolism. In patients with untreated acromegaly the increased circulating growth hormone is paralleled with enhanced serum copper level and urinary zinc excretion; however, serum zinc level is reduced. In contrast, growth-hormone deficiency is associated with increased serum zinc and reduced excretion. Treatment of these patients with growth hormone lowers zinc level and increases elimination. It seems that the fact that growth hormone levels are inversely related to zinc levels may be linked with an effect on the binding of zinc to protein ligands.

Estrogen enhances plasma copper concentration mainly by inducing the synthesis of ceruloplasmin, a copper-carrier protein.[14] These changes have been observed during pregnancy and following fertility drug treatment. Under these conditions serum iron level is also increased. This may be associated with the ferroxidase activity of ceruloplasmin.

Abnormalities of adrenal corticosteroid metabolism are linked with zinc and copper metabolism. Adrenalectomy and adrenal cortical insufficiency from several causes including Addison's disease are associated with increased serum copper and zinc concentration, increased tissue retention, and decreased urinary elimination. Exogenous administration of adrenal corticosteroids reverses these changes. Elevated endogenous production of these corticosteroids occuring in Cushing's syndrome or adrenal cortical carcinoma is related to decreased plasma copper and zinc level, decreased tissue accumulation, and elevated urinary excretion. Treatments are associated with return to the normal serum level and a reverse in urinary excretion and tissue concentration.

V. EFFECTS ON INFECTIOUS DISEASES

Infections affect the metabolism and tissue content of many trace elements, particularly serum concentrations and the homeostasis of iron, copper, and zinc.[12,104,114] Other trace elements may also be affected.[20,193] Depressed serum iron levels accompany most infectious diseases, including those caused by bacteria, viruses, and parasites. Iron level starts to decline within hours of invasion by the microorganisms, in some cases even earlier, before any clinical symptoms are apparent.[15] The reduction of serum iron concentration is most marked in generalized bacterial infection especially when accompanied by a pronounced inflammatory process.

There is a slow decline in serum iron binding, but in most acute infections iron is accumulated in tissue depots, particularly in the liver, bone marrow, and the reticuloendothelial system. The movement of the serum iron into hepatic ferritin stores occurs at early stages of infection, whereas the storage in hemosiderin in the liver, bone marrow, and spleen is associated with late phases.[93] The transfer from serum-binding proteins is facilitated by receptors located on the cell surfaces of reticulocytes. This process is energy dependent and requries ATP and ascorbic acid. Ascorbic acid is necessary to reduce the transferrin-bound ferric ion to the ferrous form.

Parallel with the decrease of serum iron concentrations, the amount of zinc is also diminished during generalized infections and during inflammatory reactions.[6,108,109] Zinc is mostly accumulated in the liver, although no storage forms are found comparable to those of iron. In contrast to these two elements, serum copper values are increased in infectious or inflammatory states, mainly due to the presence of the cuproprotein ceruloplasmin. Actually, the acute infection or inflammation stimulates the hepatic synthesis and release of ceruloplasmin. The increase of serum copper starts at the early stages of the infection but lags behind the changes of serum iron and zinc. Moreover, the increased serum copper and ceruloplasmin levels persist longer during the convalescence period than zinc and iron levels.

Changes occur in the association between these ions and their specific and nonspecific binding proteins.[122] Ceruloplasmin synthesis is raised parallel with the increased copper concentration, but there is no change in iron- and copper-binding proteins during infection. The accumulation of trace elements in the inflammatory areas and loss from necrotic cells are connected with the pathological changes. However, it has been suggested that the accelerated output of copper-ceruloplasmin complex from the liver and the flux of iron and zinc into the depot organ represent a physiologic response triggered by the action of leukocyte endogenous mediator.[65,66,107] The leukocyte endogenous mediator exerts its major effect on the liver cell. It is a heat-stable metalloprotein of small molecular weight and probably identical with the endogenous pyrogen. The endogenous pyrogen is derived from phagocytozing leukocytes or from macrophages, monocytes, and polymorphonuclear leukocytes and is released into the serum.[86,106] By acting on the temperature-regulating centers in the hypothalamus, it initiates the fever response. In response to this release, elevated amounts of leukocytic endogenous mediator circulate in the blood during febrile infectious or inflammatory conditions.

Leukocytic endogenous mediator exerts two actions on hepatocytes (Figure 2). Initially, on the transport sites it stimulates the cellular uptake of zinc, iron, and large amounts of amino acids from the serum, then enhances the synthesis of nuclear and ribosomal RNA and increases the production of copper-ceruloplasmin complex and other acute phase reactant proteins by the endoplasmic reticulum. These proteins are released from hepatocytes into the serum as a response to inflammation. The effect of leukocytic endogenous mediator on the liver is similar to a hormone action; the effect on protein synthesis is dependent on the presence of adrenal glucocorticoids. These effects on acceleration of protein synthesis are probably mediated by a mechanism involved with both the hepatocyte nucleus and rough endoplasmic reticulum.

VI. EFFECTS ON CELLULAR MEMBRANES

Zinc ions are important in preserving the integrity of the cells and tissues.[143] Zinc acts as a nonspecific mitogen in lymphocytes; the enhanced mitosis is related to the increased activity

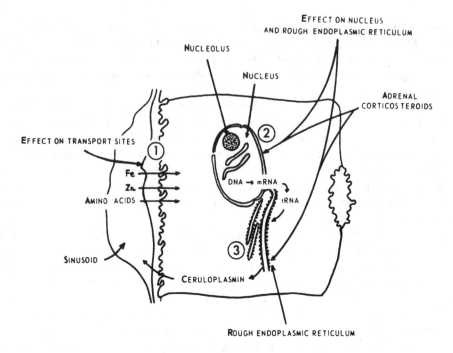

FIGURE 2.   The effect of leukocytic endogenous mediator on hepatocytes. Following acute infectious diseases this factor alters the serum concentration and metabolic homeostasis of several trace elements. The initial action of the leukocytic endogenous mediator on the membrane transport site is an increased influx of iron, zinc, and several amino acids into the cell (1). A subsequent effect is a stimulation of nuclear and ribosomal RNA synthesis (2), which subsequently affects the endoplasmic reticulum (3) and enhances the synthesis and release of ceruloplasmin and various other proteins into the serum. Adrenal glucocorticoids probably play a role in the protein-synthesizing action of the leukocytic endogenous mediator.

of DNA polymerase and reverse-transcriptase, which are zinc-containing enzymes.[126] Several trace elements, such as manganese and copper, elicit inhibitory action on the cadmium, nickel, and cobalt in lymphocyte culture system.

The disruption of mast cells by antigen-antibody reaction of lecithinase A treatment can be inhibited by zinc. This ion inhibits the spontaneous release of histamine from isolated mast cells, probably by acting on the cell membrane. A similar mechanism has been suggested for the stabilization of the membranes of hepatic lysosomal vacuoles by zinc.

The function of the platelets is affected by zinc ions. Other ions are also important: for example, calcium is required for contraction and release mechanisms. Manganese ions inhibit platelet aggregation, probably by interacting with calcium ions. Magnesium is also a competitor of calcium action and is important in the process of deaggregation. The viability and functional activity of macrophages, polymorphonuclear leukocytes, and probably other cells are dependent on zinc ions.

Although these various cells show morphologic, functional, and genetic heterogeneity, zinc ions may act by a common mechanism (Figure 3). They may interact with some functional groups of the plasma membrane such as sulfhydryl groups of proteins, carboxyl groups of sialic acid, and proteins and phosphate groups of phospholipids.[89] This binding alters membrane fluidity or stabilization. The activity of several enzymes which control membrane structure and function, such as adenosine triphosphatase, 5'-nucleotidase, and phospholipase A_2, is influenced by zinc. Several receptor sites may exist on plasma membrane which function as gates for transmitting the information to the intracellular space and where histamine- or serotonin-releasing agents elicit their action. Zinc inhibits the oxidation of

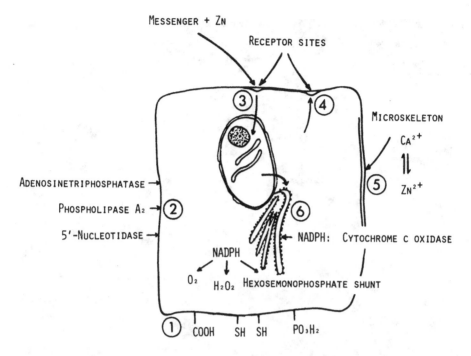

FIGURE 3. Site of action of zinc on the activities of various cells. Zinc can interfere with cellular functions by binding to components of the plasma membrane (1), or by inhibiting the activity of essential enzymes (2). At special receptor sites it may initiate the transmission of extracellular information into the cell (3) and the release of granular content into the extracellular space (4). It may influence the function of the contractile elements of the microskeleton by competing with the essential calcium ions (5) and may modify NADPH oxidation and related functions of the endoplasmic reticulum (6).

NADPH by the endoplasmic reticulum. This may represent the actual mechanism by which zinc ions inactivate various functions of macrophages, monocytes, and polymorphonuclear leukocytes.

It has been shown that contractile elements of microtubules and microfilaments are in some way responsible for the mobility of cellular organelles and transport of granules to the membrane, as well as excitability of the plasma membrane itself.[27] In these functions calcium ions play an essential role with phosphatidylserine suggested as a binding site. Chelation of these ions interferes with the function of the cellular microskeleton. Moreover, displacement of calcium by zinc may cause a change in the molecular structure of plasma membrane, resulting in deterioration of related functions.[7,155] Aluminum can impair bone mineralization.[31,40,56,117]

VII. DISEASES ASSOCIATED WITH SELECTED ELEMENTS

A. Iron

This element is essential in the function of respiratory pigments including hemoglobin.[5,42] The level of the latter provides the most useful test for the iron status in man. Lack of iron causes deficiency diseases, most commonly anemia, in man. This is largely due to nutritional inadequacies, to the lack of biological availability of food iron, and to the choice and preparation of diet.[30,83] In addition, physiological loss of iron during menstruation may be a contributing factor. In children anemia is usually an iron responsive disorder. In adults anemia may occur at different hemoglobin levels and may arise from other causes such as copper or some vitamin B deficiency. Iron can be utilized for the production of respiratory pigments only if these is adequate dietary copper.

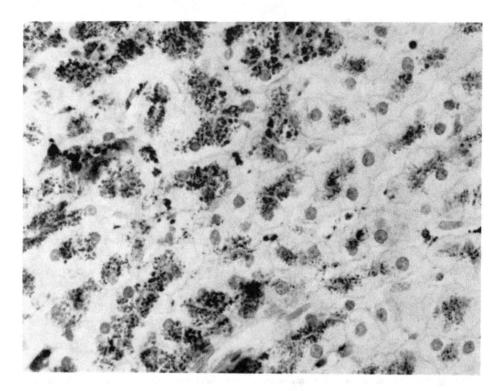

PLATE 2   Hemochromatosis in liver. Iron particles are deposited in the reticuloendothelial cells and hepatocytes.

In the body virtually all iron is present in bound form. It is stored in tissues as a mixture of ferritin and hemosiderin.[62,88] Increased digestion of iron increases the hemosiderin proportion of the store, but excess intake can cause toxic action resulting even in tissue damage. These effects are associated with familial hemochromatosis occuring in persons who use iron cooking utensils for food preparation (especially for fermenting beverages) or in people who consume a diet high in iron, largely due to contamination. The latter condition is frequent in Ethiopia and is linked with the consumption of the staple cereal teff. Excess deposition of iron particles causes hemochromatosis in the liver (Plate 2). This condition may be associated with cirrhosis (Plate 3).

B. Copper

Copper is an essential micronutrient and there are many important copper proteins in various organs.[16,44,48,143] These are copper-dependent enzymes, and most of them catalyze reactions of physiologic significance; therefore, copper deficiency leads to many abnormal functions resulting in specific pathologic lesions. Some cuproenzymes have physiologic importance but cannot be linked with specific pathological changes. The function of several others has not been established. The catalytic activity of cuproenzymes is associated with oxidation and they are localized in various tissues such as albocuprein I and II in the brain, ceruloplasmin, spermine oxidase in the plasma, diamine oxidase in the kidney, uricase in liver and kidney, cytochrome *c* oxidase, ferroxidase II, lysyl oxidase, tyrosinase, dopamine-β-hydroxylase, monoamine oxidase in mitochondria and other cellular organelles of many organs such as brain, erythrocytes, heart, aorta and cartilage, adrenal gland, melanomas, and skin.[76,149] Some copper-containing proteins occur in the cytoplasm such as cerebro-, erythro-, and hepatocuprein.

Since almost any diet contains the necessary daily requirement, copper deficiency is a rare disorder. Infant hypocupremia has only been found in a marked number of cases in

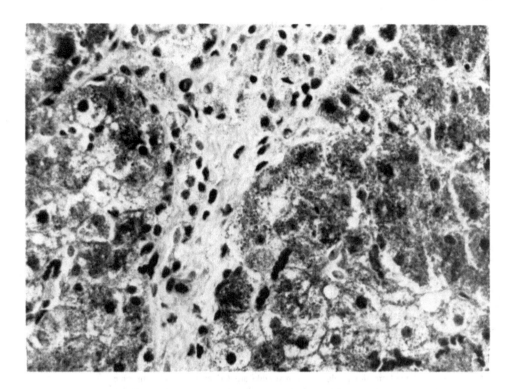

PLATE 3   Cirrhosis of the liver in a patient with hemochromatosis. A wide fibrosis band separates nodules of regenerating hepatocytes that contain dense deposits of iron particles.

Peru. Due to chronic diarrhea or malabsorption the clinical symptoms linked with copper deficiency in man are anemia or inherited conditions associated with Menkes' syndrome.[32,94,113] Most cases of copper-deficiency anemia occur in children, but prolonged intravenous alimentation can lead to the development of anemia in adults because of the lack of copper in the perfusate. The cause of anemia is the inadequate production of a copper protein, ceruloplasmin, which possesses ferroxidase activity. Iron is absorbed as ferrous ion and must be converted to the ferric ion to be transported by transferrin, the plasma iron-binding protein. The condition of anemia can be corrected by restoring normocupremia. Another cuproprotein, ferroxidase II, has been suggested as being responsible for the maintenance of normal absorption and transport of iron since the blood of some patients with Wilson's disease contains a very low amount of ceruloplasmin and has normal hemoglobin production.[48,129,131,145,152,156]

Menkes' steely hair syndrome is a sex-linked, inherited progressive fatal brain disease in infants.[32] The encephalopathy involves most of the central nervous system. Hypothermia and rapid mental deterioration cause early death. Due to the failure of copper absorption and transport, the hypocupremia can be detected by the very low levels of serum copper and ceruloplasmin and in spite of adequate dietary copper, the nervous tissue is almost devoid of the cuproenzyme cytochrome c oxidase. The steely structure of the hair is related to the presence of more free sulfhydryl groups and less disulfide bonds. The metabolic relationship is complex since the severe deficiency, abnormalities of the hair, and deficiencies of the copper-protein production apparently cannot be prevented by copper administration.

Lack of copper is also associated with spontaneous fractures and cardiac failure in domestic and experimental animals. This is related to an impairment of collagen cross-linking by copper deficiency.[105] The integrity of the vascular system, particularly of the large arteries, depends largely on the amount and quality of collagen and elastin present in the wall of the

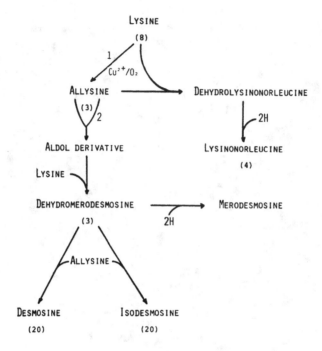

FIGURE 4.   Biosynthesis of elastin and collagen cross-linkages. In
the formation of elastin all cross-links are derived from lysine. In some
collagen, cross-links with hydroxylysine also occur; this amino acid is
absent in elastin. Lysyl oxidase (1) in the presence of molecular oxygen
converts lysine or hydroxylysine residue to an aldehyde derivative
allysine or hydroxyallysine. Condensation (2) with further lysine or
allysine residues leads to the formation of desmosine or isodesmo-
sine.[160,164] The numbers in brackets indicate the number of lysine residues involved in the cross-linkages per peptide units containing 1000
amino acids.

vessels. Elasticity and tensile strengths of these connective-tissue proteins are, however,
determined by intra- and intermolecular cross-linkages of the constituent polypeptide chains.
The presence of copper is essential in cross-linking, and is markedly reduced by copper
deficiency.[140] The cross-links in collagen and elastin are derived from lysine and involve
the oxydation of peptide-bound lysine to an aldehyde derivative, allysine, catalyzed by lysyl
oxidase, a copper-dependent enzyme (Figure 4). Further steps via condensation and reduction
procedure cross-linkages are formed containing the pyridinium ring (Figure 5). Lysyl oxidase
activity can be inhibited by β-imino-propionitrile, the lathyrogen found in sweet pea seeds.
This explains the production of pathological connective tissue by lathyrogens.

Atherogenesis is affected by copper deficiency.[98,99,103] The unsaturation of dietary fat and,
consequently, fatty acid composition are altered by copper level. Hypercholesterolemia can
be produced by an increase of the ratio zinc to copper in the diet;[71] the relative ratio of
tissue copper to lead content also shows some association with ischemic heart disease.

The toxicity of copper is relatively common in animals due to the very narrow threshold
between the toxic and required levels — 2:1 — as compared to that of some other essential
elements where it is greater than 100:1. Accidental or suicidal ingestion of copper causes
vomiting and gastritis. If the intake increases beyond complete protein binding that is greater
than 1 g, then shock, hepatic necrosis, renal toxicity, hemolysis, coma, and finally death
may occur.[139] Copper toxicity has been seen in patients with Wilson's disease.[158] At birth
children have low serum concentration of ceruloplasmin and high hepatic concentration of
copper. In normal children a physiologic increase of ceruloplasmin and a reduction of liver

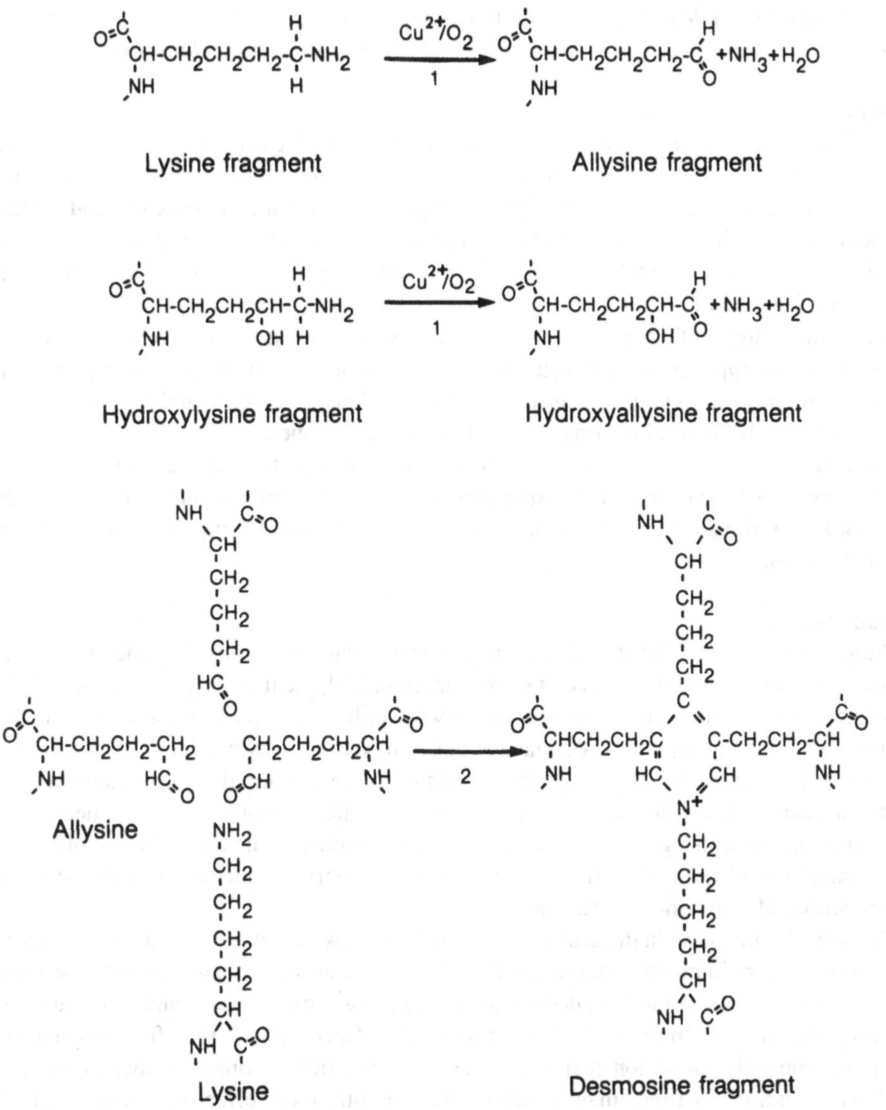

FIGURE 5. Molecular mechanisms of desmosine formation. Lysyl residue is oxydized to allysyl residue catalyzed by lysyl oxidase (1). One lysine and three allysine residues are further converted to desmosine, a pyridinium derivative, through an oxidative step, condensation, and reduction reaction (2).[165,167]

copper content occur. However, infants with Wilson's disease maintain the abnormal concentrations throughout childhood and adult life due to the inherited defect in homeostatic mechanism.[3,116] Usually, the excess fraction of dietary copper is excreted through hepatic lysosomes into the bile; but in Wilson's disease this route and mode of excretion are defective. Subsequently, excessive accumulations of copper produce inflammatory changes, fatty deposition, cellular necrosis, and postnecrotic cirrhosis. In association with the structural changes, functional abnormalities become evident, and generally serum transaminase levels are pronounced. The binding capacity of the liver for copper is exhausted and if it suddenly leaves the liver, the high concentration of copper which is free or only loosely bound to albumin causes hemolytic anemia. If the release is more gradual, it is deposited throughout the body and produces toxic effects in susceptible cells such as cerebral cortical grey matter, basal ganglia, and spinal cord. The consequence of copper manifestations in the nervous system will be described in Volume 3. Deposits of copper in the kidney cause abnormalities of glomerular and tubular function which occasionally lead to nephrocalcinosis.

C. Zinc

Zinc is an essential element for the growth and development of most living cells.[16,51-53,55,75,114,120,141,147,157] Many zinc enzymes are known which regulate a variety of metabolic processes including steps in protein, lipid, carbohydrate, and nucleic acid synthesis and degradation. Zinc is not only an integral part of the catalytic property of these enzymes but also a structure stabilizing factor.[4,24,25] It is present in DNA and RNA, probably as a structural component.

The clinical signs of zinc deficiency occur most prominently in growing children and including loss of appetite, poor growth rate, loss of taste and smell senses, and hypogonadism in males. In adults the symptoms of lack of zinc are loss of appetite and loss of the senses of taste and smell. Wound healing is retarded by zinc deficiency.[104]

The action of excess zinc in man is not known. The ingestion of great amounts of zinc causes poor growth and anemia in experimental animals. Accidental intake is accompanied by nausea, vomiting, and diarrhea. Corrosive zinc compounds bring about ulceration and severe necrosis.

D. Manganese

Manganese is widely distributed in trace amounts in various tissues.[16,130] Manganese plays an essential role in the maintenance of some important biological functions. Lack of sufficient manganese results in an impaired rate of growth, failure to reproduce, and shortened life span in experimental animals. Manganese deficiency is also associated with postural and skeletal defects (porosis, ataxia), probably related to an abnormality of collagen and mucopolysaccharide formation in connective tissues. Other symptoms include neurological impairments, abnormal glucose metabolism in the pancreas, impaired blood clotting, and mitochondrial oxidation. Mitochondrial damage and abnormal function are well-established consequences of manganese deficiency.

The role of manganese in these disorders is probably linked to its association with enzymes. Manganese occurs in many metalloproteins and may act as a dissociable cofactor of various enzymes (phosphatases, kinases, decarboxylases, glycosyl transferases, and some peptidases and dehydrogenases). In most of these enzymes the effects are not specific for manganese; this is not the only metal ion that stimulates these reactions. Sometime manganese is the most active, but many times magnesium is an apparently more effective activator. In other cases nickel, cobalt, and zinc can replace manganese. In contrast to these enzymes, manganese is built rather firmly into several proteins, including a transport protein in the serum (manganese-transferrin) and enzymes occuring in the liver of many species (pyruvate carboxylase, superoxide dismutase, and avimanganin). Adenylcyclase is also activated by manganese. In pyruvate carboxylase this metal ion can be replaced by magnesium; however, it may play a specific role in superoxide dismutase. Superoxide dismutases are very commonly distributed enzymes; some function with copper and zinc. These enzymes remove superoxide radical or closely related substances which can cause widespread damage to a variety of cells and intracellular organelles.

The importance of manganese ions in the biosynthesis of mucopolysaccharides has been implied. They have an essential activator role in glycosyltransferases associated with the biosynthesis and metabolism of chondroitin sulfate (Figure 6), and probably other similar substances. During manganese deficiency, defects in connective tissues are apparent. Glycosyltransferases are involved in the synthesis of glycoproteins in a wide variety of conditions related to many biological functions including blood coagulation, hormone action, and membrane function. Several impairments of these functions have been seen during manganese deficiency.

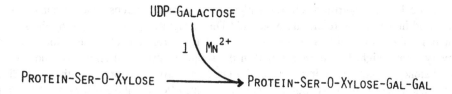

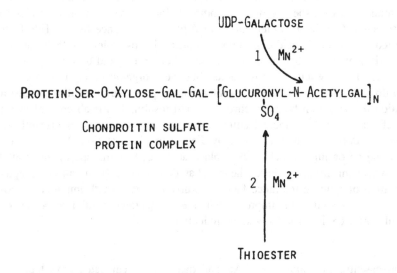

FIGURE 6. Biosynthesis of chondroitin sulfate. Manganese is very effective in promoting the incorporation of carbohydrates into the mucopolysaccharide structure. The synthesis involves the stepwise binding of two galactose molecules catalyzed by galactosyl-transferase (1) followed by the addition of the main carbohydrate portion, a repeating unit of glucuronate and N-acetylgalactosamine joined together by manganese-activated transferase (1). The sulfate moiety of this part of the structure is attached by thioesterase (2) in conjunction with manganese ions as enzyme activators.[159,162,163,166,168]

Chronic manganese poisoning causes some effects on the nervous system, indicating that the brain is susceptible to both manganese deficiency and excess. In man toxic effects have developed in workers exposed to manganese ores or fumes. The symptoms begin with a psychiatric disturbance resembling schizophrenia; the later neurologic phase is similar to Parkinson's disease. Levodopa treatment actually brings about improvement, which suggests the involvement of the dopaminergic apparatus. In experimental animals chronic manganese intoxication causes low striatal dopamine levels. These experiments have also shown that the brain of the neonate is enriched with manganese which is linked with a significant increase of cerebral dopamine.

E. Chromium

Chromium is essential for the maintenance of normal glucose tolerance.[148] A mild degree of chromium deficiency results in a lowering of glucose assimilation. Autopsy samples have shown a steady decline of chromium with age, in marked contrast to the known accumulation of other elements, particularly certain heavy metals from enviromental contaminations. Based on the analysis of the hair chromium content, repeated pregnancies are associated with significant decrease of chromium stores compared to nulliparae. Juvenile diabetics have also been found to have signicantly lower chromium level in hair than normal controls.

Protein-calorie malnutrition is linked with severely impaired glucose tolerance in children, as reported in Turkey, Jordan, and Nigeria.[50] This is significantly improved by the supplementation of the diet with a single dose of 250 μg chromium chloride. Moreover, the recovery from weight loss is more rapid in the supplemented children than in those given protein alone, indicating the importance of a marginal chromium content in the food. Since the animal protein is the best source of chromium, insufficient intake can cause chromium-deficiency symptoms.

In the U.S. adults with impaired glucose tolerance responded to chromium supplementation with a considerable lag period. The lag phase represents the time necessary to convert inorganic chromium salts to a specific organic complex. Incubation with yeast accelerated formation of the active form of this element, named glucose tolerance factor. This factor has been identified as a nicotinic acid-chromium complex. Glucose tolerance factor activity is present in a variety of foods such as meat, cheese, whole grain, and brewer's yeast.

Glucose tolerance factor elicits improvements, but the inorganic chromium cannot be considered as a drug. Its action is always dependent on the presence of insulin, suggesting a close association in the action between chromium and insulin. In the absence of insulin chromium is ineffective. In experimental animals the insulin response is significantly depressed in chromium deficiency and increased by dietary supplementation. The interaction between insulin and chromium probably takes place at the cell membrane of the insulin-sensitive tissue, where chromium increases the initial association of insulin with its receptor. The supplementation of glucose tolerance factor to human subjects with impaired glucose tolerance brings about a significant improvement; the exaggerated insulin response to a glucose load and serum cholesterol show some reduction.

F. Selenium

Selenium is an essential micronutrient.[144] An inadequate supply can cause growth defects, liver necrosis, muscular dystrophy, pancreas atrophy, and calcifications.[77] Selenium deficiency probably plays an important role in the development of kwashiokor in children suffering from protein malnutrition. Many of its effects are related to vitamin E, but nutritional studies have shown that selenium supplementation can reverse the pathologic changes independent of the action of tocopherol.[29,134]

Dietary liver necrosis develops in the absence of selenium, due to the reduction of hepatic glutathione peroxidase levels. Selenium constitutes the active site of this enzyme. A single injection of 50 μg selenite per 100 g body weight is sufficient to restore normal enzyme level within 1 to 2 days. The time lag shows that selenium is not simply incorporated, but new enzyme is synthesized *de novo*. In dietary hepatic necrosis the action of tocopherol is related to that of selenium; they are probably essential substances for alternate pathways of intermediary peroxide metabolism. These factors function probably as part of glutathione peroxidase which has a great importance in the metabolism of the kidney, erythrocytes, lens of the eye, and endothelial lining of blood vessels. This enzyme utilizes peroxide to oxidize reduced glutathione.

A number of enzyme systems are involved in the normal production of hydrogen peroxide.[124] Peroxides may play an active role in the facilitation of oxygen transport. Cytochrome P-450 occurs in microsomes in peroxide form and its level depends on the amount of selenium present in the diet.[17] Selenium deficiency causes an inhibition of hepatic drug metabolism, perhaps through the regulation of cytochrome P-450 level. Glutathione peroxidase is also an important enzyme in the biosynthesis of prostaglandins. It converts prostaglandin G, which contains a hydroperoxide group, to protaglandin E and F. These prostaglandins are essential for the maintenance of normotension. Glutathione peroxidase activity is disturbed, and prostaglandin conversion is impaired in high blood pressure. If peroxides are not metabolized readily, highly reactive free hydroxy radical and singlet oxygen are formed which may destroy tissue membranes and other components. There is a theory that aging is related to damage caused by free radicals; therefore, the functions of selenium and vitamin E may be linked with an increase of the cellular life span.

In areas where selenium levels are high in the soil, plants can accumulate large amounts in the form of selenocysteine, methylselenocysteine, and selenocystathionine. These selenium-amino-acids are incorporated into the proteins of grazing animals and cause the so-called "alkali disease".

G. Mercury

The presence of mercury as a toxic metal is mainly due to the conversion of its salts by microorganisms to methylmercury which goes through the food chain and appears in our diet.[10,72,119] Mercury salts form complexes with sulfhydryl groups and inactivate many essential biological substances such as cysteine and glutathione, inhibit several enzymes, and elicit deleterious action on the function of cellular membranes.

Metallic mercury is also toxic since it vaporizes at room temperature. Vapors cause pneumonitis if the exposure is intense. Mercury chloride is corrosive to gastrointestinal mucosa. Mercury salts and organic mercury compounds are usually concentrated in the kidney.[10] Some of them have been used as diuretics. Consumption of grain treated with pesticide containing mercury has been associated with the outbreak of mercury poisoning in man. An alkylmercury content of some fish has been suggested to be responsible for neurological damage characteristic of Minamata disease.[138]

The absorption of methylmercury is about double that of inorganic mercury salts. The biological half-life of alkylmercury in man is more than 2 months: its effect is therefore more pronounced than that of inorganic mercury. Children born to women exposed to mercury may have severe neural damage.[74]

H. Lead

Lead is a trace element that is highly toxic to biological systems. Acute or chronic lead toxicity causes damage to the kidney and to the nervous and hematopoietic systems.[23,72,98,112] Brain damage may show edematous swelling and the toxic action of lead may specifically manifest in the transmission between nerve and muscle. Lead impairs renal tubular transport and induces anemia.[115] Prolonged exposure leads to irreversible kidney insufficiency. The anemia is related to an abnormal porphyrin synthesis; chronic action of lead causes increased urinary output of δ-aminolevulinic acid.

The symptoms of lead poisoning are headache and nervous disorders including convulsions, acute abdominal colic pain, and coma in severe cases. Children who have ingested paints containing lead display irritability and fatigue, vomiting, and apathy. Intense exposure to lead elicits mental retardation.[69] The action of lead is enhanced by low-calcium-containing diets. In the mechanism of lead toxicity, probably the blocking of sulfhydryl groups is essential. Irreversible binding of lead to proteins interacts with the function of copper and iron essential in the activity of specific enzymes which play important roles in biological functions.

I. Cobalt

The metal cobalt is involved in nitrogen fixation in some plants and is required by man as a component of vitamin B_{12}. Deficiency of vitamin B_{12} in man is associated with pernicious anemia. Ruminants do not need this vitamin since microorganisms in their digestive tract can synthesize it if cobalt is present in the feed. However, they develop anemia grazing on cobalt-deficient soil. This syndrome is called wasting disease or enzootic marasmus and is noted by loss of appetite. The occurrence of this syndrome in New Zealand, Australia, Florida, and Michigan is due to cobalt deficiency and can be alleviated by direct supplementation. Excess cobalt affects iodine metabolism, and may cause polycythemia and associated hyperplastic marrow.[136] The addition of cobalt to beer to enhance foam production has been reponsible for the cardiomyopathy of beer drinkers.

J. Nickel

The toxic nature of nickel compounds was established long ago. Several enzyme systems function with nickel but can substitute other metals.[78] Nickel is consistently present in RNA and probably plays a role in the formation of membranes through phospholipid synthesis. Deficiency in experimental animals results in dermatitis, reduced pigmentation, and lower hepatic lipid content. Supplementation of nickel reverses these symptoms and lowers serum cholesterol.

Industrial exposure to dust or electroplating materials containing nickel may cause toxicosis. Lesions include a dermatitis, known as "nickel itch". Some nickel compounds may be carcinogenic; nickel carbonyl induces pulmonary cancer.

K. Cadmium

Cadmium is a ubiquitous element resulting in about 30 mg accumulation during a lifetime. It has a unique tendency to concentrate in the human kidney.[1,81,110] Renal cadmium concentrations are abnormally high in hypertensive human beings and the low-level long-term exposure to cadmium may be associated with essential hypertension. Despite many efforts to elucidate the pathogenesis, the mechanism by which cadmium raises the arterial blood pressure still remains uncertain; it can be related to renal sodium retention or to a direct effect on vascular smooth muscle.[20] In cadmium-induced hypertension mercury contamination also may be involved.

Heavy industrial exposure to cadmium brings about both acute and chronic toxicosis in man.[8,21,46,72] An acute effect of ingestion is connected with severe gastroenteritis; inhalation is associated with severe, sometimes fatal pulmonary edema. Prolonged exposure in combination with low calcium and vitamin D intake can produce moderate anemia, renal dysfunction characterized by proteinuria, aminoaciduria, glycosuria, and osteomalacia.

There are some data that involve cadmium in the pathogenesis of emphysema. The pulmonary and hepatic cadmium content is increased in emphysematous subjects as compared with subjects having normal lungs.

L. Molybdenum

This trace element is considered essential since it is part of a number of metalloenzymes including xanthine oxidase. Molybdenum deficiency in man has not yet been demonstrated, but this may provide the basis for the occurrence of xanthine-containing kidney stones in experimental animals kept on a molybdenum-deficient diet.

REFERENCES

1. **Adams, R. G., Harrison, J. F., and Scott, P.,** The development of cadmium-induced proteinuria impaired renal function and osteomalacia in alkaline battery workers, *Q. J. Med.*, 38, 425, 1969.
2. **Agarwal, B. N. and Agarwal, P.,** Magnesium deficiency in clinical medicine — a review, *J. Am. Med. Women's Assoc.*, 31, 72, 1976.
3. **Aisen, P., Schorr, J. B., Morell, A. G., Gold, R. Z., and Scheinberg, I. H.,** A rapid screening test for deficiency of plasma ceruloplasmin and its value in the diagnosis of Wilson's disease, *Am. J. Med.*, 28, 550, 1960.
4. **Aitken, J. M.,** Factors affecting the distribution of zinc in the human skeleton, *Calcif. Tissue Res.*, 20, 23, 1976.

5. **Alevizou-Terzaki, V., Gyftaki, H., and Vrettos, A. S.**, Serum iron-binding capacity in pregnancy, *Act Haematol.*, 45, 232, 1971.
6. **Askari, A., Long, C. L., and Blakemore, W. S.**, Urinary zinc, copper, nitrogen and potassium losses in response to trauma, *J. Parenteral Enteral Nutr.* 3, 157, 1979.
7. **Barlow, P. J., Sylvester, P. E., and Dickerson, J. W.**, Hair trace metal levels in Down syndrome patients, *J. Ment. Defic. Res.*, 25, 161, 1981.
8. **Bernard, A., Buchet, J. P., Roels, H., Masson, P., and Lauwerys, R.**, Renal excretion of proteins and enzymes in workers exposed to cadmium, *Eur. J. Clin. Invest.*, 9, 11, 1979.
9. **Bhat, K. R., Arunachalam, J., and Yegna Subramanian, S.**, Trace elements in hair and environmental exposure, *Sci. Total Environ.*, 22, 169, 1982.
10. **Bogden, J. D., Kemp, F. W., Troiano, R. A., Jortner, B. S., and Timpine, C.**, Effect of mercuric chloride and methylmercury chloride exposure on tissue concentrations of six essential minerals, *Environ. Res.*, 21, 350, 1980.
11. **Borhani, N. O.**, Exposure to trace elements and cardiovascular disease, *Circulation*, 63, 260A, 1981.
12. **Bragt, P. C. and Bonta, I. L.**, Role of trace elements in the pathogenesis and treatment of inflammation, *Agents Actions Suppl.*, 8, 231, 1981.
13. **Bremner, I. and Mills, C. F.**, Absorption, transport and tissue storge of essential trace elements, *Philos. Trans. R. Soc. London (Biol.)*, 294, 75, 1981.
14. **Briggs, M., Austin, J., and Staniford, M.**, Oral contraceptives and copper metabolism, *Nature (London)*, 225, 81, 1970.
15. **Bullen, J. J., Rogers, H. J., and Griffiths, E.**, Iron binding proteins and infection, *Br. J. Haematol.*, 23, 389, 1972.
16. **Burch, R. E., Hahn, H. K. J., and Sullivan, J. F.**, Newer aspects of the roles of zinc, manganese and copper in human nutrition, *Clin. Chem.*, 21, 501, 1975.
17. **Burk, R. F. and Masters, B. S.**, Some effects of selenium deficiency on the hepatic microsomal cytochrome P-450 system in the rat, *Arch. Biochem. Biophys.*, 170, 124, 1975.
18. **Capel, I. D., Pinnock, M. H., Williams, D. C., and Hanham, I. W.**, The serum levels of some trace and bulk elements in cancer patients, *Oncology*, 39, 38, 1982.
19. **Caprio, R. J., Margulis, H. L., and Joselow, M. M.**, Lead absorption in children and its relationship to urban traffic densities, *Arch. Environ. Health*, 28, 195, 1974.
20. **Caprino, L., Togna, G., and Togna, A. R.**, Cadmium-induced platelet hypersensibility to aggregating agents, *Pharmacol. Res. Commun.*, 11, 731, 1979.
21. **Carroll, R. E.**, The relationship of cadmium in the air to cardiovascular disease death rates, *J. Am. Med. Assoc.*, 198, 177, 1966.
22. **Chang, C. C., Tatum, H. J., and Kincl, F. A.**, The effects of intrauterine copper and other metals on implantation in rats and hamsters, *Fertil. Steril.*, 21, 274, 1970.
23. **Chisolm, J. J.**, Aminoaciduria as a manifestation of renal tubular injury in lead intoxication and a comparison with patterns of aminoaciduria seen in other diseases, *J. Pediatr.*, 60, 1, 1962.
24. **Chvapil, M., Ryan, J. N., and Elias, S. L.**, Protective effect of zinc on carbon tetrachloride-induced liver injury in rats, *Exp. Mol. Pathol.*, 19, 186, 1973.
25. **Chvapil, M., Sipes, I. G., and Ludwig, J. C.**, Inhibition of NADPH oxidation and oxidative metabolism of drugs in liver microsomes by zinc, *Biochem. Pharmacol.*, 24, 917, 1975.
26. **Clayton, B. E.**, Clinical chemistry of trace elements, *Adv. Clin. Chem.*, 21, 147, 1980.
27. **Cohn, D. V., MacGregor, R. R., Chu, L. L. H., Huang, D. W. Y., Anast, C. S., and Hamilton, J. W.**, Biosynthesis of proparathyroid hormone and parathyroid hormone, Chemistry, physiology and role of calcium in regulation, *Am. J. Med.*, 56, 767, 1974.
28. **Combs, D. K., Goodrich, R. D., and Meiske, J. C.**, Mineral concentrations in hair as indications of mineral status: a review, *J. Anim. Sci.*, 54, 391, 1982.
29. **Combs, G. F., Jr., Noguchi, T., and Scott, M. L.**, Mechanism of action of selenium and vitamin E in protection of biological membranes, *Fed. Proc.*, 34(11), 2090, 1975.
30. **Cook, J. D.**, Absorption of food iron, *Fed. Proc.*, 36, 2028, 1977.
31. **Cournot-Witner, G., Zingraff, J., Plachot, J. J., Escaig, F., and Lefevre, R.**, Aluminum localization in bone from hemodialyzed patients, *Kidney Int.*, 20, 375, 1981.
32. **Danks, D. M., Campbell, P. E., and Steven, B. J.**, Menkes' kinky hair syndrome. An inherited defect in copper absorption with widespread effects, *Pediatrics*, 50, 188, 1972.
33. **Deftos, L. J., Roos, B. A., and Parthemore, J. G.**, Calcium and skeletal metabolism, *West. J. Med.*, 123, 447, 1975.
34. **Deur, C. J., Stone, M. J., and Brandt, H. P.**, Trace metals in hematopoiesis, *Am. J. Hematol.*, 11, 309, 1981.
35. **Dickerson, J. W.**, Vitamins and trace elements in the seriously ill patient, *Acta Clin. Scand. Suppl.*, 507, 144, 1981.

36. **Di Paolo, R. V., Kanfer, J. N., and Newberne, P. M.,** Copper deficiency and central nervous system. Myelination in the rat: morphological and biochemical studies, *J. Neuropathol. Exp. Neurol.,* 33, 226, 1974.
37. **Doyle, J. J.,** Toxic and essential elements in bone — a review, *J. Anim. Sci.,* 49, 482, 1979.
38. **Dresner, D. L., Ibrahim, N. G., Mascarenhas, B. R., and Revere, R. D.,** Modulation of bone marrow, heme and protein synthesis by trace elements, *Environ. Res.,* 28, 55, 1982.
39. **Dunford, D. E.,** Pica and mineral status in the mentally retarded, *Am. J. Clin. Nutr.,* 35, 958, 1982.
40. **Ellis, H. A., McCarthy, J. H., and Herrington, J.,** Bone aluminum in haemodialysed patients and in rats injected with aluminum chloride: relationship to impaired bone mineralization, *Clin. Pathol.,* 32, 832, 1979.
41. **Evans, G. W.,** Copper homeostasis in the mammalian system, *Physiol. Rev.,* 53, 535, 1973.
42. **Fairbanks, V. F., Fahey, J. L., and Beutler, E.,** *Clinical Disorders of Iron Metabolism,* Grune & Stratton, New York, 1971.
43. **Favier, A.,** Role due foie dans le metabolisme des oligo-éléments, *Med. Chir. Dig.,* 10, 295, 1981.
44. **Gallagher, C. H., Reeve, V. E., and Wright , R.,** Copper deficiency in the rat. Effect on the ultrastructure of hepatocytes, *Aust. J. Exp. Biol. Med. Sci.,* 51, 181, 1973.
45. **Garnica, A. D.,** Trace metals and hemoglobin metabolism, *Ann. Clin. Lab. Sci.,* 11, 220, 1981.
46. **Glauser, S. C., Bello, C. T., and Glauser, E. M.,** Blood cadmium levels in normotensive and untreated hypertensive humans, *Lancet,* 1, 717, 1976.
47. **Golden, M. H. and Golden, B. E.,** Trace elements. Potential importance in human nutrition with particular reference to zinc and vanadium, *Br. Med. Bull.,* 37, 31, 1981.
48. **Goldstein, N. P. and Owen, C. A.,** Symposium on copper metabolism and Wilson's disease, *Mayo Clin. Proc.,* 49, 363, 1974.
49. **Gonsior, B. and Szirmai, E.,** Role of trace elements in pharmaceutical medication, *Agressologie,* 22, 143, 1981.
50. **Gürson, C. T. and Saner, G.,** Effects of chromium supplementation on growth in marasmic protein-calorie malnutrition, *Am. J. Clin. Nutr.,* 26, 988, 1973.
51. **Halsted, J. A.,** Zinc deficiency and congenital malformations, *Lancet,* 1, 1323, 1973.
52. **Halsted, J. A., Smith, J. C., Jr., and Irwin, M. I.,** A conspectus of research on zinc requirements of man, *J. Nutr.,* 104, 345, 1974.
53. **Hambridge, K. M.,** The role of zinc and other trace metals in pediatric nutrition and health, *Pediatr. Clin. North Am.,* 24, 95, 1977.
54. **Hambridge, K. M.,** Trace-element nutrition, *Pediatr. Ann.,* 10, 53, 1981.
55. **Henkin, R. I., Marshall, J. R., and Meret, S.,** Maternal fetal metabolism of copper and zinc at term, *Am. J. Obstet Gynecol.,* 110, 131, 1971.
56. **Hodsman, A. B., Sherrard, D. J., Brickman, A. S., Alfrey, A. C., Goodman, W. G., Maloney, N., Lee, D. B. N., and Coburn, J. W.,** Bone aluminum osteomalacic renal osteodystrophy correlation with excess osteoid, *Kidney Int.,* 19, 127, 1981.
57. **Holzbecher, J. and Ryan, D. E.,** Some observations on the interpretation of liver analysis data, *Clin. Biochem.,* 15, 80, 1982.
58. **Hurley, L. S.,** Teratogenic aspects of manganese, zinc and copper nutrition, *Physiol. Rev.,* 61, 249, 1981.
59. **Ingeberg, S., Deding, A., and Jensen, M. K.,** Bone mineral content in myelomatosis, *Acta Med. Scand.,* 21, 19, 1982.
60. **Iseri, I. T., Freed, J., and Bures, A. R.,** Magnesium deficiency and cardiac disorders, *Am. J. Med.,* 58, 837, 1975.
61. **Iyengar, G. V.,** Human health and trace element research: problems and propects, *Sci. Total Environ.,* 19, 105, 1981.
62. **Jacobs, A.,** Serum ferritin and iron stores, *Fed. Proc.,* 36, 2024, 1977.
63. **Johnson, M. A., Baier, M. J., and Gregor, J. L.,** Effects of dietary tin, copper, iron, manganese and magnesium metabolism of adult males, *Am. J. Clin. Nutr.,* 35, 1332, 1982.
64. **Juswigg, T., Batres, R., Solomons, N. W., Pineda, O., and Milne, D. B.,** Effect of temporary venous occlusion of trace mineral concentrations in plasma, *Am. J. Clin. Nutr.,* 36, 354, 1982.
65. **Kampschmidt, R. F., Upchurch, H. F., and Eddington, C. L.,** Multiple biological activities of a partially purified leucocytic endogenous mediator, *Am. J. Physiol.,* 224, 530, 1973.
66. **Kampschmidt, R. F. and Pulliam, L. A.,** Effect of delayed hypersensitivity on plasma iron and zinc concentration and blood leukocytes, *Proc. Soc. Exp. Biol. Med.,* 147, 242, 1974.
67. **Kirchgessner, M., Reichlamyr-Lais, A. M., and Schwartz, F. J.,** Interactions of trace elements in human metabolism, *Prog. Clin. Biol. Res.,* 77, 189, 1981.
68. **Kjellin, K. G.,** Trace elements in the cerebrospinal fluid in neurological diseases, *Clin. Toxicol.,* 18, 1237, 1981.
69. **Klein, R.,** The pediatricians and the prevention of lead poisoning in children. *Pediatr. Clin. North Am.,* 21, 277, 1974.

70. **Klein, R. and Haddow, J. E.,** Trace elements in health and disease, *Practitioner,* 201, 314, 1968.
71. **Klevay, L. M.,** Hypercholesterolemia in rats produced by an increase in the ratio of zinc to copper ingested, *Am. J. Clin. Nutr.,* 26, 1060, 1973.
72. **Koller, L. D.,** Immunosuppression produced by lead, cadmium and mercury, *Am. J. Vet. Res.,* 34, 1457, 1973.
73. **Kontula, K., Janne, O., and Luukkainen, T.,** Progesterone-binding proteins in human myometrium. Influence of metal ions on binding, *J. Clin. Endocrinol. Metab.,* 38, 500, 1974.
74. **Koos, B. J. and Longo, L. D.,** Mercury toxicity in the pregnant woman, fetus and newborn infant, *Am. J. Obstet. Gynecol.,* 126, 390, 1976.
75. **Kozma, M. and Szerdahelyi, P.,** Zinc-deficiency-induced trace element concentration and localization changes in the central nervous system of the albino rat, *Acta Histochem.,* 70, 54, 1982.
76. **Krishnamachari, K. A. V. R.,** Some aspects of copper metabolism in pellagra, *Am. J. Clin. Nutri.,* 27, 108, 1974.
77. **Kumei, Y. and Sato, A.,** Effects of sodium selenite on the cytotoxicity of dental amalgam, *Toxicol. Appl. Pharmacol.,* 59, 257, 1981.
78. **La Bella, F. S., Dular, R., and Leman, P.,** Prolactin secretion is specifically inhibited by nickel, *Nature (London),* 245, 330, 1973.
79. **La Bella, F., Dular, R., and Vivian, S.,** Pituitary hormone releasing or inhibiting activity of metal ions present in hypothalamic extracts, *Biochem. Biophys. Res. Commun.,* 52, 786, 1973.
80. **Laker, M.,** On determining trace element levels in man: the uses of blood and hair, *Lancet,* 2, 260, 1982.
81. **Lauwerys, R., Bernard, A., Buchet, J. P., and Roels, H.,** Dose-response relationship for the nephrotoxic action of cadmium in man, in Int. Conf. Management and Control of Heavy Metals in the Environment, London, 1979, pp 19.
82. **Leichter, I., Margulies, J. Y., Weinreb, A., Mizrahi, J., and Robin, G. C.,** The relationship between bone density, mineral content and mechanical strength in the femoral neck, *Clin. Orthop.,* 163, 272, 1982.
83. **Linder, M. C. and Munro, H. N.,** The mechanism of iron absorption and its regulation, *Fed. Proc.,* 36, 2017, 1977.
84. **Lombeck, I.,** The clinical significance of trace elements in childhood, *Ergeb. Inn. Med. Kinderheilkd.,* 44, 1, 1980.
85. **Lu, S. H. and Lin, P.,** Recent research on the etiology of esophageal cancer, *Z. Gastroenterol.,* 20, 361, 1982.
86. **Mangal, P. C. and Sharma, P.,** Effect of leukaemia on the concentration of some trace elements in human whole blood, *Indian J. Med. Res.,* 74, 559, 1981.
87. **McClain, C. J.,** Trace metal abnormalities in adults during hyperalimentation, *J. Parenter. Nutr.,* 5, 424, 1981.
88. **McFarlane, H., Reddy, S., and Cooke, A.,** Immunoglobulins, transferrin, caeruloplasmin and heterophile antibodies in kwashiorkor, *Trop. Geogr. Med.,* 22, 61, 1970.
89. **McLaughlin, S. G. A., Szabo, G., and Eisenmann, G.,** Divalent ions and the surface potential of charged phospholipid membranes, *J. Gen. Physiol.,* 58, 667, 1971.
90. **Mertz, W.,** The scientific and practical importance of trace elements, *Philos. Trans. R. Soc. London,* 294, 9, 1981.
91. **Mertz, W.,** Mineral elements: new perspectives, *J. Am. Diet. Assoc.,* 77, 258, 1980.
92. **Mertz, W.,** The essential trace elements, *Science,* 218, 1332, 1981.
93. **Menard, H. A., Baretti, M., Lamoureux, G., Leconte, R., and Paradis, P.,** Trace elements and acute phase reactants in gold treated rheumatoid arthritis patients, *J. Rheumatol. Suppl.,* 6, 143, 1979.
94. **Menkes, J. H.,** Kinky hair disease, *Pediatrics,* 50, 181, 1972.
95. **Miller, P. D., Dubovsky, S. L., McDonald, K. M., Arnaud, C., and Schrier, R. W.,** Hypocalciuric effect of lithium in man, *Miner. Electrol. Metab.,* 1, 3, 1978.
96. **Mills, C. F.,** Metabolic interactions of copper with other trace elements, in *Biological Roles of Copper,* Ciba Foundation Symposium, Ser. 79, Excerpta Medica, Amsterdam, 1981, 46.
97. **Mills, C. F.,** Some outstanding problems in the detection of trace element deficiency diseases, *Philos. Trans. R. Soc. London (Biol.)* 294, 199, 1981.
98. **Morgan, J. M.,** Tissue copper and lead content in ischemic heart disease, *Arch. Environ. Health,* 25, 26, 1972.
99. **Moses, H. A.,** Trace elements: an association with cardiovascular diseases and hypertension, *J. Natl. Med. Assoc.,* 71, 226, 1979.
100. **Newberne, P. M.,** Influence of pharmacological experiments of chemicals and other factors in diets of laboratory animals, *Fed. Proc.,* 34, 209, 1975.
101. **Newberne, P. M.,** Disease states and tissue mineral elements in man, *Fed. Proc.,* 40, 2134, 1981.
102. **Nielsen, F. H., Hunt, C. D., and Uthus, E. O.,** Interactions between essential trace and ultratrace elements, *Ann. N.Y. Acad. Sci.,* 355, 282, 1980.

103. **Nuutinen, L. S., Ryhänen, P., Hollmen, A., and Tyrvainen, L.,** The levels of zinc, copper, calcium and magnesium in serum and urine after heart valve replacement, *Infusionsther. Klin. Ernahr.,* 8, 214, 1981.
104. **Oberleas, D., Seymour, J. K., and Lenaghan, R.,** Effect of zinc deficiency on wound-healing in rats, *Am. J. Surg.,* 121, 566, 1971.
105. **Partridge, S. M.,** Elastin, biosynthesis and structure, *Gerontologia,* 15, 85, 1969.
106. **Patriarca, P., Cramer, R., and Moncalvo, S.,** Enzymatic basis of metabolic stimulation in leucocytes during phagocytosis: the role of activated NADPH oxidase, *Arch. Biochem. Biophys.,* 145, 255, 1971.
107. **Pekarek, R. S., Powanda, M. C., and Wannemacher, R. W., Jr.,** The effect of leukocytic endogenous mediator (LEM) on serum copper and ceruloplasmin concentrations in the rat, *Proc. Soc. Exp. Biol. Med.,* 141, 1029, 1972.
108. **Pekarek, R. S., Wannemacher, R. W., Jr., and Beisel, W. R.,** The effect of leukocytic endogenous mediator (LEM) on the tissue distribution of zinc and iron, *Proc. Soc. Exp. Biol. Med.,* 140, 685, 1972.
109. **Pekarek, R., Wannemacher, R., and Powanda, M.,** Further evidence that leukocytic endogenous mediator (LEM) is not endotoxic, *Life Sci.,* 14, 1765, 1974.
110. **Perry, H. M., Jr. and Erlanger, M. W.,** Metal-induced hypertension following chronic feeding of low doses of cadmium and mercury, *J. Lab. Clin. Med.,* 83, 541, 1974.
111. **Phillips, G. D. and Garnys, V. P.,** Parenteral administration of trace elements to critically ill patients, *Anaesth. Intensive Care,* 9, 221, 1981.
112. **Pounds, J. G., Wright, R., Morrison, D., and Casciano, D.,** Effect of lead on cellular calcium homeostasis in the isolated rat hepatocyte, *Toxicol. Appl. Pharmacol.,* 63, 389, 1982.
113. **Prohaska, J. R. and Wells, W. W.,** Copper deficiency in the developing rat brain: a possible model for Menkes' steely-hair disease, *J. Neurochem.,* 23, 91, 1974.
114. **Powanda, M. C., Cockerell, G. L., and Pekarek, R. S.,** Amino acid and zinc movement in relation to protein synthesis early in inflammation, *Am. J. Physiol.,* 225, 399, 1973.
115. **Rabinowitz, M. B., Wetherill, G. W., and Koppel, J. D.,** Studies of human lead metabolism by use of stable isotope tracers, *Environ. Health Perspect.,* 7, 145, 1974.
116. **Ravin, H. A.,** Rapid test for hepatolenticular degeneration, *Lancet,* 1, 726, 1956.
117. **Recker, R. R., Blotcky, A. J., Leffler, J. A., and Rack, E. P.,** Evidence for aluminum absorption from the gastrointestinal tract and bone deposition by aluminum carbonate ingestion with normal renal function, *J. Lab. Clin. Med.,* 80, 810, 1977.
118. **Retnam, V. J. and Bhandarkar, S.,** Trace elements in diabetes mellitus, *J. Postgrad. Med.,* 27, 129, 1981.
119. **Reuhl, K. R. and Chang, L. W.,** Effects of methylmercury on the development of the nervous system: a review, *Neurotoxicology,* 1, 21, 1979.
120. **Riordan, J. R. and Richards, V.,** Human fetal liver contains both zinc and copper-rich forms of metal-lothionein, *J. Biol. Chem.,* 255, 5380, 1980.
121. **Riordou, J. F. and Vallee, B. L.,** The functional roles of metals in metalloenzymes, *Adv. Exp. Med. Biol.,* 48, 33, 1974.
122. **Rosenberg, I. H. and Solomons, N. W.,** Biological availability of minerals and trace elements: a nutritional overview, *Am. J. Clin. Nutri.,* 35, 781, 1982.
123. **Golden, M. H.,** Trace elements in human nutrition, *Hum. Nutr. Clin. Nutr.,* 36, 185, 1982.
124. **Rotruck, J. T., Pope, A. L., and Ganther, H. E.,** Selenium: biochemical role as a component of glutathione peroxidase, *Science,* 179, 588, 1973.
125. **Ruberg, R. L. and Mirtallo, J.,** Vitamin and trace element requirements in parenteral nutrition: an update, *Ohio State Med. J.,* 77, 725, 1981.
126. **Rubin, H.,** Inhibition of DNA synthesis in animal cells by ethylene diamine tetraacetate and its reversal by zinc, *Proc. Natl. Acad. Sci. U.S.A.,* 69, 712, 1972.
127. **Sanstead, H. H.,** Trace elements in uremia and hemodialysis, *Am. J. Clin. Nutr.,* 33, 1501, 1980.
128. **Sanstead, H. H.,** Trace element interactions, *J. Lab. Clin. Med.,* 98, 457, 1981.
129. **Sass-Kortsak, A.,** Wilson's disease — a treatable liver disease in children, *Pediatr. Clin. North Am.,* 22, 963, 1975.
130. **Scheuhammer, A. M. and Cherian, M. G.,** The influence of manganese on the distribution of essential trace elements, *Toxicol. Appl. Pharmacol.,* 61, 227, 1981.
131. **Scheinberg, I. H. and Sternlieb, I.,** Wilson's disease, *Annu. Rev. Med.,* 16, 119, 1965.
132. **Schenker, J. G., Jungreis, E., and Polishuk, W. Z.,** Oral contraceptives and serum copper concentration, *Obstet. Gynecol.,* 37, 233, 1971.
133. **Schroeder, H. A. and Mitchener, M.,** Toxic effects of trace elements on the reproduction of mice and rats, *Arch. Environ. Health,* 23, 102, 1971.
134. **Schwarz, K., Porter, L. A., and Fredga, A.,** Some regularities in the structure-function relationship of organo-selenium compounds effective against dietary liver necrosis, *Ann. N.Y. Acad. Sci.,* 192, 200, 1972.

135. **Sever, L. E. and Emanuel, T.,** Is there a connection between maternal zinc deficiency and congenital malformation of the central nervous system in man?, *Teratology*, 7, 117, 1973.
136. **Sesame, H. A. and Boyd, M. R.,** Paradoxical effects of cobaltous chloride and salts of other divalent metals on tissue levels of reduced glutathione and microsomal mixed-function oxidase components, *J. Pharmacol. Exp. Ther.*, 205, 718, 1978.
137. **Shah, B. G.,** Bioavailability of trace elements in human nutrition, *Prog. Clin. Biol. Res.*, 77, 199, 1981.
138. **Shiraki, H. and Takeuchi, T.,** Minamata disease, in *Pathology of the Nervous System*, Minckler, J., Ed., McGraw-Hill, New York, 1971, 1651.
139. **Shokeir, M. H. K. and Schreffler, D. C.,** Cytochrome oxidase deficiency in Wilson's disease, *Proc. Natl. Acad. Sci. U.S.A.*, 62, 867, 1969.
140. **Siegel, R. C. and Martin, G. R.,** Collagen cross-linking. Enzymatic synthesis of lysine-derived aldehydes and the production of cross-linked components, *J. Biol. Chem.*, 245, 1653, 1970.
141. **Smith, J. C., Jr., McDaniel, E. G., Fan, F. F., and Halstead, J. A.,** Zinc: a trace element essential in vitamin A metabolism, *Science*, 181, 954, 1973.
142. **Smythe, W. R., Alfrey, A. C., Craswell, P. W., Crouch, C. A., Ibels, L. S. and Rudolph, H.,** Trace element abnormalities in chronic uremia, *Ann. Intern. Med.*, 96, 302, 1982.
143. **Sinha, S. N. and Gabrieli, E. R.,** Serum copper and zinc levels in various pathologic conditions, *Am. J. Clin. Pathol.*, 54, 570, 1970.
144. **Sosa-Lucero, J. C., de la Iglesia, F. A., Lumb, G., and Feuer, G.,** Nutritional influences on drug metabolism, *Rev. Can. Biol.*, 32 (Suppl.), 69, 1973.
145. **Sternlieb, I. and Scheinberg, I. H.,** The diagnosis of Wilson's disease in asymptomatic patients, *JAMA*, 183, 747, 1963.
146. **Subramanian, K. S. and Meranger, J. C.,** Simultaneous determination of 20 elements in some human kidney and liver autopsy samples, *Sci. Total Environ.*, 24, 147, 1982.
147. **Thind, G. S. and Fischer, G. M.,** Relationship of plasma zinc to human hypertension, *Clin. Sci. Mol. Med.*, 46, 137, 1974.
148. **Toepfer, E. W., Mertz, W., and Roginski, E. E.,** Chromium in foods in relation to biological activity, *J. Agric. Food. Chem.*, 21, 69, 1973.
149. **Topham, R. W., Sung, C. S., and Morgan, F. G.,** Functional significance of the copper and lipid components of human ferroxidase-II, *Arch. Biochem. Biophys.*, 167, 129, 1975.
150. **Underwood, E. J.,** The incidence of trace element deficiency diseases, *Philos. Trans. R. Soc. London (Biol.)*, 294, 3, 1981.
151. **Underwood, E. J.,** Trace metals in human and animal health, *J. Hum. Nutr.*, 35, 37, 1981.
152. **Uzman, L. L., Iber, F. L., and Chalmers, T. C.,** Mechanism of copper deposition in the liver in hepatolenticular degeneration, *Am. J. Med. Sci.*, 231, 511, 1956.
153. **Van Hook, R. I.,** Potential health and environmental effects of trace elements and radionuclides from increased coal utilization, *Environ. Health Perspect.*, 33, 227, 1979.
154. **Wacker, W. E. C. and Parisi, A. F.,** Magnesium metabolism, *N. Engl. J. Med.*, 278, 658, 712, 772, 1968.
155. **Wacker, W. E. C., Ulmer, D. D., and Vallee, B. L.,** Metalloenzymes and myocardial infarction II. Malic and lactic dehydrogenase activities and zinc concentrations in serum, *N. Engl. J. Med.*, 255, 449, 1956.
156. **Walshe, J. M.,** The physiology of copper in man and its relation to Wilson's disease, *Brain*, 90, 149, 1967.
157. **Williams, R. O. and Loeb, L. A.,** Zinc requirement for DNA replication in stimulated human lymphocytes, *J. Cell. Biol.*, 58, 594, 1973.
158. **Wilson, S. A. K.,** Progressive lenticular degeneration: a familial nervous disease with cirrhosis of the liver, *Brain*, 34, 295, 1912.
159. **Baker, A. P. and Hillegass, L. M.,** Enhancement of UDP-galactose: mucin galactosyltransferase activity by spermine, *Arch. Biochem. Biophys.*, 124, 218, 1974.
160. **Gallop, P. M., Blumenfeld, O. O., and Seifter, S.,** Structure and metabolism and connective tissue proteins, *Annu. Rev. Biochem.*, 41, 617, 1972.
161. **Klein, R. and Haddon, J. E.** Trace elements in health and disease, *Practitioner*, 201, 314, 1968.
162. **Leach, R. M., Jr.,** Role of manganese in mucopolysaccharide metabolism, *Fed. Proc.*, 30, 991, 1971.
163. **Leach, R. M., Jr., Muenster, A. M., and Wien, E. M.,** Studies on the role of manganese in bone formation. II. Effect upon chondroitin sulfate synthesis in chick epiphyseal cartilage, *Arch. Biochem. Biophys.*, 133, 22, 1969.
164. **Partridge, S. M.,** Elastin, biosynthesis and structure, *Gerontologia*, 15, 85, 1969.
165. **Pinnell, S. R. and Martin, G. R.,** The cross-linking of collagen and elastin: enzymatic conversion of lysine in peptide linkage to alpha-aminoadipic-delta-semialdehyde (allysine) by an extract from bone, *Proc. Natl. Acad. Sci. U.S.A.*, 61, 708, 1968.

166. **Robinson, H. C., Telser, A. and Dorfman, A.**, Studies on biosynthesis on the linkage region of chondroitin sulfate-protein complex, *Proc. Natl. Acad. Sci. U.S.A.*, 56, 1859, 1966.

167. **Siegel, R. C. and Martin, G. R.**, Collagen cross-linking. Enzymatic synthesis of lysine-derived aldehydes and the production of cross-linked components, *J. Biol. Chem.*, 245, 1653, 1970.

168. **Telser, A., Robinson, H. C., and Dorfman, A.** The biosynthesis of chrondroitin-sulfate protein complex, *Proc. Natl. Acad. Sci. U.S.A.*, 54, 912, 1965.

FURTHER READINGS

Burch, R. E. and Sullivan, J. F., *Trace Elements*, Medical Clinics of North America, W. B. Saunders, Philadelphia, 1976.

Burrows, D., Ed., *Chromium: Metabolism and Toxicity*, CRC Press, Boca Raton, Fla., 1983.

Fribert, L. T., Piscator, M., Nordberg, G., and Kjellstrom, T., *Cadmium in the Environment*, CRC Press, Boca Raton, Fla., 1974.

Newberne, P. M., Ed., *Trace Substances and Health*, Marcel Dekker, New York, 1976.

Pfeiffer, C. C., Ed., *Neurobiology of the Trace Metals Zinc and Copper*, Academic Press, New York, 1972.

Prasad, A. S., Ed., *Trace Elements in Human Health and Disease*, Academic Press, New York, 1976.

Sarkar, B., Ed., *Biological Aspects of Metals and Metal-Related Diseases*, Raven Press, New York, 1983.

Underwood, E. J., *Trace Elements in Human and Animal Nutrition*, Academic Press, New York, 1971.

Chapter 3

ABNORMALITIES OF CELLULAR ORGANIZATION

I. INTRODUCTION

Various structural elements of the cell represent specialized functions. However, their concerted action is necessary to maintain normal cellular activity associated with normal conditions. There are great structural and functional differences between various cell types and variations between subcellular organelles in different tissues.[275,429,430,460] There are also limitations; not all cells contain all organelles, and certain structures are found only in particular cell types.

The membrane constituents of the cell are complex, with many varied functions.[505] Specificities play a major role in differentiating the various types of membranes, although all share some common structural features. All membranes contain phospholipid bilayers as their basic structural framework, with proteins absorbed on the surface or penetrating the lipid layer. The various membrane components, proteins, lipids, and carbohydrates, are in asymetric disposition across the lipid bilayer. The various membranes represent a dynamic structure;[504] the mobility of structural components is affected by membrane fluidity and interactions with various protein components on the surface. The chemical composition of the membranes is also not static; they are synthesized, repaired, or modified during the lifetime of the cell. These processes are necessary in order that the subcellular membranes can adapt to the ever-changing environment and maintain their function. When this adaptation fails, impairment and disease conditions become apparent.[153,448]

Many aspects of intra- and extracellular organization are important in health and disease.[316,501] Certain features of this organization will be dealt with in specialized sections. These are (1) discussions on changes involving specialized cells which represent special functions, such as nerve, muscle, and various glandular tissues, and their interrelationship with disease; (2) the association with disease of several extracellular elements which are vital to tissue organization, such as basement membrane, cartilage, collagen, and bone; and (3) structural changes of the cytoplasm which occur concomitant with physiological functions, and impairments of these changes related to various disorders, such as phagocytosis and pinocytosis, secretion, cell movement, and contraction. In this chapter we are not going to discuss the contribution of any special structural changes to disease; rather, reference will be made primarily to the hepatic parenchyma cell as a prototype.

Studies on the intimate morphological and biochemical interrelations of intact cells have demonstrated that any injury affecting one subcellular organelle or impairing one biochemical reaction may directly cause a disease and may result in a rapid disorganization of many other cellular and noncellular components, triggering chain reactions and modifying many interrelated biochemical processes. In the etiology of most diseases, the structural and molecular bases are still obscure. Recently, the abnormal events at a structural level that give rise to some diseases have been elucidated. In some syndromes, the structural changes are only clarified in general terms, without precise definition of the specific alteration involved.

II. MEMBRANE STRUCTURE

A. General

Normal homeostasis of the cell is associated with the integrity of the various membranes.[53] These membranes exhibit selective permeability, regulate the composition of the environment within and outside the cell, and are responsible for an active communication between these

Table 1
DISTRIBUTION OF PROTEIN AND
LIPID IN VARIOUS CELLULAR
AND SUBCELLULAR MEMBRANES

| Membrane | Percentage | | Ref.[a] |
	Protein	Lipid	
Erythrocyte	60	40	1
Retina receptor cell	40	60	2
outer segment	61	39	3
Microsomes	68	32	4
Mitochondria	74	26	2
Myelin	20	80	5

[a] 1. **Maddy, A. H. and Malcolm, B. R.,** *Science,* 150, 1616, 1965.
2. **Fleischer, S. and McConnel, D. G.,** *Nature (London),* 212, 1366, 1966.
3. **Sjostrand, F. S.,** *Ergeb. Biol.,* 21, 125, 1959.
4. **Spiro, M. J. and McKibbin, J. M.,** *J. Biol. Chem.,* 219, 643, 1956.
5. **O'Brien, J. S. and Sampson, E. L.,** *J. Lipid Res.,* 6, 537, 1965.

compartments. Some of these membranes, such as myelin or plasma membranes of blood cells, behave primarily as barriers separating various cells within one organ from each other or from another part of the body. These various membranes contain proteins and are generally rich in lipids (glycolipids and phospholipids), mainly responsible for passive permeability characteristics. In the various cellular membranes some diversity exists in protein to lipid ratio (Table 1). There are some membranes which are the functional sites of organized enzyme activity, such as the inner mitochondrial membrane which contains a high proportion of strongly bound proteins and is poor in lipids. This distinction is rather broad, since most membranes act to some extent as permeability barriers and exhibit several enzyme activities. Nevertheless, structural abnormalities may reflect minor changes in permeability to moderate effects on cellular organization leading even to impairment of the normal function of the cell.

The various facets of membrane structure and function are interrelated. The essential features of this organization have been described in many models.[275,504,505] According to the latest hypothesis, the membrane forms a mosaic structure, consisting mainly of a fluid phospholipid bilayer in which solid bodies of globular proteins are placed randomly (Figure 1). Within the double layer, cholesterol and glycolipids are also found. The fluidity changes with the lipid composition; the presence of saturated fatty acids and increased cholesterol concentration increases rigidity. The bilayer is also responsible for the stability, integrity, and impermeability of the cell millieu. Proteins may be attached to the surface of the membrane, may be partially embedded into the outer lipid bilayer with a part emerging at the surface, or may penetrate into the lipid bilayer to varying degrees. Many proteins are bound by hydrophobic bonds and are accessible from only one side of the bilayer, while some go through the entire membrane. Proteins are an integral part of the membranes, but they are not bound rigidly and therefore are capable of some lateral movement within the lipid bilayer. This movement of proteins on the surface is important in interactions with antibodies, hormones, and receptors. Proteins are responsible for the catalytic function of the membranes which depends on the presence of the lipids such as microsomal enzymes. Proteins may be located through the membranes with a hydrophobic end outside and hy-

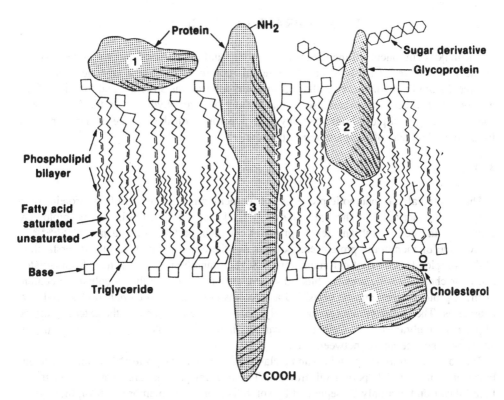

FIGURE 1. Organization of cellular membranes. There are diverse relationships between the lipid matrix and proteins. (1) Some proteins or glycoproteins are absorbed or attached with bonds to the surface of the membrane; (2) some are embedded into the lipid bilayer but parts are projected into the surrounding medium or the interior of the cell; and (3) others pass through the entire membrane. The organization of amino and carboxylic terminals provides an amphipatic character to the molecule, responsible for unidirectional movement of water and electrolytes.

drophilic end inside. This localization represents water binding on the inside and repellance on the outside. The amphipatic character of these proteins may be responsible for water and electrolyte movements. If any change occurs in the molecular organization of these membranes, disease processes may be initiated.

B. Cytoskeletal Membrane System

It is generally accepted that cells are structured with a cytoskeletal network.[53,109,451] This has been established in erythrocyte membranes[315] and indirectly proven in other membranes by the effects of various drugs which disrupt the organization of microtubules and microfilaments with subsequent alteration of receptor distribution.[275]

In the erythrocyte membrane, the spectrin-actin cytoskeletal network controls the shape and membrane changes result in deformation.[85,315,326,494,548] Genetic defects and metabolic changes modify the cell shape. ATP depletion and Ca^{2+} accumulation leads to discocyte-spheroechinocyte transformation.[493,576] Mg-ATP also induces shape changes,[492] although these are related to phosphorylation of spectrin.[315] In senescent or nonfunctional erythrocytes, phagocytosis is connected with an alteration of the normal cell surface which can bind IgG autoantibodies and recognize phagocytes.[261] In fibroblasts and lymphoid cells, there is transmembrane control of cell-surface receptor distribution, connected with the actin-myosin structure.[356] The cytoskeletal actin-myosin filaments may play a role in transmembrane control of many cell-surface proteins such as collagen and other matrix proteins.[235]

III. MEMBRANE COMPOSITION

The major components of membranes are proteins and lipids. Some proteins are embedded in the lipid bilayer, and these are the integral proteins.[505,524] These moieties cannot be separated from the membranes without completely altering the structure and function of the membrane. The cell membrane also contains about 5% carbohydrates, in the form of glycoproteins or glycolipids. Some carbohydrates are found on the surface of the membrane firmly bound as part of the membrane structure.

A. Proteins

1. General

The functional and metabolic properties of biomembranes are derived from proteins (and enzymes) interacting with the lipid bilayer.[204,422] Thus, the specificities of the biological actions of the cell membrane reside primarily in their protein constituents.[85,154,326,436,518] Knowledge of membrane protein components are therefore essential in our understanding of the complex functions of cell membranes. Several membrane proteins have been isolated and their chemical structures established. There are differences in the carbohydrate content of these proteins; they can be classified as proteins containing no carbohydrates, and glycoproteins. The carbohydrate residues in glycoproteins are localized on the external surface of plasma membranes.[178,380] Cell-surface carbohydrates play a role in receptor function, cell adhesion, and interaction between cells.

The various membrane proteins can be classified into two categories:[504] (1) integral membrane proteins, and (2) peripheral proteins. The integral proteins are firmly bound in the lipid bilayer and can only be separated by solubilization using membrane-disrupting agents such as lipid solvents, detergents, and chaotropic agents. The peripheral proteins are mainly localized outside the lipid framework of the membrane. These proteins are probably attached to the membrane by hydrogen or ionic bonding through the lipid head groups or integral proteins. The binding of peripheral proteins to the membrane is relatively weak, and they can be separated easily by chelating agents or by changing the ionic strength or pH of the medium. Under these circumstances, the lipid matrix remains unchanged and integral proteins are still bound to the lipid bilayer. Some peripheral proteins may be connected with the structural network and they cannot be detached easily from the membrane, such as erythrocyte spectrin and actin. These proteins form aggregates under certain conditions.[51] The membrane network in a hepatocyte is illustrated in Plate 1.

The nature and types of proteins found in various membranes show wide variations, reflecting differences in the functions of these membranes. The dissimilarities in enzyme composition are most marked among intracellular membranes.[224] In some instances, parts of the same subcellular organelle contain different enzyme systems such as the outer and inner membrane of mitochondria. The differences in enzyme activities in various cell membranes can be biochemically characterized, leading to the identification of "marker" enzymes.

Although the protein composition of various membranes is complex, there is a general pattern concerning of protein structure. The integral membrane proteins are generally amphipathic molecules which exhibit bimodal or trimodal structure and contain zones of hydrophilic and hydrophobic amino acid residues. Integral proteins also contain a large number of amino acids with apolar residues.[80] The hydrophilic zones are exposed to the aqueous milieu; in contrast, the hydrophobic regions are buried in the lipid bilayer. Most integral membrane proteins are glycoproteins; some also contain proteolipid components, such as esterified fatty acids, as in the myelin membrane[436,528] and sarcoplasmic reticulum.[317-319] The amino acid compositions of peripheral membrane components are similar to those of the soluble cytoplasmic proteins.[148]

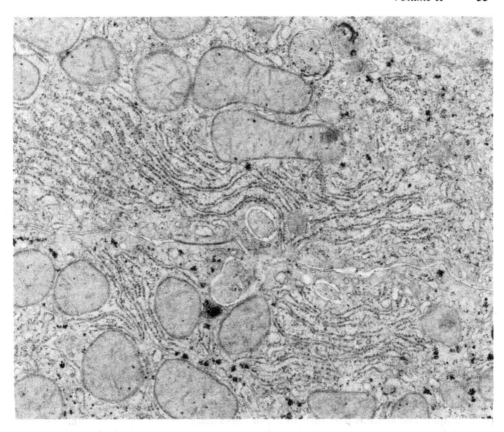

PLATE 1. Membrane network in the hepatocyte cytoplasm represented by regularly stacked membranes of the endoplasmic reticulum.

2. Molecular Organization of Membrane Proteins

There is some relationship (1) between various membrane proteins and the lipid bilayer, and (2) between various protein components.[296,496,590] Protein components are asymmetrically organized across the membrane, and their relation to the lipid bilayer varies. Some proteins may be partially embedded from the cytoplasmic side, and some from the extracellular side. Some proteins completely traverse the lipid bilayer and others may be superficially bound to membrane surfaces[504] (Figure 1). Some cell-surface glycoproteins span the lipid bilayer and represent the routes of communication between the internal and external cell environment.[246] Several important glycoproteins are present in cell membranes, including myelin,[436] the visual pigment rhodopsin,[351] and calsequestrin and Ca^{2+}-ATPase in the sarcoplasmic reticulum.[320] Injection of human myelin membrane into experimental animals induces the synthesis of antibodies which can cause allergic encephalomyelitis, a disease condition resembling multiple sclerosis.[145]

The asymmetric distribution of proteins within the membrane represents specialization in membrane function. Different cell membranes are characterized by differences in the type and ratio of the various membrane components. The chemical composition of a particular membrane may show variations due to metabolic turnover and degradation.[500,537] Changes can also occur as a result of differentiation,[180] growth[335] virus infection,[291] malignancy,[178,234] hormones,[81] drug administration,[398] intense physical activity,[529] or variations in nutritional status.[231,507]

Table 2
LIPID COMPOSITION OF CELLULAR AND SUBCELLULAR MEMBRANES

| Membrane | Cholesterol | Percentage | | | Ref.[a] |
		Glycerophosphatides	Sphingolipids	Others	
Erythrocyte	25	55	18	2	1
Retina receptor	all				
outer segment	2	81	5	12	2
Microsomes	6	94	0	0	3
Mitochondria	6	93	0	1	3
Myelin	25	32	31	12	4

[a] 1. **Ways, P., Reed, C. F., and Hanahan, D. J.,** *Clin. Invest.,* 42, 1248, 1963.
2. **Sjostrand, F. S.,** *Ergeb. Biol.,* 21, 128, 1959.
3. **Spiro, M. J. and McKibbin, J. M.,** *J. Biol. Chem.,* 219, 643, 1956.
4. **O'Brien, J. S. and Sampson, E. L.,** *J. Lipid Res.,* 6, 537, 1965.

B. Lipids
1. General

Lipids represent a heterogeneous population of molecules. Their only common characteristic is water insolubility. They show amphipathic characteristics and when dispersed in aqueous solutions, the hydrophobic or apolar parts aggregate spontaneously to form micelles or bilamellar structures. The hydrophylic or polar parts are oriented towards the aqueous phase of the environment. Among several types of lipids found in biological membranes, phospholipids and steroids play roles of major importance.[54,89,100,182,513,518,613] Various other types of lipid, such as cerebrosides, gangliosides, or others, are usually found in smaller amounts, with the exception of brain structures which are rich in sphingolipids.[171,388,574] Some membranes contain triglycerides, glycolipids, and sulfolipids, but these components usually do not play an important structural or functional part.[112,210,532] The relative proportion of various lipids varies considerably in different types of membranes (Table 2).

The most essential structural steroid is cholesterol.[60,190] Its concentration in the membranes shows variations, the highest being 25% of total lipids. Some membranes contain bound steroid hormones, such as estradiol, testosterone, or progesterone[6,41,137,226,415,450,608] (Figure 2). Phospholipids, on the other hand, consist of 60 to 80% of the lipid composition, indicating that these substances represent the most important structural units. Phospholipids are molecular derivatives of triglycerides. They are composed of glycerol and two fatty acids, and a fatty acid residue of the triglyceride is replaced by a phosphate group alone or with the addition of various nitrogen containing bases.[12,545]

Differences in the skeleton and base give rise to various phospholipid classes: phosphatidic acid, phosphatidylethanolamine, phosphatidylserine, phosphatidylcholine, phosphatidyl-inositol, and sphingomyelin (Figures 3 to 5). In addition, wide variation can occur in the degree of unsaturation of the hydrocarbon chain of fatty acids attached to the glycerol backbone. As a result of these variations, it has been estimated that most membranes contain many chemically different phospholipid molecules. The erythrocyte membrane is probably composed of 150 to 200 different units,[498,499] and this number is considerably greater in the various membranes present in complex cells, such as hepatocytes. This high degree of variability may be related to constituent fatty acid synthesis which exceeds that of any other internal organ. Sphingomyelin constitutes part of gangliosides, which are important constituents of brain membrane lipids (Figure 6).

Minor lipid components are plasmalogens which contain α, β-unsaturated hydrocarbon residues joined to the glycerol carbon one by ether linkage, lysophosphatides which contain

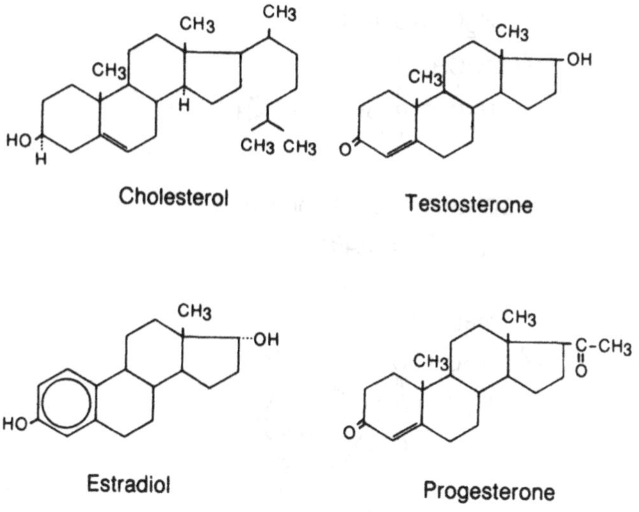

FIGURE 2. Structure of steroids occurring in cellular membranes.

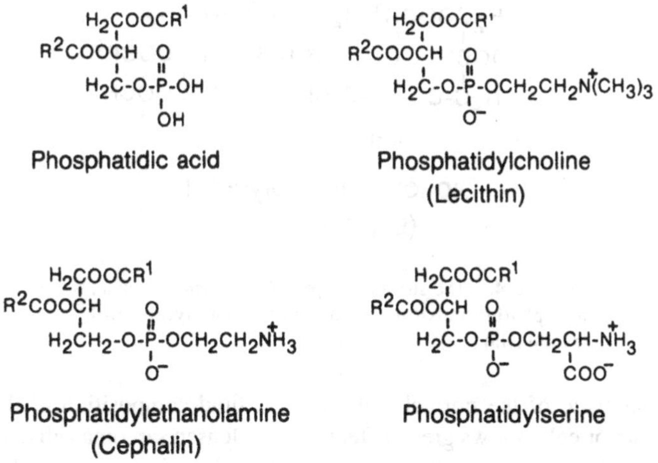

FIGURE 3. The structure of glycerophosphatides. R^1 and R^2 represent hydrocarbon chains.

only one fatty acyl residue, and cardiolipin (Figure 4). These are occasionally found in various cellular membranes.

2. Molecular Organization of Membrane Lipids

a. Phospholipids

There are some specific differences in the lipid composition of mem-branes.[147,188,263,401,460,503] The inner membrane of mitochondria is rich in cardiolipin and these components are associated with more unsaturated fatty acids than the outer mitochondrial membrane. The inner membrane is poor in sphingomyelin and contains no cholesterol and little or no phosphatidylinositol. In addition, this specificity of mitochondrial lipid com-

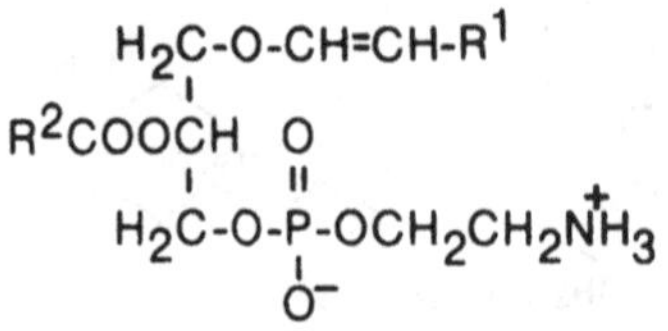

$$H_2C-O-CH=CH-R^1$$
$$R^2COOCH \quad O$$
$$H_2C-O-\overset{\overset{\displaystyle O}{\|}}{P}-OCH_2CH_2\overset{+}{N}H_3$$
$$O^-$$

Phosphatidal ethanolamine
(Plasmalogen)

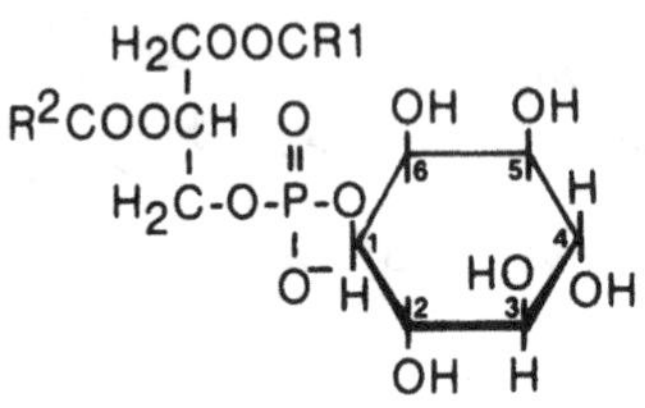

$$H_2COOCR1$$
$$R^2COOCH \quad O$$
$$H_2C-O-\overset{\|}{P}-O$$
$$O^-$$

Phosphatidylinositol

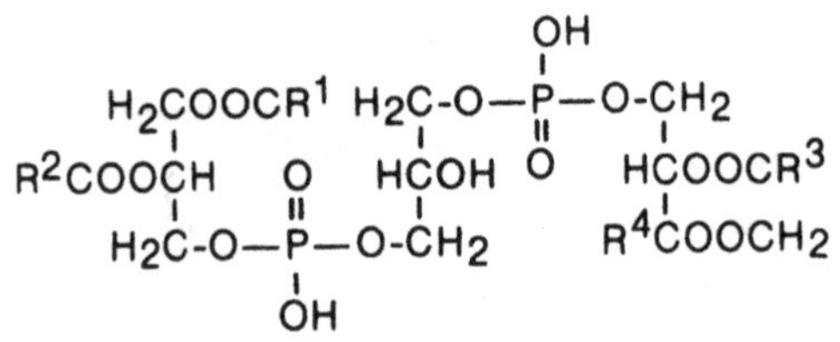

$$H_2COOCR^1 \quad H_2C-O-\overset{OH}{\overset{|}{P}}-O-CH_2$$
$$R^2COOCH \quad O \quad HCOH \quad O \quad HCOOCR^3$$
$$H_2C-O-\overset{\|}{P}-O-CH_2 \quad R^4COOCH_2$$
$$OH$$

Diphosphatidylglycerol
(Cardiolipin)

FIGURE 4. The structure of phosphatidyl ethanolamine, phosphatidylinositol, and diphosphatidylglycerol (cardiolipin).

position is characteristic of the normal cell, whereas lipid composition of this subcellular cell organelle in tumor cells shows great differences. In Jensen sarcoma cells, sphingomyelin is found in elevated amounts in mitochondria and cardiolipin occurs in the microsomal fraction; this phospholipid is virtually absent in the normal endoplasmic reticulum. Similar abnormalities have been described in the poorly differentiated, fast-growing hepatoma, but not in a minimal-deviation hepatoma or in regenerating liver cells.[34,35,328,561] The peculiarities of phospholipid composition of the individual membranes probably reflect the effect of protein composition and enzyme activity.[430,458,518] In human erythrocyte ghosts, phosphatidylserine is closely associated with membrane protein and cannot be freely exchanged with phosphatidylethanolamine,[136,498] and these moieties appear to be located in the inner surface. Conversely, the outer erythrocyte membrane consists predominantly of choline-containing phospholipids, lecithin, and sphingomyelin, and only few phosphatidylserine and ethanolamine residues have been located there.[513,518,600,613]

The asymmetric distribution of phospholipids on both sides of the erythrocytic membranes is probably not unique.[175,197] Since the overall phospholipid composition is similar, it is possible that the plasma membrane of other cells is constructed on a similar asymmetric basis. It is also interesting that the distribution of saturated/unsaturated fatty acids reveals asymmetry. About one third of the total of double-bond-containing components are found

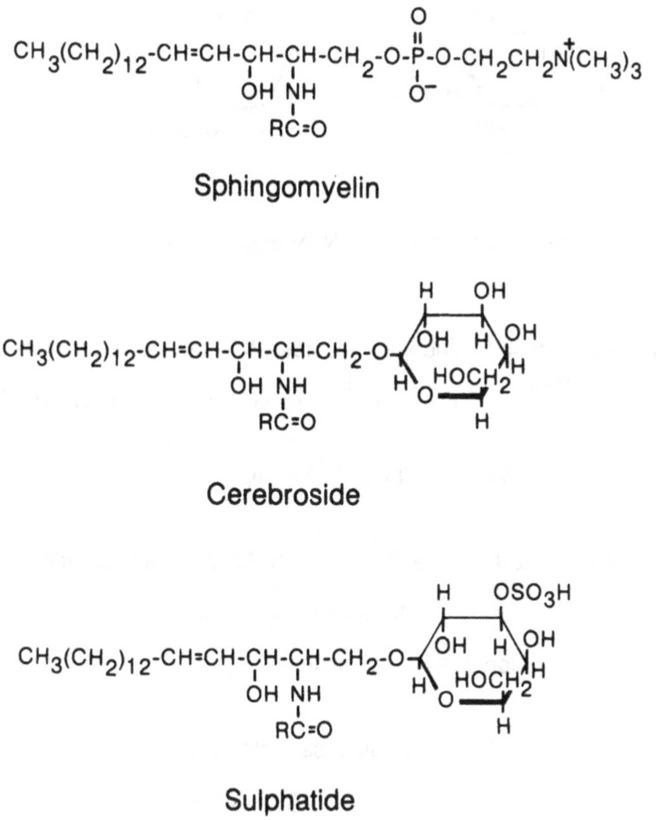

$$CH_3(CH_2)_{12}-CH=CH-CH-CH-CH_2-O-\overset{\overset{O}{\|}}{\underset{\underset{O^-}{|}}{P}}-O-CH_2CH_2\overset{+}{N}(CH_3)_3$$

$$\underset{\underset{RC=O}{|}}{\underset{OH\ \ NH}{}}$$

Sphingomyelin

$$CH_3(CH_2)_{12}-CH=CH-CH-CH-CH_2-O-$$

Cerebroside

$$CH_3(CH_2)_{12}-CH=CH-CH-CH-CH_2-O-$$

Sulphatide

FIGURE 5. The structure of sphingomyelin, cerebroside, and sulfatide.

in the outer phospholipids and in phosphatidylcholine and spingomyelin fractions. The remaining two thirds occur in phosphatidyl-ethanolamine, -serine, and -inositol, which contain more polyunsaturated species. Differences in the assymetrical distribution of phospholipids and in particular the uneven localization of unsaturated fatty acids influences the physical properties of the biological membranes.[560,568,613,614] Studies on the thermal properties of synthetic membranes using liposomes from pure phospholipids or from phospholipid-cholesterol mixtures revealed that, by raising the temperature, the organized structure was converted to a disorganized one. The temperature at which the transition occurs from crystal to liquid-crystal is dependent on the chemical characteristics of the phospholipids, namely on the nature of the polar head group, the chain length of the fatty acid hydrocarbon, hydration rate, and degree and type of unsaturation of the side chain. Liposomes made of phosphatidylcholine have a lower transition temperature than those composed of phosphatidylethanolamine. If the head group is identical and the extent of hydration is the same, phospholipids containing more unsaturated or longer fatty acids have transition temperatures reduced, as compared to more saturated or shorter side chains. Shorter chains lower the temperature, while *cis*-unsaturated derivatives decrease, and *trans*-unsaturated chains elevate the transition point. These data suggest that the phospholipid composition of biological membranes influences the transition temperature, with attendant changes in membrane fluidity. In turn, the fluidity change affects the movement of protein components, their catalytic activity, and the space between the constituents associated with the diffusion of smaller molecules through the membranes.[499] Essentially, asymmetrically distributed phospholipids bring about also asymmetrical membrane fluidity which may account for the major regulatory

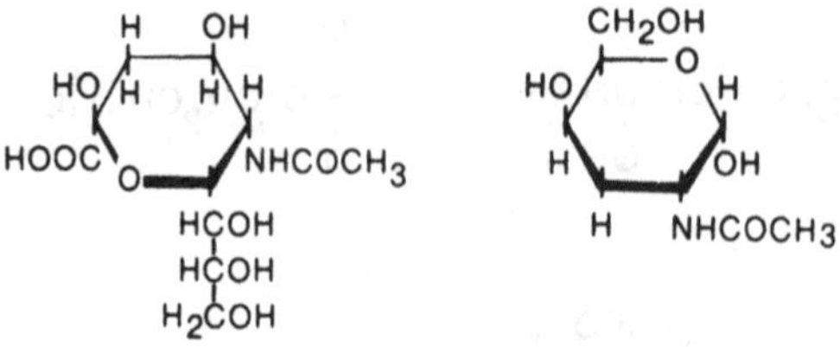

N-Acetylneuraminic acid   N-Acetygalactosamine

Acyl-Sphingosine-Glucose-Galactose-N-Acetylgalactosamine
(ceramide) | |
              N-Acetylneuraminic acid   Galactose

Monosialoganglioside, G_{M1}

Ceramide-Glucose-Galactose-N-Acetylgalactosamine
                    |
                Sialic acid

Monosialoganglioside, G_{M2}

Ceramide-Glucose-Galactose
                    |
                Sialic acid

Monosialoganglioside, G_{M3}

FIGURE 6.   The structure of ganglioside constituents and major monosialo-gangliosides in human brain.

factor in the unidirectional transport process.[197,257,546] Possible simultaneous coexistence of fluid and crystalline (gel) regions in phospholipid membranes have also been suggested.[52,147]

The hepatic endoplasmic reticulum represents another important cellular organelle concerning the organization of membrane lipid in relation to membrane function. These membranes contain about 40% of the total phospholipid present in hepatocytes.[89,100,132,182] The action of foreign compounds on these cell organelles is associated with changes in phospholipid content. Furthermore, the capability of a drug to induce or inhibit the function of the endoplasmic reticulum is related to its action on phospholipid structural patterns. Inducers raise the *de novo* synthesis of membrane-bound phosphatidylcholine from ethanolamine by step-wise methylation and lysophosphatidylcholine content, probably by an increased degradation of the overflow of phosphatidylcholine. Hepatotoxic compounds inhibit phospholipid synthesis and reduce their content in hepatic microsomes.[15,107,165-167,229,565,610,611] This contrasting action is also manifest in alterations of the total quantity of microsomal fatty acids and shows a shift in the distribution of various individual fatty acids bound to phosphatidyl-choline and -ethanolamine. Drugs enhance the production of phospholipids rich in unsaturated fatty acids and low in saturated chains. Conversely, toxic compounds reduce the amount of unsaturated and increase the amount of saturated fatty acids. The increased unsaturated and polyunsaturated chains in the phospholipid bilayer alter the fluidity of the crystalline membrane structure. This effect modifies the function of endoplasmic reticulum membrane-bound processes, favoring proliferation of membranes. In contrast, the decrease of these essential components by increasing saturation bring about more gel-like behavior, which appears to be unsatisfactory for the normal function of membranes and triggers off an abnormal response leading to biochemical lesion and cellular damage.[90,128,310]

Abnormalities of phospholipid synthesis are connected with various diseases.[48,551,567,568] Some of them are related to the lack of phosphatidylcholine production due to inadequate intake of methyl donors. In experimental animals, dietary choline deficiency leads to the onset of a variety of pathological changes in several organs including the development of fatty liver, hypertension, vascular changes, hemorrhagic kidney, and other hepatic and renal lesions including hepatocarcinoma[203,208,216,282,305,308,352,378,583] The inadequate nutritional status in man may result in phospholipid-precursor deficiency thus leading to the development of various diseases. In contrast to reduced phospholipid levels, there are conditions which cause lipid infiltration into the cell and elevate to the phospholipid content, probably by blockage of metabolism and elimination. These conditions are consequences of the action of hepatotoxicants or drug side effects.[157,304,321,459,461]

Many disorders of phospholipid biosynthesis are connected with the lack of specific enzymes responsible for the formation or metabolism of a particular phospholipid in the brain, liver, and other organs.[453] Various drugs affect liposomes, the artificial phospholipid bilayers, and probably elicit similar action on the phospholipid moieties in natural membranes. Viral infections can induce changes in membrane fluidity.[68] Lipid-soluble drugs and other compounds lower the transition temperature and widen the temperature range in which the crystal and fluid structures of the membranes occur. This dual effect is subsequent to the action of anesthetics and similar lipophilic compounds which increase fluidity and raise disorganization and mobility of the membrane bilayers. These changes can modulate protein function in the hepatic response to the administration of foreign compounds.[86,240]

b. Steroids

These constitute part of neutral lipids found in various biological membranes. Cholesterol is the major steroid in plasma membranes, and cholesteryl esters and triglycerides are present occasionally.[185,287,363,423,606] Among the intracellular membranes, the nuclear membrane and the Golgi apparatus have high cholesterol content. Cholesteryl ester content of nuclear envelopes is also high. The presence of cholesterol may be important in structural development and membrane fluidity.[106,363,449] An interaction between cholesterol and phospholipid fatty acid chains regulates the fluidity of biomembranes and maintains the membrane in an intermediate fluid state. The role of cholesterol is dependent upon temperature changes. Below the transition temperature, cholesterol disrupts the gel-like structure of the phospholipid hydrocarbon chains and allows increased movement. Above the transition temperature, on the other hand, cholesterol gives rigidity to the acyl chains and reduces fluidity. This condition contributes to the strength of various membranes. Moreover, culture studies have shown the cells cannot survive in the absence of cholesterol either provided from an exogenous source or synthesized within the cell.[60]

Recent investigations revealed that various steroid hormones are specifically bound to biological membranes, and through this binding they modulate target cell metabolism.[28,298] The generally accepted mechanism of action is that steroid hormones bind initially to soluble cytosolic receptor proteins. The steroid-receptor complex is translocated to the nucleus and by an interaction with the chromatin it activates the genes and modifies protein and RNA synthesis.[243,391] Steroid hormones also bind significantly to various biomembranes such as plasma membranes,[6,415] lysosomes,[226] and microsomes.[41,137,450,608] In these cases, steroid receptor binding is not connected with direct gene activations, but the steroid hormone-membrane receptor interaction may be involved in the regulation of the firing rates in synaptic membranes,[333] in intracellular Ca^{2+} distribution,[26] in controlling the capacity of the endoplasmic reticulum to bind to ribosomes,[41] or in the proliferation of these membranes and their enzyme activity.[137]

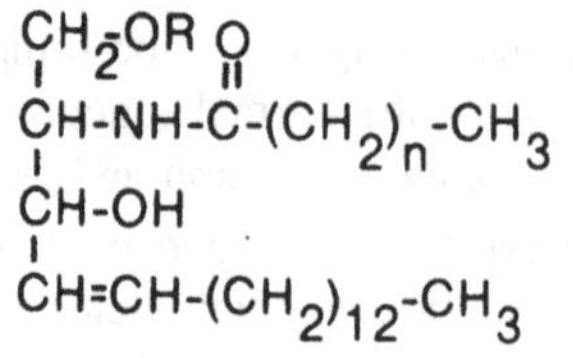

$$\begin{array}{l} \text{CH}_2\text{OR} \quad \text{O} \\ | \qquad\quad || \\ \text{CH-NH-C-(CH}_2)_n\text{-CH}_3 \\ | \\ \text{CH-OH} \\ | \\ \text{CH=CH-(CH}_2)_{12}\text{-CH}_3 \end{array}$$

FIGURE 7. Structure of glycolipids. *R* represents mono- or
oligosacharides, where n = 14—
24.

c. Glycolipids

Two major groups of glycolipids are found in mammalian membranes: glycoglycerolipids and glycosphingolipids. Glycoglycerolipids contain glycerol, fatty acid or fatty ether, and carbohydrate. Glycosphingolipids contain sphingosine, fatty acid, and carbohydrate. The basic structure of these lipids is shown in Figure 7. Glycoglycerolipids are found in the brain and sperm cells, and in small amounts in the kidney; glycosphingolipids are present in all tissues.[532]

There are considerable variations in the fatty acid composition of glycolipids. In erythrocytes there is a variation of chain length between C_{16} and C_{25}, and the degree of unsaturation is also variable. Carbohydrate residues may vary widely depending on the glycolipid type. Neutral glycososphingolipids contain neutral sugar moieties; gangliosides contain sialic acid; sulfatoglycosphingolipids have sulfate ester groups attached to the carbohydrate residue.

Glycolipids and glycoproteins represent the blood group substances of the erythrocyte membranes.[531] Changes in the glycolipid pattern in membranes are connected with many pathological conditions, such as malignancy, tumorigenic virus transformation, and other diseases.[209,210,519] These moieties are important in growth regulation and in interactions between cells.[112] Glycolipids also function on the membranes as receptor sites for hormones, toxins, and interferon.[112,171]

IV. MEMBRANE ABNORMALITIES IN CELL INJURY

Lethal cell injury develops on account of any type of injurious agents. Injurious agents may originate either inside or outside the cell. Injuries of endogenous origin are mainly genetic alterations. Exogenous injurious agents are (1) microbiological agents or cellular parasites and their products (bacteria, viruses, protozoa, metazoa, fungi, etc); (2) drugs and other foreign compounds; (3) deprivation of oxygen and/or various substrates; (4) trauma, extreme temperatures, extreme pH, or radiation; and (5) various immunological phenomena. Following lethal injury, the cell reactions can be grouped into two phases: (1) a reversible phase which precedes cell death and (2) the irreversible phase, consisting of those changes occurring after the death of the cell. In the latter phase changes are irreversible even if the injurious agent is removed and the cell is returned to a normal environment.[553] Various cell activities form an integrated functional unit. Even after the cell is dead, individual cellular organelles such as the mitochondria or fragments of the endoplasmic reticulum still continue functioning in vitro in suitable media for long periods of time. These organelles, however, are unable to restore the complete cell functions.

During both the reversible and irreversible phases many changes occur in the function and structure of cell membranes.[389,533] The reversible phase is characterized by alterations in membrane functions such as changes in permeability, ATP and protein synthesis, water and electrolyte movements, changes in metabolic intermediates, and redistribution of cytoplasmic proteins. During the irreversible phase, autolysis and cellular degradation by proteolytic and hydrolytic enzymes and denaturation occur. Following sublethal injury, the cell can adapt to even a continuously present injurious stimulus by changing its physiological status within its homeostatic adaptability. Such adaptive changes include fatty changes, hypertrophy, atrophy, increased formation of cellular organelles (mainly lysosomes), aging, or neoplastic transformation. These changes may represent the increased formation of existing organelles such as proliferation of endoplasmic reticulum; formation of secondary lysosomes by autophagy; megamitochondria; peroxisome proliferation; and the accumulation of various abnormal products in the cell, such as hyaline bodies in chronic alcoholism, or abnormal phospholipid granules in chemically induced phospholipidosis.[128] These adaptations associated with structural and functional changes are important in many instances. These include the complex reactions in fatty liver production involving changes in the synthesis of triglycerides, phospholipids, cholesterol, and carrier proteins; increased membrane movements connected with enhanced lysosome formation by autophagocytosis; significant increases of autophagic vacuoles and membrane breakdown in atrophy; increased synthesis of endoplasmic reticulum membranes in the liver brought about by induction with barbiturates and other inducers; and the marked changes in antigenic properties and electric changes of the cell surface in malignant transformation.[69]

Acute lethal injury can result from two types of interactions with cell homeostasis: (1) conditions that inhibit ATP synthesis, and (2) conditions that alter directly plasma membrane permeability and transport systems. There is a thorough interrelation between these interactions. Ischemic cell injury occurs by cessation of blood flow which results in the combined effects of anoxia and lack of substrates.[153] In the kidney and liver, various changes are seen within 5 min which include clumping of nuclear chromatin and disappearance of the mitochondrial matrix. Mitochondria cease to function as soon as the oxygen tension becomes limiting. Total cellular ATP level falls rapidly to zero. The endoplasmic reticulum undergoes early fragmentation and dilatation correlated with the influx of water and sodium. Polysomes are lost and protein synthesis becomes defective. Lysosomal enzymes are released as the result of increased fragility.[218,551]

In sublethal injury the cells are able to adjust by attaining an altered level of homeostasis. Many types of sublethal injuries are associated with changes in cell membranes. Lysosomes are altered, they release hydrolases into tissues, and numerous autophagic vacuoles are formed, as in many diseases such as gout, rheumatoid arthritis, nephritis, and vasculitis.[554]

A. Diseases of Membrane Transport Mechanism

Transport of biologically important substances from one site in the organism to another, or passage into and out of the cell is essential for normal cellular function.[8,326,515,517,518] Disorders arise from the defects in transport mechanisms across plasma membranes or complex multicellular surfaces, mainly representing epithelial transport. Transport processes occur at different levels, including (1) material translocation within cells, (2) movement across cell membranes, and (3) integrated transport mechanisms, specifically related to an individual organ such as kidney or liver. These transport processes can be active or passive. In active transport, a connection exists between the movement of substances and energy-yielding metabolic reactions. In this case, energy is required for the transfer of a given component. In passive transport, a diffusion proceeds through membranes without the need for metabolic energy. The passive transport may represent a simple diffusion along an electrochemical potential gradient, or a facilitated transport which requires the need for a carrier molecular associated (may be transiently) with the substance translocated. The active transport mechanisms have greater physiological importance.

1. Active Transport

Components of the cell membrane function as carriers in the process of active transport and it is most likely that these carriers are linked with proteins.[31,51,312,355,399,470,516,557] However, the nature of carriers is probably varying with the nature of substances which are transported through the membrane. There are two stages in this transport. In the first stage an association is established with the substance to be transported. In the second stage, the process of migration through the membrane to the interior of the cell involves either conformational changes in the carrier molecule itself or actual inward migration of the carrier-substance complex.[63,92,228,277,295,349,428] It may be that passive exchange diffusion is also carrier-mediated and a stimulus triggers the conformational change on the carrier. Physical changes such as temperature variations can initiate such changes in the membrane. In the active transport the change is thoroughly connected with an energy-producing process such as the synthesis of an energy-rich bond by phosphorylation. This provides the binding force for the carrier-substance association and induces the conformational change. The utilization of the energy results in the actual translocation of the substance.

In the cellular transport mechanism, structural proteins play an important part. They are associated with the cell walls. Some proteins with adenosine triphosphatase activity have been identified in liver cell membranes, red cell ghosts, platelet membranes, and other membranes.[39,51,161] The enzyme activity could be separated into components. One shows actomyosin-like contractile protein characteristics — activated by Ca^{2+} and Mg^{2+} ions and unaffected by cardiac glycosides. The role of this structural protein probably concerns structural modification of the cell surface by contraction, and calcium and magnesium transport. This enzyme especially has considerable importance in blood platelets during coagulation. It is called thrombosthenin and takes part in the viscosity changes and subsequent clot retraction processes exhibited by the platelets during coagulation. The other adenosine triphosphatase activity is stimulated by Na^+ and K^+ ions and inhibited by cardiac glycosides. This enzyme is intricately connected with the transport of monovalent cations and responsible for the function of the sodium pump.

The concentration difference on the opposite sides of the cell membrane between intracellular K^+ and extracellular Na^+ is very strictly maintained. Since both Na^+ and K^+ ions migrate freely through membranes and the uneven maintenance requires energy, the transport of sodium is essential in many cellular processes: (1) in the regulation of amino acid and sugar uptake by the epithelial and other cells, (2) in the action of insulin in the cell, (3) in the control of absorption and secretory control of kidney tubular cells, and (4) in the excitation and conduction of nerve cells. However, the active transport system makes the cell selectively permeable to sodium and potassium and therefore concentration gradient is produced for these ions across the cell membrane. This difference develops from the net transport of cations as the result of two basic processes: (1) the active transport of potassium into and sodium out of the cell, and (2) the passive diffusion along respective electrochemical processes. Water freely penetrates through the cell membrane; therefore, changes in the rate of cation movement are connected with either water uptake or loss by or from the cell.

The mechanism of translocation of ions, the so-called sodium or potassium pump, has considerable importance in secretory structures such as renal tubular epithelium and in hormone-producing glands. Adenosine triphosphatase enzymes located in the cell membrane probably play an essential role. Inhibition of these enzymes is connected with abnormalities. Impairment of the sodium pump represents an inhibition of the active sodium transport. Subsequently sodium ions migrate freely into the cell from the extracellular compartments with an accompanying excess of water passage resulting in cellular edema and increased volume. Various hormones participate in the control of the active electrolyte transport, including antidiuretic hormone and various steroids which influence salt and water turnover in the cell. These are capable of modifying cation transport and fluid distribution between various compartments.[298,454]

There are other active transport systems which affect the translocation of amino acids. These are energy dependent and are influenced by the length of chain, location, and number of amino groups. Several distinct amino acid transport systems function in the kidney tubules. Some are highly specific and only translocate a single amino acid, while others transport only dibasic amino acids, and a third system is associated with many different types of amino acids.

Glucose and other simple sugars are also actively transported by specific, energy-dependent systems which utilize ATP as an energy donor and require Na^+ ions to function. This system is very specific in the kidney. Glucose is passively filtered at the glomeruli and actively reabsorbed in the proximal convoluted tubules. In contrast, the membrane of the liver cell is freely permeable for glucose and therefore hepatic enzymes play a significant role in the regulation of plasma glucose levels.

Specialized areas of transport occur in several organs with highly developed absorptive surfaces. These are associated with the selective uptake and selective conservation or retention of essential substances. These specialized organs include the intestinal mucosa for selective uptake. Selective retention occurs in the concentration of water and essential solutes in the kidney tubular system.

2. Epithelial Transport

Cells with similar structure and function form complicated multicellular tissues by joining together. In turn, one type of multicellular organization can be associated with other cellular complexes to organize various organs. For instance, connective tissue and muscle cells from vessel walls in epithelial tissues take part in the specific function of these structures.

A wide variety of disorders are connected with disturbed multicellular organization.[551] Most of them are relatively infrequent but their investigations and clinical significance represent important assets in understanding the genetic control and basic mechanism of transport process in the human body. These diseases derive usually from a genetic defect or from abnormalities, dependent upon the nature of the biological membrane involved, the altered membrane function, or membrane organization. Many hormones, vitamins, and drugs affect these molecular mechanisms. The mode of action or how these components influence transport processes are not yet completely understood.

a. Diseases of Disturbed Transport

Many diseases are connected with disturbances of active transport systems, including amino acid and phosphate transport defects, renal glycosuria, tubular acidosis, vitamin B_{12} malabsorption, cystic fibrosis, and pseudohypoparathyroidism[131,411,457] (Table 3 and see Plate 2.) Among the various amino acid transport defects, cystinuria and Hartnup disease are the most important. Cystinuria is a hereditary defect of cystine transport by the kidney.[221,345] In this disease, the clearance of cystine is close to the maximum glomerular filtration rate. Considerable amounts of cystine, arginine, ornithine, and lysine are elminated in the urine; the amount excreted often exceeds 1 g daily. The excretion of dibasic amino acids follows different patterns. Occasionally, cystine stones are formed in the kidney, due to relative insolubility of this amino acid.

In Hartnup disease, neutral and aromatic amino acids, such as alanine, serine, valine, threonine, asparagine, glutamine, phenylalanine, tyrosine, tryptophan, and histidine,[346,495] are excreted in excessive amounts. Dibasic and dicarboxylic amino acids are excreted normally. The defect causing this abnormality is present in the kidney tubules and in the intestine; both tubular reabsorption and intestinal absorption are defective. Clinical signs of this condition include skin rash resembling pellagra and cerebellar-type ataxia, probably originating from an inhibited transformation of tryptophan to nicotinamide. Abnormalities in glycoprotein secretions are characteristic of cystic fibrosis (Plate 2).

b. Hypophosphatemia

Hypophosphatemia is characterized by excessive urinary excretion of phosphate and cor-

Table 3
CELLULAR TRANSPORT DEFECTS OF VARIOUS BODY CONSTITUENTS CAUSING DISEASES

Disease	Site of defect	Affected substance	Major clinical symptoms
Cystinuria	Proximal convoluted tubule, intestinal mucosa	Cystine, arginine, ornithine, lysine	Kidney stones
Hartnup disease	Intestinal mucosa, proximal convoluted tubule	Monoamino acids	Cutaneous lesions, ataxia
de Toni-Fanconi syndrome	Proximal and distal convoluted tubules	Multiple amino acids, glucose, water, PO_4^{-3}	Acidosis, osteomalacia
Renal glycosuria	Proximal convoluted tubule	Glucose	Symptomless, glycosuria
Glucose-galactose malabsorption	Intestinal mucosa, ileum	Glucose, galactose	Diarrhea
Renal tubular acidosis	Distal convoluted tubule	H^+	Nephrocalcinosis, systemic acidosis
Vasopressin-resistant diabetes	Distal convoluted tubule	Water	Polydipsia, polyuria
Hypophosphatemic rickets	Intestinal mucosa, proximal convoluted tubule	Ca^{+2}, PO_4^{-3}	Osteomalacia
Pseudohypoparathyroidism	Intestinal mucosa, proximal convoluted tubule, skeleton	Ca^{+2}, PO_4^{-3}	Cataract, nephrocalcinosis
Cystic fibrosis	Eccrine glands and associated organs, respiratory mucosa, pancreas, sweat glands	Na^+, mucins	Malabsorption syndrome, pulmonary infections, pulmonary and pancreatic fibrosis, hyponatremia, heat intolerance
Hereditary spherocytosis	Erythrocyte membrane	Na^+	Hemolytic anemia
Erythrocyte ATPase deficiency	Erythrocyte membrane	Na^+, K^+, and other cations	Hemolytic anemia
Vitamin B_{12} malabsorption, intestinal type	Intestinal mucosa, acceptor sites for B_{12} complex entry	Vitamin B_{12}	Megaloblastic anemia

responding low serum level. This condition is the consequence of reduced tubular reabsorption and decreased uptake of phosphate from the intestine. The absorption of calcium from the intestine is also low, resulting in the clinical conditions of osteomalacia or rickets. There are some indications that the reduced intestinal uptake is the primary cause of the disease leading to secondary stimulation of the parathyroid gland causing decreased phosphate reabsorption by the kidney tubules. Subsequently, an increased elimination of phosphate occurs in the urine and hypophosphatemia sets in. The consequence of the impairment of phosphate and calcium metabolic balance is a reduced formation of the mineral matrix of the bones and partial loss of skeletal calcification.

c. Tubular Reabsorption Defects

Renal glycosuria is a rare disease connected with a defect of renal tubular reabsorption of glucose.[281,604] In this condition the blood glucose level is normal, the glucose tolerance test is normal, and ketosis is absent, but glucose appears in the urine in greater amounts. The reason for this disorder is that the nephrons do not function uniformly. In some nephrons, the carrier sites are easily saturated, the concentration of the tubular fluid exceeds the threshold level and, consequently, glycosuria occurs.

Renal tubular acidosis is a complex condition involving coordinated impairment in the functions of the kidney and respiratory system.[211,483,520] It is characterized by low plasma

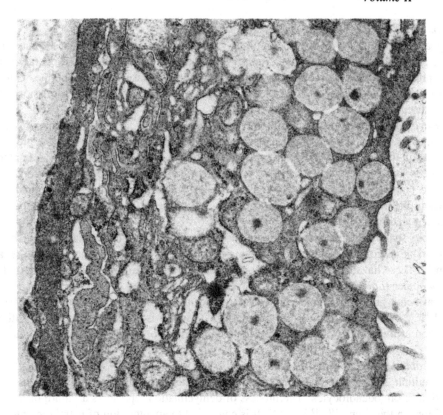

PLATE 2. Transmission electron micrograph of mucus-secreting cells from the lung in cystic fibrosis. Hypertrophy and hypersecretion of mucus-secreting cells is accompanied by alterations in glycoprotein secretions and electroytes. (Courtesy of Dr. J. M. Sturgess, Warner-Lambert/Parke-Davis Research Institute, Sheridan Park, Ontario, Canada.)

bicarbonate and high chloride levels and the inability to excrete urinary acid. The etiology of this syndrome is complex. Major observations indicate a genetically determined autosomal pattern of inheritance, but symptoms of renal tubular acidosis are present in heavy metal poisoning and cystinosis. Many other complex defects, which often affect several secretory and reabsorptive functions, impair kidney tubules. These severe disturbances are associated with complex aminoaciduria, glycosuria, hypercalciuria, hyperphosphaturia, and excessive uricosuria, as in the Toni-Fanconi syndrome.[476] In a variety of metabolic abnormalities complex or special transport defects develop: copper metabolism in Wilson's disease, galactose-1-phosphate metabolism in galactosemia, and cystine metabolism in cystinosis. Heavy metal poisoning and toxic organic compounds such as Lysol can damage the renal tubular epithelium, causing abnormal deposition and transport defects similar to the conditions found in the Toni-Fanconi syndrome.

B. Membrane Abnormalities in Neurological Disorders

Many important properties of the cells in the nervous system are connected with membrane structure. Moreover, the elimination of endogenous products or the effect of exogenous substances are also associated with membrane-related mechanisms. Alterations of membrane composition, structure, receptor function, and transmitter release can therefore cause significant neurological dysfunction. In the pathogenesis of many neurological disorders neuronal membrane changes are involved.[91,248]

1. Synaptic Transmission Abnormalities

Synaptic transmission is disturbed by many toxic actions. Intoxication can be caused by toxins produced by *Clostridium botulinum*.[113,259] Botulinus toxins bind to peripheral cholinergic presynaptic terminals of the axons and block the release of acetylcholine. Clinical manifestations of botulism thus represent a blockade of cholinergic neurotransmission at the neuromuscular junctions.

Clostridium tetani produces an exotoxin which is responsible for the clinical manifestation of tetanus.[425] This condition is connected with the uptake of the toxin by cholinergic nerve terminals in the muscle. This is then transported along the axons to the perikarya of anterior horn cells and further across the plasmalemma to comparatively low glycine-containing synaptic boutons on the surface of the cell body.[426] There the tetanus toxin impairs the release of the inhibitory neurotransmitter glycine, resulting in the manifestation of spasm of the innervated muscles.

Various snake and spider venoms contain toxic components which bind to acetylcholine receptors at neuromuscular junctions leading to weakness through respiratory paralysis and death.[176] These hazards include the venom of the cobra and banded krait snakes and black widow spider (*Lactrodectus mactans*).

Myasthenia gravis is connected with a significant disorder of neuromuscular transmission.[114,158,502] The major clinical symptom of this disorder is weakness, connected with the reduction of the number of functional acetylcholine receptors at neuromuscular junctions.

2. Excitable Membrane Abnormalities

In multiple sclerosis the clusters of axons are deficient of myelin, probably due to destruction of oligodendrocytes. The process of demyelination represents a failure of impulse conditions in the axons.[436,442,573,578,588] It is connected to an alteration of membrane proteins so that current flow is often insufficient to initiate an action potential.[82] In some demyelinative neuropathies, such as the Landry-Guillain-Barré syndrome, the demyelination process is followed by infiltration of immunocompetent cells.[544] Clinical manifestations reflect the lack of impulse condition in the demyelinated axons. Due to remyelination of the Schwann cells in the peripheral nervous system, the disease may be completely reversible.

In many demyelinating diseases such as multiple sclerosis, myasthenia gravis, and Landry-Guillain-Barré syndrome, an immune response to membrane-associated antigens plays an important role in development. This may also be reflected in the pathogenic mechanism of viral disease in the central nervous system.

Seizure disorders reflect a repetitive neuronal firing which spreads along the fiber path.[427] The exact etiology of neuronal firing has not been established. The fact that most anticonvulsants act at membrane level, either by interfering with membrane permeability or by altering the affinity of the receptors for certain neurotransmitters, indicates that membrane phenomena is connected with seizure disorders.[149]

C. Membrane Abnormalities in Metabolic Disorders

Several inborn errors of metabolism involve the excess accumulation of membrane lipids, or are connected with an inability to generate normal membrane components. Most of the disorders are caused by lysosomal enzyme defects.[512] These enzymes are serving normal lipid metabolism and three distinct processes are associated with the impairments: (1) the defect may be directly related to the synthesis of a vital membrane component, (2) an accumulation of substrates due to metabolic block leads to insertion of these molecules into the membrane with subsequent alteration of its properties, and (3) accumulating substrates may interfere generally with cellular metabolism resulting in specific membrane abnormalities. These disorders include Gaucher's disease, a sphingolipid storage disorder caused by altered β-glucosidase activity and resulting in accumulation of glucocerebrosides. Most of the increased glucocerebrosides are stored in the reticuloendothelial cells of the liver, spleen, and bone marrow. Niemann-Pick disease is associated with a deficiency of sphingomyelinase activity, resulting in abnormal storage of sphingomyelin. Krabbe's disease, or globoid cell leukodystrophy, is caused by a deficiency of galactocerebroside-β-galactosidase activity and characterized by an enhanced production of globoid cells in the central nervous system. Fabry's disease involves α-galactosidase deficiency connected with the accumulation of ceramidetrihexoside or globoside. Tay-Sachs disease is associated with an almost complete lack of hexosaminidase A activity.

The clinical manifestations of lipid storage diseases vary considerably from mild bone changes and enlargement of the liver, spleen, and viscera, kidney failure, and, finally, severe mental retardation. When the disease manifests at an early age, the prognosis is most severe. Attempts of treatments by applying enzyme replacement therapy or organ transplant have so far been unsuccessful.

1. Brush Border Membrane Disorders

A large number of functions reside in the brush border, and primary or secondary effects that modify the structure of these membranes or their composition can lead to various disorders.[53,99,267,322,323,485,486] Primary diseases are relatively rare. The deficiency or lack of an essential membrane enzyme in the small intestine usually causes malabsorption and osmotic diarrhea. In some cases, a parallel defect of the same enzyme activity occurs in the proximal tubular membrane associated with abnormalities in the kidney transport mechanism.

a. Carbohydrate-Related Abnormalities

Abnormal absorption of carbohydrates can occur as a consequence of a defective step of carbohydrate digestion or transfer into the cell.[110,117,486] The unabsorbed sugars are degraded by bacterial action in the lower ileum and colon resulting in a mixture of monosaccharides and metabolites such as acetic, propionic, butyric, and lactic acids. The unabsorbed and nonhydrolyzed carbohydrates may cause osmotic diarrhea.[412]

The most prevalent disorder is dietary lactose intolerance, due to low lactase levels in the intestinal brush border.[173,230] Hereditary lactase deficiency has been reported; however, the lactase deficit is more frequently acquired and manifests itself in later life. Lactose intolerance occurs between 2 and 10 years of age, although young children are generally tolerant to lactose even in the affected population. In the acquired deficiency, dietary lactose is not hydrolyzed completely due to low enzyme levels.

Glucose-galactose malabsorption is a hereditary disorder.[146,200,336] Symptoms start from the beginning of neonatal life and include severe watery diarrhea. Low-rate renal glucosuria also occurs in these cases. The disorder is due to a reduction in the number of glucose transport carriers in the brush border. Fructose is the only carbohydrate tolerated by these patients.

Sucrose-isomaltase deficiency is a hereditary disorder, characterized by a total lack of sucrose activity and only traces of isomaltase activity in the brush border.[65,173,199,271,424,485] Glucose, lactose, and maltose tolerance is normal, and intake of starch, dextrin, or sucrose by these patients causes diarrhea.

b. Amino Acid- and Protein-Related Abnormalities

Amino acids are actively absorbed from the small intestine and the kidneys by membrane transport mechanisms.[337,338] These mechanisms involve various transporters which fall into four broad categories: neutral amino acids, dibasic amino acids, dicarboxylic amino acids, and glycine-amino acid transporters.[59,143,265,482,587] Intestinal transport defects of a specific amino acid category usually run parallel with a similar defect in renal tubular reabsorption. Amino acid malabsorption is therefore often associated with aminoaciduria.[116,345,480] Exceptions may occur in isolated cases, such as methionine malabsorption or tryptophan malabsorption. In methionine-transport failure, abnormal amounts are eliminated in the feces. An oral load of methionine evokes diarrhea and increases excretion of α-aminobutyric acid and fecal loss of many neutral amino acids due to competition. In tryptophan malabsorption the intestinal tryptophan is converted to indican and other indolyl compounds through bacterial degradation. These substances are responsible for the characteristic color of the urine. This disorder is also called as "blue diaper syndrome".[140]

Hartnup disease is characterized by constant amino aciduria, pellagra-like skin rash, and cerebellar ataxia, and is in some instances connected with mental retardation.[346,495] In these cases tryptophan absorption from the gut is defective, resulting in an increased urinary excretion of indoles as a consequence of bacterial action on tryptophan in the gut. The clearance of other neutral amino acids such as phenylalanine, valine, threonine, isoleucine, and histidine is increased in the kidney, and absorption is reduced from the intestine. The pellagra-like symptoms are related to the lack of tryptophan, which is an essential amino acid. These can be cured by therapy with nicotinamide, which is synthesized normally from tryptophan.

Cystinuria is entirely due to a defect in renal reabsorption.[21,345] Often cystinuria is combined with dibasic aminoaciduria, caused by a defect of transport of these amino acids in the small intestine and kidney. The renal clearances of cystine, arginine, lysine, and ornithine are high, and since cystine is poorly soluble, kidney and bladder calculi are formed, leading to renal colic, infections, and eventual renal failure and death.

Dicarboxylic aminoaciduria is an autosomal-recessive transport disorder which also affects the small intestine and the kidney[542,587] with concurrent high renal clearance of aspartic and glutamic acid. Hereditary iminoglycinuria is characterized by a renal membrane defect leading to enhanced elimination of glycine, proline, and hydroxyproline.[194] This disorder is asymptomatic, since the amino acids involved are not essential. Lysinuria represents an inherited transport disorder of dibasic amino acids.[59,400]

c. Renal Tubular Abnormalities

Renal tubular acidosis is characterized by persistent metabolic acidosis, low serum bicarbonate, and elevated chloride under near normal glomerular filtration rates.[211,483,520] The defect is connected with the inability to achieve a net tubular hydrogen ion excretion, probably caused by either an impaired bicarbonate reabsorption in the proximate tubule or by the lack of sustained transepithelial acid-base gradient in the collecting ducts. The consequence of these abnormalities is that normal acid-base equilibrium is not maintained in the body. Large amounts of bicarbonate are lost in the urine; this loss is often connected with glucosuria, aminoaciduria, and hyperphosphaturia. In several clinical conditions water transport in the intestines shows abnormalities.[262] Congenital alkalosis is also connected with diarrhea.[37]

The renal proximal tubular reabsorption of phosphate is a sodium-dependent process located in the brush border membrane and is under the influence of parathyrin and dietary phosphate.[277,526] Parathyrin inhibits phosphate reabsorption whereas low phosphate diet or parathryoidectomy increases it. When the reabsorption of filtered phosphate is defective, hypophosphatemia occurs. This is a familial condition, with diminished renal tubular reabsorption of phosphate, known as vitamin-D-resistant rickets osteomalacia. The small-intestinal absorption of phosphate and calcium is defective and cannot be cured with vitamin D treatment.

In cystinosis, very high concentrations of free cystine are found in lysosomes, resulting in crystal depositions in various tissues.[476] The cause of cystine accumulation is not known. Clinically in the most severe forms of the disease, polyuria, polydipsia, glomerular damage, and uremia are seen, caused by tubular and glomerular dysfunction, leading to death at an early age.

In Fanconi's syndrome tubular defects occur first, and subsequently tubular reabsorption of amino acids, glucose, water, and electrolytes is severely reduced.[476] Cystinosis is the most common cause of Fanconi's syndrome, but other inherited diseases can also lead to nonspecific renal proximal tubular dysfunction.

d. Secondary Brush Border Membrane Disorders

Several disease conditions are connected with generalized dysfunction of the brush border membrane.[325,334] This is particularly important in the small intestine, leading to malabsorption and malnutrition. In these disorders the mucosal cells are functionally and structurally abnormal, resulting in abnormal digestion and absorption of almost all food.[283] The microvilli are shortened and sometimes almost flat and reduced in number. Consequently, the functions of the entire cell are modified, and intracellular organelles and cytoplasmic activity are impaired.

Celiac sprue is also described as celiac disease, idiopathic steatorrhea, and gluten enteropathy. It affects primarily the intestinal mucosa.[57,272,547] Cereal grains (such as wheat, rye, oats, and barley) containing gluten are involved in the pathogenesis. Celiac sprue is characterized by malabsorption of all nutrients, absence of normal intestinal villi, and degenerative changes in mucosal cytoplasm and organelles. The digestive brush border enzymes are decreased and the absorption of most nutrients becomes insufficient. When gluten is removed from the diet the conditions quickly return to normal.

The disease is connected with a low-molecular-weight acidic polypeptide, a toxic substance in gliadin which is responsible for the lesion.[57] How the toxicity manifests in the intestinal cells is not known, and it is probable that immunologic mechanisms are involved. The lesion usually occurs in part of the intestine and only in severe cases, is all the length of the small intestine affected. The extent of the intestinal damage usually indicates the intensity of the symptoms.

There are some other conditions with associated intestinal mucosal changes. In Kwashiorkor, a disease in children brought about by protein-calorie malnutrition, the absorption of various nutrients is decreased.[417] The height of the intestinal villi is reduced, representing less-efficient absorption surfaces and diminished activity of digestive enzymes in the brush border. In the Zollinger-Ellison syndrome, a massive hypersecretion of gastric acid is associated with tumors of non-β islet cells of the pancreas which produce and secrete gastrin.[325,334] The enhanced production of this agent is responsible for the enhanced acid secretion. The primary manifestation of this disease is ulceration of the upper gastrointestinal tract. Secondary malabsorption is due to intestinal mucosal damage by gastric acid.

2. Platelet Membranes

Platelets exert an important role in hemostasis. The nonactivated circulating platelets have an oblate shape; when they are activated, their shape may show changes. These include invagination of the plasma membrane, development of pseudopodia, and adherence to other platelets and to subendothelial structures. Intraplatelet granules are released during activation. The internal platelet reactions are triggered by an increase of Ca^{2+}. After activation, O_2 consumption is elevated, ATP content is lowered, and arachidonic acid is liberated from membrane phospholipids and oxidized to prostaglandin endoperoxides and thromboxane. Various substances, such as ADP, serotonin, and thromboxane A_2, which are secreted from activated platelets, are able to activate other platelets. The secreted substances enhance or modify the initial activation by probably interacting at the site of the external plasma membrane. Inherited and acquired defects of platelet activation are associated with abnormalities in membrane integrity.[161]

Exposure of platelets to an activating agent modifies the number and location of receptor sites on the cell surfaces. When the shape of platelets is altered, their surface area is increased by invagination of membranes and larger amounts of glycoproteins become available. In pathological conditions such as the Bernard-Soulier syndrome, characterized by giant platelets, the distribution of certain membrane glycoproteins is abnormal.[385] Prostaglandin E_1 reversibly inhibits platelet activation, including shape changes, by increasing platelet cAMP. Increased cAMP levels alter the pattern of protein phosphorylation in the plasma membrane.[217] Although these changes are complex, platelet plasma membranes represent the site for many unique functions such as receptor binding of ADP, thrombin, epinephrine, and collagen, as well as binding and transport of serotonin. Platelet membranes are the structural sites for storage and release of arachidonic acid, and the sites for the formation of fibrin in hemostasis. The special surface of platelet membranes is also essential in the process of adhesion to other platelets and to subendothelial surfaces.

Genetic dysfunctions are associated with specific defects in the glycoprotein profile of the platelet membrane. These include (1) defects in sugar intake by the cell, (2) defective glycosylating system, and (3) decreased levels of precursors needed for the synthesis of glycoprotein oligosaccharides. Patients with the Bernard-Soulier syndrome and Glanzmann thrombasthenia show a significant reduction in sugar transferases, probably connected with an inborn error in some steps of glycoprotein biosynthesis.[97,386]

3. Membrane Surface Receptors
a. Regulation and Disease

Cell surface receptors are membrane proteins and they are important in intracellular communication processes.[436] These receptors translate the binding of hormones and transmitters to changes in gene expression, cellular metabolism, or activities which just simply modify electrolyte distribution. Cell surface receptors are under regulatory mechanisms which include some basic stages: (1) binding of the hormone or transmitter, (2) movement of the receptors within the surface membrane and the formation of aggregates or clusters, (3) internalization of receptor-ligand complexes with production of intracellular vesicles, and (4) lysosomal fusion of vesicles, a step which then further processes internalized molecules. The regulatory processes can be modified, including stimulation of the cell surface receptors by hormones or neurotransmitters which alters the concentration and location of these receptors. Experimental denervation causes changes in surface receptors leading to target cell supersensitivity.[79] Loss of receptor sites by receptor internalization modulates the sensitivity of the cell to further stimulation. Recent investigations characterized acetylcholine receptors on muscle cells, insulin receptors, and β-adrenergic receptors on the surface of many cells. The mechanism of internalization of low density lipoprotein (LDL) through receptor sites has also been described.[276]

There is some evidence that changes in cell surface receptors are associated with disease. In myasthenia gravis patients, the nicotinic-acetyl choline receptor is reduced due to antibody-mediated autoimmune processes that accelerate the degradation of the surface receptor.[341] In hyperthyroidism the development of supersensitivity mediates many physiological variations, including cardiac hyperresponsiveness to catecholamines. In contrast, thyroxine or triiodothyronine enchance receptor levels in the myocardium.[266,591] Endogeneous catecholamines can regulate the diurnal variation in the number of β-adrenergic receptors in the pineal gland. The synthesis of *N*-acetyl transferase, an enzyme involved in melatonin synthesis, is regulated in the pineal by catecholamine levels. The circadian rhythm of *N*-acetyl transferase activity is connected with the amount of norepinephrine released by the sympathetic fibers which innervate the pineal gland.[262] The development of melanoma associated with increased pineal production of melatonin may be connected with changes in surface receptors.[270]

The internalization of LDL is connected with receptor-mediated endocytosis regulated by peptide hormones.[61] LDL transports cholesterol from the liver to the other cells of the body.

Although all cells require cholesterol, excess amounts of this substance in the blood can result in its deposition in the arterial wall leading to atherosclerosis. During the binding process the target cell receptor recognizes the protein component of the LDL complex. The internalized LDL particle is then fused with lysosomes, where the protein components are metabolized and cholesterol esters are degraded to free cholesterol. Cholesterol is important as a membrane constituent and as precursor in the synthesis of steroid hormones and bile acids. The role of cholesterol transport by LDL, the action of intracellular receptors, and further processing of cholesterol represent efficient ways to avert the development of atherosclerotic lesions.

The regulatory role of insulin in glucose metabolism is connected with its binding to cell surface receptors; in diabetes this association may become impaired. Insulin was one of the first hormones demonstrated to regulate its own receptor. Fasting or impaired pituitary function, which lowers blood insulin levels, brings about an increase of insulin receptor concentration leading to hypersensitivity.[13] On the other hand, loss of insulin receptors in liver and adipose tissue is characterized by insulin resistance and chronic hyperinsulinemia associated with obesity and increased secretion of growth hormone or glucocorticoids.

b. Role in Carcinogenesis

Cell surface constituents play a role in transformation and tumorigenesis. When normal cell growth is compared with tumorous or transformed cell growth in culture, the normal cells have limited life span, contact inhibition of growth, and are highly organized with normal morphology. Tumor cells in culture have unlimited life span, grow at high densities, and form multilayers. These differences may be related to lipid composition of the cell surface. Human leukemic lymphocyte cell membranes show greater fluidity than lymphocytes from healthy donors.[413] This is related to an enhanced molar ratio of cholesterol to phospholipids. The increase of membrane fluidity is directly related to increased cell proliferation. In leukemic patients serum lipid composition is altered. When isolated leukemic lymphocytes are incubated with LDL, the membrane viscosity is increased, indicating that membrane LDL receptors are functional and bind these lipoproteins. The difference in fluidity between leukemic and normal lymphocytes is probably due to a defect in absorption of serum lipids and modified cell surface constituents.

V. SUBCELLULAR ORGANELLES

A. Nucleus

The nucleus is the prime controller of cellular activity.[390] The chromosomes play the essential role in inheritance. Research in genetics and molecular biology of microorganisms revealed that the information relating to molecular structure is contained in the genome and is transcribed and translated for use in protein biosynthesis.[179,601] The genome is organized into chromosomes in which the DNA is closely associated with protein; ribosomes are almost entirely found in the cytoplasm. Although the messenger hypothesis and the genetic code have been described for microbial systems, including steps such as the synthesis of DNA, the transcription of DNA, and the translation of RNA, it is probable that the same system also operates in eukaryotic cells. It may be that there are mechanisms in animal cells that are not found in microorganisms. The coordinated growth of a cell requires close association between nuclear and cytoplasmic activities.[95] There is a continuous exchange of information in both directions crossing the nuclear membranes. We are concerned with the nature of these interactions between the nucleus and cytoplasm.

The nucleus represents an integrated structure where interactions occur between various macromolecules, and these interactions play an important part in its normal function. The fully differentiated nucleus responds to cytoplasmic and extracellular stimulations with great sensitivity. Controlling factors involved in nuclear functions are reversible changes in DNA and intranuclear electrolyte concentrations, and the availability of histone and nonhistone proteins and metabolites.[214,592] In the cell, the entry of cytoplasmic proteins initiates RNA and DNA synthesis, probably regulated by nonhistone protein repression.[71] Histone proteins are specifically associated with DNA in the fully differentiated cell. Various enzyme changes influenced or promoted by cytoplasmic factors can modify the microstructure of many histones. Although some of these changes may occur concomitantly, it has been suggested that the alteration of histone microstructure is controlled by nonhistone proteins by specific derepression processes.

Some characteristics of the nucleo-cytoplasmic interactions are determined by the nucleus and some derived from the cytoplasm. The nucleus is the main site of the genetic material. The information which specifies the structure of cellular proteins resides in the DNA of the genome in the form of specific sequences of purine and pyrimidine bases. The synthesis of proteins is directly controlled by nucleic acid genes. Many of these proteins are structural proteins and many of them function as enzymes, which are necessary to centralize various biochemical reactions. The absence of any of these essential enzymes can lead to abnormal function and disease. The synthesis of RNA in the nucleus and its translocation into the cytoplasm represents an important aspect of the nuclear interactions.[191]

Normally the process of protein and hence enzyme synthesis is remarkably efficient. Occasionally, the essential enzymes or proteins are not produced, or they are produced in a different or inactive form. These changes are genetically determined and derive from some DNA abnormalities in the nucleus. The decreased synthesis or an overall defect in the production of these enzymes may not always cause disease. Some enzymes are not vital and, if they are not formed, alternate pathways can compensate for them and take over the process they would not ordinarily catalyze. Some enzymes are, however, vital and their absence is connected with disease conditions. In these cases, the alternate pathways do not compensate satisfactorily for the loss of essential enzyme activity. When such important enzymes or proteins are missing or are defectively produced, functional and often pathological changes occur.

Those enzyme defects are connected with inborn errors of metabolism. Some of these defects are extremely severe, causing early death; others are moderate and some are even latent. Probably all individuals have a variable biochemical makeup. Individual differences exist in the enzyme content of various cells and tissues. Some borderline inborn error of metabolism is likely to exist in most individuals and it is also likely that not every healthy individual has the same quantity or quality of enzymes as their siblings. The variation in cellular proteins, including antibodies and enzymes, influences the susceptibility of the individual to disease.

The cytoplasm in the nucleo-cytoplasmic interactions has not been studied very intensely.[11] The nucleus is ultimately the source of basic information that determines the ultimate potential of the cell. This potential represents, however, a complex and often subtle interaction with the cytoplasm. Once the nucleus has initiated a particular sequence of events and passed it to the cytoplasm, the subsequent events are controlled by the cytoplasm with moderate influence by the nucleus. The cytoplasmic contribution to the interactions includes the appropriate processing of macromolecules which contain specific information. Proteins and RNA belong to this group of substances. The different types of RNA determine through the sequences of purine and pyrimidine bases the amino acid structure of a particular protein.[471] It is likely that some of these highly specific proteins are involved at least in some part in the regulation of gene function.[111] Further, macromolecules, and particularly proteins, carry a broad range of information which controls the overall direction of cell metabolism. The rapid exchange of substances between the nucleus and cytoplasm include several classes of proteins. DNA polymerases represent an important example of these regulators of cellular metabolism.[205]

Small molecules and electrolytes may exert profound action on cell structures and are responsible for the initiation of certain cellular processes. For instance, Mg^{2+} and Ca^{2+} affect the conditions of chromatin aggregation; hormones and cyclic AMP influence many biosynthetic activities of the cell;[10,254] thyroid hormones modulate various cellular oxidation reactions;[27] and pituitary hormones are involved in the regulation of hormones produced by and released from peripheral glands. Sex hormones derived from the cytoplasm promote nuclear reorganization, initiate the nuclear function, and trigger off protein synthesis and cell multiplication.[87,88,454,586] In the function of the nerve cells many small molecules are involved such as various types of neurotransmitters and memory peptides.[541] Some examples of nucleo-cytoplasmic interactions, such as the changes brought about by various chemical carcinogens on nuclear macromolecules and structures and the role of the endoplasmic reticulum in the activation of procarcinogens and its nuclear consequences, are discussed later.

B. Mitochondria

These organelles are present in the cytoplasm of all cells of higher animals, and have characteristic shape, size, structure, and function. The mitochondrion is bounded by a double membrane and the inner membrane forms invaginations which are the cristae.[213] The matrix often contains granules that are related to the ability of mitochondria to accumulate ions. Granules are prominent in the mitochondria of cells concerned with the transport of electrolytes and water such as kidney tubule cells, epithelial cells of the small intestine, and the osteoclast.[290] The inner mitochondrial membrane is also involved in oxidative phosphorylation which provides the basis for respiratory activity essential for the survival of the cell.[410] The surface area of cristae in various types of mitochondria is roughly proportional to respiratory activity.[402] For instance, heart muscle mitochondria have a higher density of cristae and of respiratory enzymes compared to liver mitochondria. Mitochondria are responsible for many oxidation reactions and for ATP synthesis.

Interruption of oxidative phosphorylation causes changes in the shape and volume of mitochondria. These changes appear promptly in ischemia,[66] and in conditions of acute injury.[328] Loss of oxygen due to hypoxia or ischemia is associated with dilatation and disruption of mitochondria. As a result of impaired oxidative phosphorylation, cellular ATP content rapidly decreases and electrolyte transport ceases. The results of these events are manifested in loss of potassium and influx of sodium, calcium, and water into the cell from the interstitial fluid, and changes develop in membrane phospholipids and fatty acid composition. Such reactions have been observed in the liver of patients receiving antiatherosclerotic drugs.[126,258,311,514] Hypertrophy of mitochondria is found in many disease conditions, such as deficiencies of vitamins (vitamin E[455] and riboflavin[538]), essential minerals (copper, iron,[119] and manganese[233]), essential fatty acids,[598] and in chronic alcoholism.[256] The increased mitochondrial volume may be connected with fusion of several mitochondria due to defects in the cellular respiratory system.[274,455]

1. Ischemic Injury

It has been demonstrated that the contractile function of the myocardiocyte abruptly ceases in ischemic regions. Normal myocardial function depends upon a continuous flow of oxygenated blood, and upon normal subcellular function (contraction, relaxation, protein synthesis, and active transport) which is predominantly regulated by oxidative phosphorylation leading to ATP synthesis by mitochondria.[468,481,488] During ischemia the supply of oxygen to the myocardial cell is reduced and a series of intracellular events results in depletion of high-energy phosphate levels. Subsequently, water and electrolyte transport across the sarcolemma is disturbed, protein and nucleic acid synthesis is altered, fatty acid oxidation is blocked, and the rate of glycolysis is enhanced. All these changes are probably connected

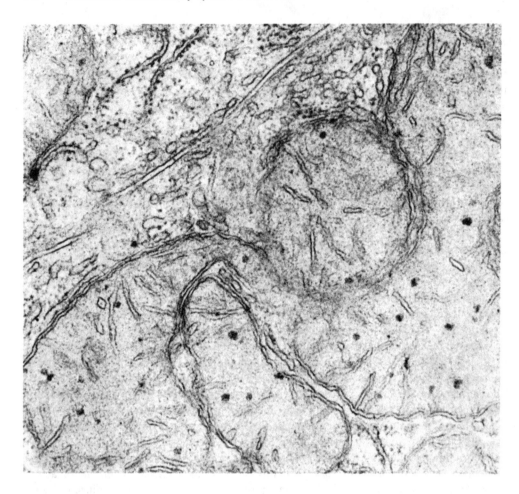

PLATE 3.   Alterations of the mitochondrial membrane in experimental acute alcoholism. The main irregularity consists of membrane coalescence.

with alterations in mitochondrial structure and function.[242,286,592] The membrane of mitochondria is damaged in acute alcoholism (Plate 3).

The precise sequence of events and the exact biochemical mechanism of the lesions involved in the irreversible cell injury are unknown. Possible factors are lowered intracellular pH due to increased intracellular acidosis, release of lysosomal enzymes, sarcolemma defects, and cell swelling. However, mitochondrial damage plays the major role in the sequence of events leading to irreversible changes. Mitochondrial damage occurs at early stages in the response of the myocardium to ischemia. Animal studies demonstrated that a 50% decrease in mitochondrial respiration occurs within 20 min after anoxia, and other subcellular organelles deteriorate more slowly. The function of the sarcoplasmic reticulum is reduced to 50% of normal at about 60 min; adenylate cyclase activity is reduced to 50% at about 90 min, and Na^+/K^+-ATP-ase activity is decreased by about 10% after 2.5 hr of ischemia. However, regardless of what is the primary defect of the consequence of ischemia, it has been established that any measure that prolongs the integrity of mitochondria increases the probability of myocardial cell survival after the ischemic episode.

Mitochondrial damage by ischemia is connected with significant changes of electrolyte composition; these are rapid loss of potassium, increased sodium and calcium, and delayed reduction of magnesium.[552] The most significant change is the accumulation of calcium in myocardial tissue damaged during ischemia. The reduction of extracellular calcium prior to ischemia can provide a significant degree of protection to contractile and mitochondrial functions of the myocardium.[288] Mitochondrial calcification during ischemic injury is probably associated with the deposition of amorphous calcium phosphate with organic material followed by progressive crystal growth.[67,206] The conversion of calcium deposits to crystalline formed in ischemic tissues may be facilitated by low levels of ATP and magnesium.[379]

Myocardial damage with myofibrillar hypercontraction and mitochondrial calcification represents a common pattern of injury that occurs in many types of toxic and metabolic lesions. These include the exposure to high levels of chemicals, or metastatic calcification associated with chronic anemia. In metastatic calcification apatite-like crystals accumulate in the myocardium and other soft tissues.[4] Progressive metastatic calcification is connected with marked muscle loss and fibrosis. It seems that abnormal calcium accumulation represents a manifestation of altered cell membrane function induced by cell injury.[159,374,552] Electrolyte balance becomes progressively impaired as the result of deficient energy metabolism. Prolonged injury and more severe membrane damage is followed by excess calcium influx to the cell. In contrast in hepatocellular injury calcium influx may precede the evolution of damage.[159]

2. Toxic Injury

Impairment of various mitochondrial oxidation processes and phosphorylation is one of the more common manifestations of toxic lead injury. Frequently, more than one enzyme of the electron transport system is affected. Lead toxicity involves mainly three organ systems: the hematopoietic system, the central nervous system, and the kidneys.[198] The effect of lead is associated with binding to mitochondrial membranes and changes in the ultrastructure and function in mitochondria from hepatocytes, renal proximal tubule, and bone marrow cells. Lead toxicity is also manifest in the central nervous system; however, a relationship between mitochondrial actions of lead and the central nervous system pathology has not yet been established.

The effect of lead poisoning on mitochondria of bone marrow reticulocytes is an impairment of heme synthesis. This is probably the cause for the subsequent anemia. Lead inhibits δ-aminolevulinic acid dehydratase, resulting in a block in the utilization of δ-aminolevulinic acid. δ-Aminolevulinic acid synthetase is depressed and ferrochelatase, an enzyme localized on the inner membrane of mitochondria, is inhibited.[252] In acute lead poisoning, particularly in children, the urinary excretion of amino acids, glucose, and phosphate is extensive. Whether the reduced reabsorption of these substances is related to reduced energy production by the impaired mitochondria or is the consequence of the action of lead on other cellular membranes is not well known. In man, lead poisoning due to excessive industrial exposure causes reduced sodium reabsorption.[469] This defect and aminoaciduria are reversible by treatment with chelating agents.

Poisoning with inorganic or organic mercurials also affects mitochondrial functions.[174] The effect on heart mitochondria is related to structural and functional changes by complexing thiol sites. This binding inactivates phosphate transport by mitochondrial membranes.[556] In mercury poisoning renal cell mitochondria also become irreversibly injured showing reduced activity of many enzymes, including succinic dehydrogenase and cytochrome c oxidase.[563]

There are many drugs and toxicants which selectively act on mitochondrial function. It appears, however, that physiological factors could lead to changes and it is often difficult to distinguish from pathological alterations. The various changes include enlargement, changes in matrix density, and loss of cristae. Abnormalities are paralleled with ATP reduction in many instances. Chemical injury causes nonspecific distortion and fragmentation of mitochondria. Swelling may also occur in response to a variety of conditions and some of them are physiological or adaptive. Some changes show selectivity and may be related to the

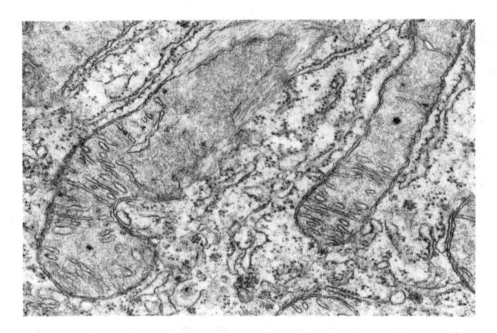

PLATE 4.   Fibrillar degeneration of mitochondrial membrane under conditions of chronic exposure to alcohol.

chemical structure of the injurious agent.[331,420] Salicylate-induced mitochondrial changes resemble the mitochondrial injury associated with the Reye's syndrome.[609] Extremely enlarged mitochondria, termed "megamitochondria", are characteristic of chronic alcoholism.[245,543] Paracrystalline inclusions are reported in patients with adverse drug reactions or other hepatic injuries such as alcoholism, but may represent some normal or senescent variations.[554] Foreign chemicals cause toxic damage of mitochondrial membranes. Examples include chronic alcoholism (Plate 4), and treatment with benzylidene yohimbol (Plate 5) or an antiatherosclerotic drug[126] (Plate 6).

3. Genetic Diseases

Mitochondrial mutagenesis was proposed as a first step in carcinogenesis many years ago by Warburg, based on the defects of mitochondria that are widespread in cancer cells of many types. It was difficult to define a direct relationship between respiratory deficiency and the cancerous condition, since mitochondrial aberrations lead only to a loss of energy and slowing of metabolic rate. Recent findings, however, indicate that mitochondria are more involved in cellular processes than by simply providing ATP. Mitochondria function in the modulation of the activity of certain nuclear genes which specify cell surface/plasma membrane components.[152]

The mechanism of this mitochondrial control is not yet known but the implications for carcinogenesis are clear, since changes in surface characteristics represent the first indication of the onset of the neoplastic condition. Changes in cell surface properties disturb the biochemical and physicochemical signals that pass from the external environment into the cell and direct its function. Cancer cells do not receive the proper signals and so behave abnormally, usually continuing to grow and divide instead of differentiating.

There are some inherited diseases mediated by mitochondrial aberrations. Mitochondrial cytopathy is related to abnormalities in mitochondrial structure and associated deficiencies in a variety of enzymes.[358,569] These deficiencies include reduced cytochrome b and cytochrome a_3 content and cytochrome c oxidase, ATP-ase and reduced nicotinamide adenine dinucleotide coenzyme Q reductase activities. All these biochemical parameters belong to

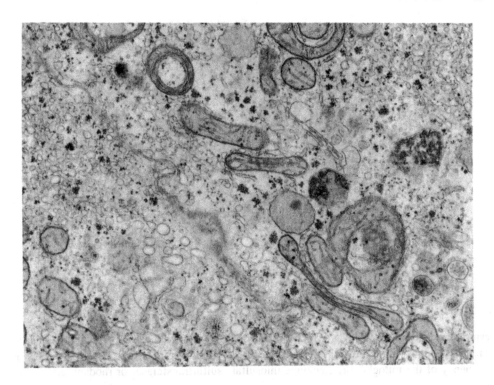

PLATE 5.   Mitochondrial alterations induced by toxic injury. Rat liver after benzylidene yohimbol treatment.

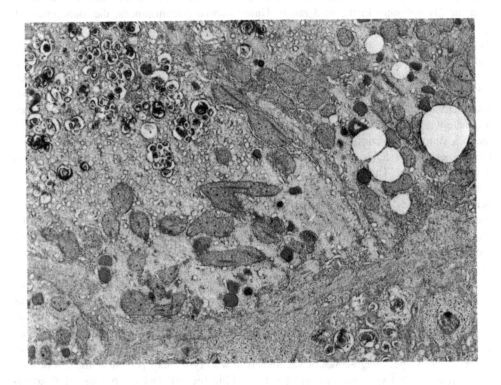

PLATE 6.   Morphological alterations in mitochondria in human liver with drug-induced phospholipidosis.

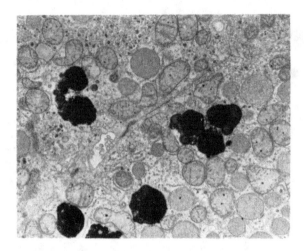

PLATE 7.   Normal appearance of lysosomes in the liver cell (dark granules). They are usually found in the vicinity of the bile canaliculus and contain hydrolases and degradation products such as lipofuscin.

respiratory enzyme complexes which are encoded by mitochondrial DNA.[144] In Laber's hereditary optic atrophy, which is connected with mitochondrial transmission, there is a deficiency of the mitochondrial enzyme thiosulfate sulfurtransferase, or rhodanase.[73]

C. Lysosomes

These cellular organelles were first described in liver cells. They contain several hydrolases with the pH optimum in the acid range. Under normal circumstances, the membranes of lysosomes are impenetrable to the substrate of these hydrolases, which are normally inactive. Activation may be by an action exerted on the lysosomal membrane by a variety of agents and physical or chemical treatments, acting either as physiological stimuli or as damaging factors. The intact lysosomes become more permeable and permit the entry of various substrates which are digested. These substrates include proteins, lipoproteins, lipids, polysaccharides, nucleic acids, and compounds of low molecular weight. These substances originate within the cytoplasm of the cell when the cell is undergoing autolysis. Lysosomes take up endogenous components synthetized by the cell in a process called endocytosis. Lysosomes exert a detoxication role when they incorporate nutritional or toxic factors from the external milieu. This action is described as phagocytosis. When cell autolysis is involved, these particles are called autophagosomes, and, in the case of extracellular substances, heterophagosomes. The effect of lysosomes is similar to the effect of monocytes and polymorphonuclear blood cells or free macrophages of the reticuloendothelial system, which incorporate large bodies such as bacteria and viruses. Lysosomes thus generally function as the intracellular digestive apparatus.[207] An electron micrograph of normal hepatic lysosomes is shown (Plate 7).

The cell absorbs large molecules through the mechanism of endocytosis. The cell membrane forms an invagination which surrounds, encloses, and finally isolates various molecules or particles, whereby this process produces intracytoplasmic vesicles or vacuoles of various sizes. At this stage the vesicles or vacuoles elicit no acid hydrolase activity, although they possess cell membrane enzymes such as alkaline phosphatase and adenosine triphosphatase. In an ulterior stage, phagosomes or vacuoles fuse with lysosomes present in the cell or with Golgi vesicles.[232] Lysosomes travel very fast within the cell and the process of fusion is very rapid. Subsequently, acid hydrolases are activated and the digestion process is completed.

The function of the lysosomes is fairly uniform; however, there is a variation in morphological appearance in various tissues throughout the body. This variation depends on the

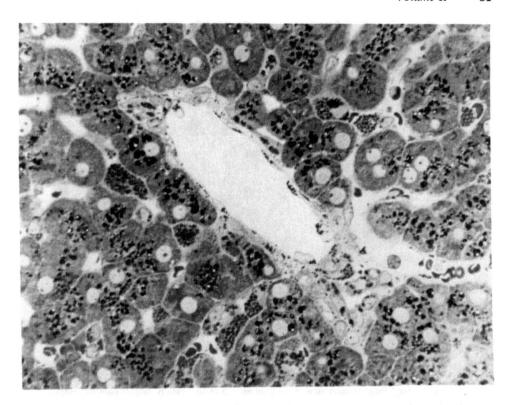

PLATE 8. Light microscopy aspect of liver in drug-induced experimental phospholipidosis with Coralgil. Note the accumulation of dark bodies in every cell type, including the vascular endothelium.

tissue and is probably related to cell types and function of each tissue. The chemical constituents probably reflect the morphological variability. The biochemical composition of lysosomes includes structural proteins and acid hydrolases, probably specific and tightly associated with structural elements. Lysosomal lipids appear to be similar to those in mitochondria. However, a strongly acidic phospholipid component was isolated from kidney lysosomes which binds lipophilic cations. Lysobisphosphatidic acid has been identified in secondary hepatic lysosomes following treatment with diethylamino ethoxyhexestrol. These strongly acidic lipid substances probably represent specific components that bind toxic substances and protect the cell from damage. Through this mechanism, lysosomes incorporate great amounts of drugs as well as high proportions of iron and flavins.[123,135,222,324,506,580] In drug-induced phospholipidosis, lysosomal structures in the liver are altered, as seen under light microscope (Plate 8) or electron microscope (Plate 9).

The major role of lysosomes is to form a system of cell defense in response to injurious factors attacking the cell. In some cases, however, lysosomes elicit an important physiological role, such as in thyroid hormone secretion. In the thyroid gland the epithelium absorbs the glycoprotein thyroglobulin by endocytosis. The colloid droplets are very variable; thyroid lysosomes fuse with these colloid droplets and by proteolytic activity set the hormone free from protein support and release the active form. This effect is not synchronous; therefore, the action of thyroid hormone is dependent on the stage reached by lysosomal fusion and enzyme activity. This process is not the only mechanism regulating thyroid hormone production, but it is very important in physiological circumstances. In contrast to this role, lysosomes do not participate in the disposal of protein hormones of the anterior pituitary such as somatotropin, thyrotropin, or other peptides; neither do lysosomes play a part in disposing excess insulin.

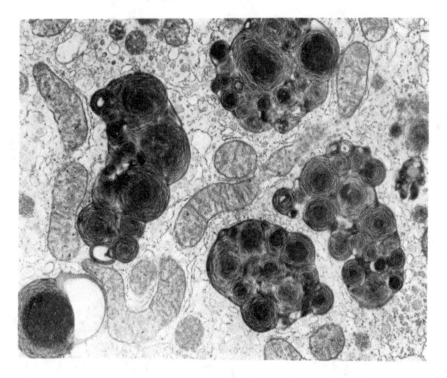

PLATE 9.   High magnification view of the lysosomes in experimentally induced phospholipidosis in mouse liver.

Steroid hormones act on the lysosomal membrane by either stabilizing or labilizing mechanisms.[382] These actions are probably unrelated to hormone activity although specific changes have been reported in lysosomes of rat prostate or castrated mice treated with estradiol. The process of activation and inhibition is probably connected with changes in the permeability of the membranes. When isolated rat liver lysosomes are incubated with cholesterol or progesterone in vitro, less or more acid phosphatase activity is released into the medium than from control lysosomes. This suggests that cholesterol stabilizes, whereas progesterone labilizes the lysosomal membrane (Figure 8). Various nonsteroid compounds also affect lysosomal membranes (Table 4). Lysosomes of adrenal cells have no effect on steroid secretion. However, the stabilizing action of cortisone and hydrocortisone is essential in reducing the inflammatory reaction in other tissues. The contrasting action of steroids on lysosomes indicates that perhaps, under physiological conditions, the variations in relative concentration regulate the activity of these subcellular organelles. This regulation is associated with several factors: (1) concentration of various steroids controlled by a balance in their synthesis, (2) metabolism and elimination, (3) feedback mechanism regulating release from organs, and (4) their availability at the site of action which sometimes may represent the hormone bound to protein, or the active free form.

The inhibition of lysosomal enzyme activity may be important in the development of some diseases. Trypan blue, a powerful inhibitor of acid hydrolase, when injected into rats is concentrated in lysosomes of phagocytic tissues and elicits teratogenic and carcinogenic action. It has been suggested that these pathological changes are related to the inhibition of lysosomal enzyme activity. On the other hand, the activation of lysosomal enzymes may represent an adaptation to correct a biochemical lesion, since the function of the lysosomes is dependent upon their enzymes. There are certain metabolic disorders where lysosomes are deficient in one or more hydrolases.[74] This deficiency may lead to severe cellular disturbances. Sometimes there is no complete loss of enzyme activity, only a great reduction,

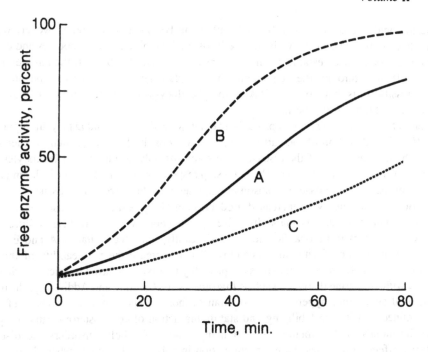

FIGURE 8. Effect of various steroids on the stability of lysosomes. Incubation of lysosomes with progesterone enhanced acid phosphatase activity whereas cholesterol caused a decrease. These effects indicate a labilization or stabilization of the membrane, respectively. Free phosphatase activities: (A) control; (B) in the presence of progesterone; and (C) in the presence of cholesterol.

Table 4
COMPOUNDS AFFECTING
LYSOSOMAL MEMBRANES

Stabilizers	Labilizers
Cholesterol	Progesterone
Cortisone	Testosterone
Prednisolone	Diethylstilbestrol
Chloroquine	Vitamin A
Antihistamines	Vitamin E
Phenothiazines	Detergents
Salicylates	Polyene antibiotics antifungal agents

and a drug which activates lysosomal enzymes may exert remedial effects on the disease.

Lysosomes are probably involved in all diseases which are connected with an increased or abnormal turnover of cells and cell constituents and where an enhanced defense is needed to restore normal circumstances. Lysosomal activity is raised in many conditions including infections, inflammations, and nutritional deficiencies.[19,75,124]

Most knowledge has been obtained on the relationship between infectious diseases and the role of lysosomes. During the defense action the intact bacteria are taken up into leukocytes by phagocytosis, the leukocyte granules then combine with the bacteria-containing phagosomes, and their hydrolases begin digestion of the bacterial wall destroying the bacterium. The leukocyte remains intact, and its nucleus and cytoplasm function normally because the acid hydrolysis was restricted within the lysosome structure. However, when bacterial toxins are released, they disrupt the lysosomal membranes of leukocytes and cause cell death. It has not been established how toxins are incorporated by cells though they may enter by diffusion through the cell membrane and damage lysosomes from the outside. Another possibility is that toxins are taken up by endocytosis and after fusion they disrupt lysosomal membranes from the inside.[207]

In rheumatoid arthritis lysosomal proteolytic enzymes cathepsin B and D play an important role.[21,579,585] The digestion of the cartilage matrix is probably the fundamental biochemical lesion in the manifestation of the disease. Lysosomes of polymorphonuclear leukocytes are involved in inflammation processes. The main step responsible for initiating the inflammatory process is probably the release of lysosomal enzymes into the surrounding tissue.

Starvation is associated with increased production of Golgi bodies and lysosomes, and β-glycerophosphatase activity is markedly raised. In patients with severe heart failure, the number of lysosomes in myocardiocytes is significantly enhanced. Vitamin A and E deficiencies and the intake of vitamin "megadoses" can increase the permeability of the lysosomal membrane. Lysosomal enzymes are probably involved in various clinical conditions associated with endocrine disorders such as thyrotoxicosis, Cushing and Addison syndromes, adrenogenital syndrome, hypersecreting ovarian adenomas, and carcinomas. These effects may be connected with the labilizing and stabilizing action of corticosteroids, thyrotropin, and sex hormones on the membranes. In many inborn metabolic disorders the disease secondarily affects lysosomes. Direct participation is only found in cases where lysosomes are deficient in enzymes which metabolize the substrate accumulating in these subcellular organelles. Often, however, the intralysosomal storage is not primarily related to lysosome action.

1. Thyroid Lysosomes in Disease

Lysosomes play a central role in the processing of thyroid hormones. In the thyroid, lysosomal digestive enzymes are directly associated with hormone secretion; this action does not occur in any other endocrine gland. The follicle cell of the thyroid contains a number of dense granules that are considered primary lysosomes and they are found dispersed throughout the cytoplasm, with higher frequency near the Golgi area. These lysosomes contain acid phosphatase, aryl sulfatase, β-glucuronidase, β-glucosamidinase, and numerous other hydrolytic enzymes.[129,602]

Following administration of thyroid-stimulating hormone (TSH), the hormone rapidly becomes bound to the surface of the thyroid cell. The binding is very firm, probably with specific receptor molecules involved. The binding can be abolished by treatment with trypsin, phospholipase, or anti-TSH antibody. The primary effect of TSH is triggering the endocytosis of thyroglobulin and also stimulation of lysosomes that migrate towards the cell apex and fuse with newly formed endocytic vacuoles and vesicles. The hydrolytic enzymes of the lysosomes split off the thyroid hormone from the large thyroglobulin molecules. At a later stage, the thyroid hormone passes into the blood from lysosomes, whereas some of the unbroken thyroglobulin molecules enter into the lymph in unchanged condition. In the endocytotic action TSH activates adenyl cyclase with the subsequent formation of cyclic AMP.[278] There is a simultaneous rise of intracellular Ca^{2+}. The increased cyclic AMP and Ca^{2+} may represent important factors in the varous steps of endocytosis.

TSH stimulates nearly all aspects of thyroid metabolism. It increases iodine and carbohydrate metabolism and respiration, enhances phospholipid turnover, and activates protein and nucleic acid synthesis. TSH causes a rise in iodide concentration in thyroid tissue, the rate of thyroglobulin iodination, and the release of thyroid hormones through endocytosis of thyroglobulin.[484] In certain conditions such as thyrotoxicosis the increased rate of synthesis and release of thyroid hormones are connected with the stimulatory action of the protein

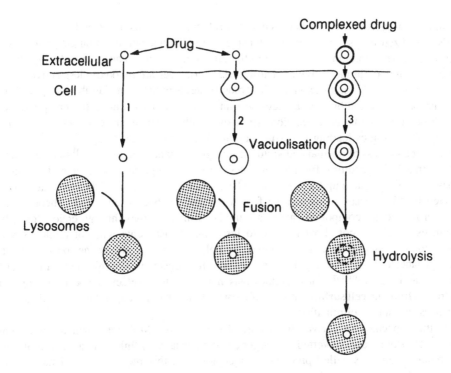

FIGURE 9. Various pathways of molecular entry into lysosomes including (1) passive permeation, (2) endocytosis, and (3) piggyback endocytosis. (Adapted from Reference 125.)

long-acting thyroid stimulator (LATS) and probably an immunoglobulin. LATS causes increased endocytosis, phospholipid synthesis, and glucose oxidation in experimental conditions.[497]

In thyrotoxicosis, characterized by an increased uptake of iodine by the thyroid gland and an increased output of thyroid hormones, there is an enhanced endocytosis and lysosomal breakdown of thyroglobulin.[484] Acute inflammation is rare in the thyroid gland. In Hashimoto's thyroiditis, however, follicle cells display poorly developed Golgi bodies, swollen mitochondria, dilated endoplasmic reticulum, and scarce colloid droplets. Increased lysosomal activity is represented by increased hydrolytic activity corresponding to autophagic vacuoles. In the interstitial tissue, phagocytes, lymphocytes, and plasmocytes appear in great number.

2. Lysosomotropic Effects

The unique properties of lysosomes to take up exogenous substances by endocytosis, their collection of bound acid hydrolases, and their internal pH between 4 and 5, provide the possibility that these cellular organelles are able to concentrate a variety of basic molecules. This process is called lysosomotropism and it has important implications in cell chemotherapy.

Molecules of the extracellular medium can enter lysosomes by three different mechanisms: permeation by diffusion, normal endocytosis, and piggy-back endocytosis[125,299,357] (Figure 9). Low-molecular-weight substances can go through membranes by permeability and become trapped in lysosomes. Most of these substances are weak organic bases. Many factors influence their intralysosomal accumulation, including the high buffering capacity of lysosome constituents. Considering this, concentration ratios of about a few hundred to a thousand times can be expected between lysosomes and the extracellular fluid. Antibiotics containing aminoglycoside groups such as streptomycin, dihydrostreptomycin, neomycin, gentamycin, and kanamycin are accumulated in lysosomes 150 to 300 times compared to the external milieu. Similarily, significant amounts of daunorubicin antibiotics of the anthracycline group are also localized in lysosomes to an extensive rate with concentrations ranging about 1000 times inside the membranes.[1,555,589] This trapping phenomenon causes a reduction in the efficiency of the antibiotic activity because these substances are markedly less active at acid than alkaline pH. This may explain why some bacteria multiply inside the living cell and are resistant to these antibiotics. Besides becoming inactive, basic antibiotics can prove very toxic through lysosomal accumulation, since overloading can severely hinder the digestive capacity of lysosomes leading to disfunction and cellular damage.[280]

During the process of normal endocytosis, foreign substances enter the cell in an enclosure of the extracellular medium, forming vacuoles derived from plasma membrane. The content of these vacuoles is taken up by lysosomes where accumulation occurs. The exact mechanism involved in this process of membrane fusion has not yet been established. Theoretically, any macromolecular compound can be accumulated in lysosomes through endocytosis, but substances which do not show affinity to lysosomal constituents will not remain stored. Stimulation of endocytosis can be achieved by polyions such as polylysine, polyglutamate, or acetyl-labeled IgG. These polymers increase the uptake of albumin and various anti-lysosome antibodies.[101,467] Thus, endocytosis may lead to a selective specificity of drug penetration from the cell surface into specific parts of the vacuolar apparatus or into lysosomes for processing and/or elimination.

The third mechanism derives from normal endocytosis. Molecules which cannot gain access to lysosomes are converted to a larger complex usually by linkage to a macromolecule. This mode of entry is called piggyback endocytosis. In this mechanism, substances may diffuse readily across membranes, although this route may be restricted by the limited availability of coupling carriers.[108,300,381,549,550] Liposomes (lipid microspheres) have been applied as substance carriers in many drug studies. In particular, in the case of antitumor drugs it seems essential to protect them by enclosure so they can effectively reach lysosomes and nuclear DNA. Adriamycin and daunorubicin can be prepared in this fashion, and the complexed drugs can enter lysosomes by the endocytosis route, with later complexing with DNA. Drugs bound in this fashion lose their pharmacological properties, but they are fully restored when DNA is hydrolyzed by lysosomal enzymes and when the free drug is released. The freed chemotherapeutic agent will then leave the lysosomes, reaching the target nucleus where it can exert cytostatic properties (Figure 10). The complexing process reduces the toxicity of many antileukemic drugs; thus, unwanted side effects on other cells such as bone marrow, kidney, and digestive tract epithelia can be minimized.

D. Endoplasmic Reticulum

The endoplasmic reticulum in the cell is a complex network of interconnected cytoplasmic membranes.[120,128,244,434,554,577] These cytoplasmic membranes are concerned with the synthesis of proteins and other cell constituents, intracellular transport, elimination of endogenous and foreign substances, and possibly sensitivity conduction. They have continuity, as membranes, with the nuclear membrane and the cell membrane, and they are distinguished from other subcellular organelles by the attachment of ribonucleoprotein particles or ribosomes. There is no preferential location; however, parallel stacks of endoplasmic reticulum membranes are often found in the vicinity of the nucleus. The linkage of the endoplasmic reticulum network with the nuclear envelope and the Golgi complex probably represents a specialized region of this subcellular structure. There are two recognized types of endoplasmic reticulum membranes. Some membranes are studded with ribosomes or ribonucleoprotein particles and these are the rough membranes. Other endoplasmic reticulum membranes are devoid of ribosomes; they constitute the smooth membranes. A continuity exists between the rough and smooth membranes, but generally the two kinds of membranes are localized in distinct regions of the cytoplasm. The Golgi apparatus is a specially organized smooth endoplasmic

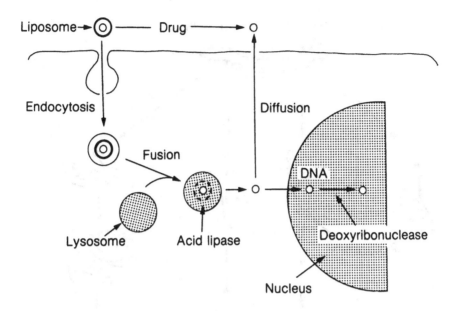

FIGURE 10. Permeation of a drug with lysosomotropic and nucleotropic characteristics into the cell. Linkage to a macromolecule initiates endocytosis, then the drug is released free from its carrier in lysosomes by enzyme or acid action and may enter the nucleus or partly diffuse out. The cellular concentration of the drug depends on the endocytotic activity of the cells. The transformation of the endocytosed drug may confer a basis for tissue selectivity. (Modified from References 621 and 622.)

reticulum usually located near the nucleus. Its probable function is to concentrate, secrete, or package substances formed elsewhere in the cell.[96,402,421,571]

Endoplasmic reticulum membranes can be isolated from disrupted liver cells by high-speed centrifugation, separating the rough and smooth microsomal fractions. These components show certain functional differences. The hepatic endoplasmic reticulum is rich in phospholipid; the membranes also contain ribonucleic acids which are not only localized in ribosomes. The membrane ribonucleic acid is essential in the structural organization of the membrane components and may play a role in the attachment of ribosomes or in the synthesis of specific membrane proteins.[440] There is a similarity between the rough and smooth endoplasmic reticulum in the amino acid content of the proteins, and in the distribution of phospholipids. Differences are in quantity; smooth membranes are richer in phospholipids than rough ones and cholesterol content is also greater in the smooth fraction, representing about 75% of total hepatic cholesterol. The molar ratio of phospholipid to cholesterol is 15 in rough membranes and 4 in smooth-surfaced membranes. The distribution of certain microsomal enzymes also shows selectivity. Fragments of rough membranes composed of polysomes attached to membranes and free polysomes are active in protein synthesis. Drug-metabolizing enzymes are localized in the smooth membranes, while various phosphatases are mainly found in the rough fraction. The wealth of functions of the hepatic endoplasmic reticulum are essential in the production of several body constituents for many physiological functions. Some enzyme activities play an important role in the liver response to disease and in the elimination of ingested exogenous substances including drugs, food additives, and environmental contaminants.

There are different critical phases in the differentiation of the liver cell involving the development of the function of the endoplasmic reticulum. Some enzyme activities of these cellular membranes are low during fetal life but there is increased activity immediately after

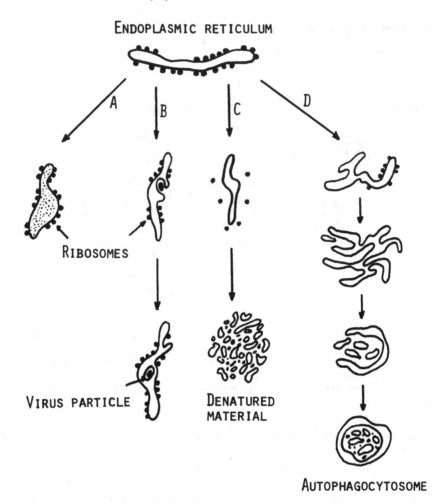

ENDOPLASMIC RETICULUM

A   B   C   D

RIBOSOMES

VIRUS PARTICLE   DENATURED
                    MATERIAL

                    AUTOPHAGOCYTOSOME

FIGURE 11.   Response of the rough endoplasmic reticulum to injury or disease. Altered environmental conditions may impair the normal structure of these cellular organelles. (A) Osmotic changes can lead to the dilation of the vesicles which may cause disruption and loss of function; (B) virus particles can be attached to the membranes and incorporated into the vesicles thus interfering with normal function; (C) toxic actions can initially degranulate ribosomes from the membranes (following the loss of ribosomes, complete breakage and disintegration destroys the entire vesicle); (D) moderate but exhaustive drug effects can also cause partial loss of ribosomes. The proliferating action may, however, extend membrane area and formation and, through structural rearrangement, this leads to the production of secondary lysosomes which tend to eliminate by autophagocytosis the drug causing these changes.

birth. It appears that the activity of microsomal constitutive enzymes is present earlier in the rough than in the smooth endoplasmic reticulum. The lipid components of these membranes are assembled in the rough endoplasmic reticulum. The delay in the development of drug metabolizing enzymes in the newborn is probably linked with the occurrence of neonatal jaundice and accounts for greater drug sensitivity.

1. Changes in Disease

Membrane changes are associated with various diseases such as autoimmune disease, virus infection, cancer, and genetic abnormalities. Dietary and various other environmental stress conditions cause alterations in the structure and function of the endoplasmic reticulum (Figure 11). Genetic defects of the membranes have not been found in man, but in a mutant mouse strain the liver lacks β-glucuronidase activity in the endoplasmic reticulum. This enzyme complex, consisting of four proteins, is missing in the mutant endoplasmic reticulum, although the hepatocytes contain the lysosomal form of the enzyme. This indicates that, in the mutant mouse strain, a structural gene is malfunctioning and a regulatory factor is missing which is required to anchor the enzyme to the endoplasmic reticulum membranes. It is probable, therefore, that the metabolism of exogenous substances including drugs, food additives, and environmental contaminants may also be defective in these animals. There are different critical phases in the differentiation of the liver cell involving the development of the function of the endoplasmic reticulum. Some enzyme activities of these cellular membranes are low during the fetal life but there is a marked surge in activity immediately after birth, probably related to either inhibitory maternal factors or to the maturation of mechanisms whereby the lipid components of these membranes are assembled in the rough endoplasmic reticulum.

The endoplasmic reticulum is the site of biosynthesis of essential constituents of cellular membranes: proteins, phospholipids, and cholesterol and various cytochromes.[307,392,403] Other nonmembrane proteins are also produced by the endoplasmic reticulum. The products are redistributed within the cell or excreted from the cell to target tissues. These processes may be associated with gross morphological changes. Membrane flow from the endoplasmic reticulum to the cell membrane is connected with the breaking off of small membrane vesicles. Cyclic morphological changes occur between the cisternae and small vesicles corresponding to the diurnal rhythm of enzyme activities of the endoplasmic reticulum membranes. Myelin-type structures, cytosomes, and membrane whorls are formed under pathological conditions such as neoplasia or chemical carcinogenesis, virus infections, autoimmune diseases, and in genetically determined storage disorders such as Niemann-Pick disease. They are also produced by exposure to toxicants, such as carbon tetrachloride poisoning, ethionine administration, or drug-induced phospholipidoses.[3,126,166,167,244,394,414]

Starvation or protein-deficient diets do not have much effect on the hepatic endoplasmic reticulum, but there is some irregularity in the parallel organization of the membranes and some vacuolization occurs in the cisternae. Some enzyme activities show changes. Differences exist in the composition and enzyme activity of the endoplasmic reticulum from normal and neoplastic liver cells.

Abnormal growth of the liver leads to the occurrence of different kinds of alteration in protein composition while the morphological structure remains apparently unaltered. Cytoplasmic disturbances affect the endoplasmic reticulum, varying from simple exhaustion of ribosomal mass linked with the excessive use of proteins, such as chromatolysis, to fairly complete blockage of ribosomal protein formation by intoxication. Conversely, any disturbance of the endoplasmic reticulum structure gives rise to local perturbation, leading to large-scale destruction when associated with the action of hepatotoxic compounds. Toxicant actions and malfunctioning enzymes can destroy the activities of the Golgi apparatus.[40,317,342,364,534,603]

It is generally accepted that a spatial and functional continuity exists in the endoplasmic reticulum. This represents the maintenance of the membrane composition within certain limits. In this homeostatic process a functional coupling is evident between the mixed function oxygenase enzyme system and phospholipase activities associated with phospholipid composition. The increase of microsomal cytochrome P-450 following the effect of inducers is necessarily connected with a corresponding increase of endoplasmic reticulum membranes. The abnormal structure is a consequence of disturbance of membrane homeostasis. The whorl formation may be due to an abnormal increase of phospholipid constituents or a segregation of membrane parts that cannot be placed into the integrated membrane structure. This imbalanced integration may cause a susceptibility to other endogenous control mechanisms influencing the effects of nutrition, hormones, or divalent cations. It is often difficult to evaluate the pathological sequencing of biomembrane alterations. In acute carbon tetrachloride poisoning, the effects on the endoplasmic reticulum may not just precede cellular necrosis but may cause cell death by inhibition of protein synthesis governed by these membranes. In other intoxications structural changes may represent an adaptation within regulatory membrane homeostasis. The homeostatic impairment of the endoplasmic reticulum involves its interdependence on various other cellular functions.

2. Protein Synthesis

The rough endoplasmic reticulum is mostly responsible for the synthesis of various proteins. These include proteins constituting an integral part of membranes, proteins which are packaged within this organelle and serve to produce new membranes in the smooth fraction, soluble proteins which are transferred to other organelles and are essential in their formation such as mitochondria, and proteins which are secreted into the circulation and serve many roles, such as blood coagulation, immune response, and transport of smaller and larger molecules into various cells.[76-78, 83, 121, 122, 223, 439]

The rough-surfaced endoplasmic reticulum, containing ribosomes, is the important source of protein synthesis, but polysomal aggregates which are either not bound to the membranes or are linked by weak forces also take part in this function. Hepatoma cells contain significantly fewer free polysomes than normal liver cells and protein synthesis is also significantly reduced in tumor tissue. Serum albumin is synthetized by bound polysomes or ribosomes attached to the outside of the membranes, but the newly formed protein passes through the membrane into the lumen of the cisternae. The peptide chain is growing further within the cisternae and released through gaps in the membranes. Glycoproteins and lipoproteins are synthetized in a manner similar to albumin and probably the endoplasmic reticulum is the site of attachment of their carbohydrate or lipid prosthetic groups. Some membrane-bound enzymes are responsible for the synthesis of phospholipids and for the addition of carbohydrate units. In the assembly and packaging of protein moieties, the smooth membranes and Golgi complex also participate. The Golgi complex is also involved in the selection and packaging of proteins prior to secretion. Some of the assembled proteins, particularly active enzymes, are discharged from the Golgi complex into lysosomes.

Protein synthesis necessary to produce membranes and subcellular constituents can be stimulated by various drugs. Many cellular components synthesize proteins but the microsome fraction is the most active. The synthesized proteins are probably enclosed by granules although there is no definitive information on how these proteins leave the liver cell. However, several proteins found packaged within an organelle probably came from the endoplasmic reticulum. Membranes of peroxisomes and zymogen granules are derived from the endoplasmic reticulum. Peroxisomes contain catalase, D-amino acid oxidase, and urate oxidase, but the initial site of synthesis of these enzymes is the rough endoplasmic reticulum and not the peroxisomes. Zymogen granules are transported from the pancreas in the form of granules. The membranes of these granules are formed in the Golgi vesicles. Several mitochondrial proteins such as cytochrome *c* and probably all soluble proteins[293] are also synthetized in the microsomal fraction. The membrane of lysosomes is also derived from the endoplasmic reticulum; the synthesis of lysosomal membranes results in organelles which store potent hydrolytic and proteolytic enzymes also produced in the endoplasmic reticulum. In fact, these very active enzymes are kept away from the cytoplasmic matrix which could be sensitive to them. In cystic fibrosis morphological changes manifest in the endoplasmic reticulum and in some other cellular structures (Plate 10).

3. Drug Metabolism

Present drug therapy is based largely on pharmacological investigations and studies on biochemical mechanisms associated with the metabolism and disposition of foreign compounds.[120] The strength and duration of the drug effect often depends on the rate at which

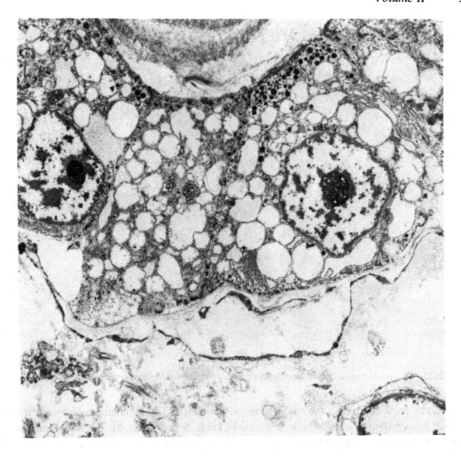

PLATE 10. Electron micrograph of exocrine pancreatic acinus in cystic fibrosis. The acinar lumen is distended with accumulated secretions. The earliest ultrastructural changes to the cystic fibrosis pancreas appear at the basal regions of the acinar cells, with dilatation of rough endoplasmic reticulum, and then loss of zymogen granules. The morphologic changes are thought to reflect an abnormality in the secretion or transport of exocrine secretions. (Courtesy of Dr. J. M. Sturgess, Warner-Lambert/Parke-Davis Research Institute, Sheridan Park, Ontario, Canada.)

drugs are activated or deactivated by various processes occurring mainly in the liver cell. The endoplasmic reticulum is the major site where drugs, foreign compounds, and some endogenous substances are metabolized. Although the endoplasmic reticulum shows morphological heterogeneity, there is evidence that drug-metabolizing enzymes are chiefly localized in the smooth membranes. The activities of these enzymes can be altered by dietary and nutritional or individual factors, such as obesity or hormonal changes, and by the ingestion of foreign chemicals.[7,441] There are several hundred compounds which bring about an excessive proliferation of smooth endoplasmic reticulum associated with increased microsomal drug-metabolizing enzyme activity and elevated microsomal cytochrome P-450 content.[55,102,445,522] The various classes of hepatic microsomal enzyme inducers include several drugs, chlorinated hydrocarbons, polyhalogenated biphenyls, and many food additives. There are several inducers which produce marked increases in some enzyme activities and changes in cytochrome P-450, without an accompanying massive proliferation of smooth membranes. These compounds include polycyclic aromatic hydrocarbons which may act through a different mechanism. Proliferation of the smooth endoplasmic reticulum in the liver cell is seen on Plate 11.

One of the major roles of drug metabolizing enzymes is to provide a primary defense mechanism for the detoxification of drugs, insecticides, pesticides, potential carcinogenic

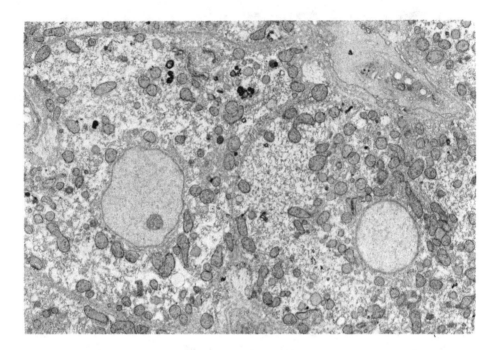

PLATE 11.   Aspect of human liver cells under conditions of smooth endoplasmic reticulum induced by drugs.

compounds, food additives, and environmental pollutants. The mechanism of detoxification involves nonsynthetic and synthetic processes (Table 5 and Table 6), the major one being the hydroxylation pathway. Through these reactions drugs are transformed into highly water-soluble compounds that can be easily excreted. The activity of the hepatic microsomal enzymes is affected by many factors which can cause either an increase or decrease in substrate biotransformation (Table 7).

The major groups of nonsynthetic metabolic reactions or Phase I processes are catalyzed by the mixed-function oxidase enzyme system. This system is found primarily in the endoplasmic reticulum of the liver and in certain other tissues such as lungs, kidney, brain, skin, and gastrointestinal mucosa. In some tissues it is bound to mitochondria, as in the adrenal cortex. This system is complex, and converts lipid-soluble compounds to hydrophilic derivatives. Mixed-function oxidases of the endoplasmic reticulum vary considerably in their specificity and rate of raction. This variation is due to the nonhomogeneous structure of these membranes, as well as to the functional status of the membrane components.

The various enzymatic processes are highly nonspecific, but it is not unusual that some show pronounced specificity to the extent that some even react only with certain stereoisomers. Mixed-function oxidases are versatile systems in that they are able to catalyze the biotransformation of many different substrates. This flexibility is possibly due to the participation of different species of cytochrome P-450 and conformational changes.[183,307]

The mixed-function oxidase enzyme system requires molecular oxygen and NADPH. During the course of the reaction, it incorporates one oxygen atom of the oxygen molecule into the compound undergoing hydroxylation and the other one into water. The complex system receives a second electron from NADH. At the end of the process, a substrate is converted into a hydroxylated metabolite (Figure 12). The hydroxylating system in the hepatic endoplasmic reticulum consists of at least two catalytic components, cytochromes P-450 and b_5, and two flavoprotein enzymes, NADPH-cytochrome P-450 reductase and NADH-cytochrome b_5 reductase, which catalyze the reduction of the cytochromes by NADPH or

Table 5
NONSYNTHETIC PROCESSES IN HEPATIC DRUG METABOLISM

Metabolic reaction	Subcellular site in most tissues
Oxidations	
Aliphatic hydroxylation	Microsomes
Alcohol oxidation	Cytosol and microsomes
Aldehyde oxidation	Cytosol
Aromatic hydroxylation	Microsomes
N-hydroxylation	Microsomes
N-oxidation	Microsomes
Sulfoxidation	Microsomes
Epoxidation	Microsomes
Steroid hydroxylation	Microsomes
Oxidative transformations	
Dealkylation	Microsomes
Dehalogenation	Microsomes
Deamination	Microsomes and mitochondria
Desulfuration	Microsomes
Reductions	
Aldehyde reduction	Cytosol
Azo reduction	Microsomes
Nitro reduction	Microsomes
Steroid reduction	Cytosol and microsomes
Hydrolysis	
Deesterification	Cytosol and microsomes
Deamidation	Cytosol and microsomes

Table 6
SYNTHETIC PROCESSES IN HEPATIC DRUG METABOLISM

Metabolic reaction	Radical	Compound
Glucuronidation	Hydroxyl	Alcoholic
		Phenolic
		Enolic
	Carboxyl	Aliphatic
		Aromatic
	Amino and related radicals	Aromatic
		Heterocyclic
		Hydroxylamine
		Carbamate
		Sulfonimide
	Sulfhydryl	Thiol
		Carbodithioic
Glucosidation	Hydroxyl	
	Purine ring N	
Other glycosidation		
Sulfation		
Methylation	*O*-Methylation	
	N-Methylation	
	S-Methylation	
Acetylation		
Amino acid conjugation	Glycine	
	Glutamine	
Mercapturic acid formation	Glutathione acetylated cystein residue	

Table 7
FACTORS INFLUENCING
HEPATIC MICROSOMAL ENZYMES

Induction
  Drugs
  Halogenated hydrocarbons
  Polycyclic aromatic hydrocarbons
  Food additives
      Antioxidants
      Coloring agents
  Nicotine and related alkaloids
  Flavones
Inhibition
  Enzyme blockers
  Protein synthesis inhibitors
  Carbon monoxide
  Hepatotoxic compounds
  Hormonal derangement
  Liver disease
  Malnutrition
  Stress
  Irradiation

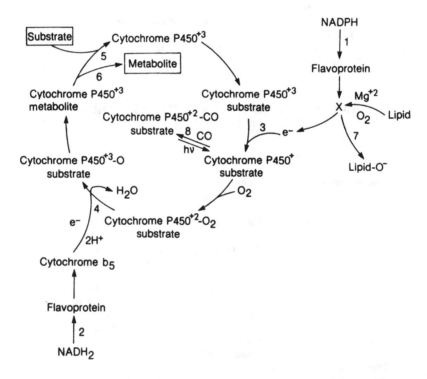

FIGURE 12.   Possible mechanism of drug hydroxylation in the drug-metabolizing enzyme system through the participation of cytochrome P450. This scheme includes the flavoprotein enzymes (1) NADPH-cytochrome c reductase and (2) NADH-cytochrome b_5 reductase. (3) NADPH-cytochrome c reductase participates in the reduction of cytochrome P450 through a possible intermediate; (4) cytochrome b_5 participates in the oxidation of cytochrome P450. (5) In the initial step the substrate becomes bound to cytochrome P450 and (6) through the cycle finally hydroxylated. (7) Lipid peroxidation may form a side reaction in this scheme. (8) Carbon monoxide binds reversibly cytochrome P450, a step related to pathological changes associated with carbon monoxide poisoning. (Adapted from References 183, 616, and 619.)

NADH, respectively. Cytochrome P-450 shows a sensitivity to carbon monoxide which binds the reduced form of the cytochrome and produces a characteristic absorption spectrum with a maximum at 450 nm. The flavoprotein NADPH-cytochrome P-450 reductase is probably closely related to NADPH-cytochrome *c* reductase, which is also found in the endoplasmic reticulum. An interaction exists between a NADH-flavoprotein-cytochrome-b_5-containing system and the cytochrome P-450-O_2-substrate complex in splitting the molecular oxygen bound to cytochrome P-450. The mitochondria-bound hydroxylating system in the adrenal cortex also functions with cytochrome P-450 but differs from the microsomal system with respect to the flavoprotein.

Recently it was found that an enzyme system converts cytochrome P-450 to cytochrome P-448 by an alternate process of oxygen transfer.[329]

Synthetic processes or Phase II metabolic reactions are catalyzed also by the mixed-function oxidase enzyme system of the endoplasmic reticulum. They have been characterized in several tissues including liver, kidney, brain, pineal gland, lungs, and gastrointestinal mucosa. Some of these reactions also occur in other cellular fractions, such as sulfotransferases in the cytoplasm from a wide variety of tissues, and methyltransferases that may be cytoplasmic or microsomal. Mercapturic acid formation is connected with the catalytic glutathione *S*-transferase action of the intracellular carrier protein ligandin.

The main difference between Phase I and Phase II reactions is that the synthetic processes need energy and the contribution of an endogenous molecule. These molecules include the formation of high-energy-containing compounds such as uridine diphospho-derivatives of sugar molecules, activation of methyl donors or amino acids by adenosine triphosphate, formation of 3'-phosphoadenosine-5'-phosphosulfonate, acetyl-coenzyme A in the acetylation reaction, and combination with the tripeptide glutathione. The latter does not involve a high-energy molecule and can occur sometimes nonenzymatically.

Using these high-energy molecules, Phase II processes compete with the routine metabolism of endogenous compounds to an extent greater than do Phase I processes. Phase II enzyme systems usually catalyze the transformation of both xenobiotic and endogeneous substrates to more water-soluble derivatives. Such a competition might exert undesirable effects on the metabolism of normal body constituents. During maturation, in advanced liver disease, or in starvation, where energy supply and the cellular content of endogeneous substances are limited, the extra load on these enzymes might exhaust their biotransformation capability. Consequently, reduced efficiency can lead to incomplete development, abnormalities, or cellular damage.

4. *Metabolism of Body Constituents*

Several normal body constituents are substrates for drug-metabolizing enzymes such as steroid hormones, cholesterol, bilirubin, and fatty acids. Accordingly, changes in the activities of drug-metabolizing enzymes bring about changes in the metabolism of these normal body constituents and, conversely, changes in the level of these body components may affect drug metabolism and disposition. In particular, some compounds that stimulate liver microsomal hydroxylation alter pathways of steroid metabolism in man.[50] Since drugs cause increased urinary excretion of 6β-hydroxycortisol, this observation led to the suggestion that the level of urinary 6β-hydroxycortisol may be useful as a measure of microsomal enzyme induction in man. The increased hydroxylation of cortisol raised the possible therapeutic value of the hepatic induction process in the treatment of Cushing's syndrome. Moreover, during late pregnancy the highly elevated production of reductive progesterone metabolites (pregnanolone and pregnanediol) may be responsible for the increased sensitivity towards drugs. These maternal steroids may be the causative factors in the delayed development in the function of the hepatic endoplasmic reticulum in the fetus and newborn.[130,164,260,375]

Microsomal metabolism of cholesterol, bile acid, and fatty acid oxidation, and bilirubin conjugation may also have relevant implications in certain diseases. Certain inducers of hepatic microsomal enzymes have been used therapeutically for the treatment of chronic intrahepatic cholestasis in adults and for the treatment of hyperbilirubinemia in jaundiced children. Increased bilirubin in the newborn may result in damaged brain function. The increase in unconjugated bilirubin in the plasma of infants could conceivably be reduced if the mothers received treatment with phenobarbital and primidone throughout pregnancy. A correlation between the degree of neonatal hyperbilirubinemia and low scores of mental or motor activity attained at 8 months of age suggests that the application of suitable inducers of drug metabolism can stimulate bilirubin metabolism allowing normal brain function and eliminating potential adverse effects on psychomotor function. It seems likely that the future application of selective inducers or inhibitors of hepatic drug-metabolizing enzymes may provide beneficial tools in the regulation of abnormal metabolism characteristic to various human diseases.

5. Response to Toxic Compounds

The toxicological importance of the endoplasmic reticulum in the liver cell is unquestionable. These membranes (1) are exposed generally to higher concentrations of drugs or toxic agents entering the body in comparison to other membranes, (2) contain sites where many toxic compounds are transformed to reactive intermediates mainly by the mixed-function oxidase enzyme system, (3) modulate the biotransformation of xenobiotics with biological consequences which are dependent on the toxic nature of the metabolism products, and (4) by virtue of their xenobiotic action, may cause changes in the structure and composition of the endoplasmic reticulum membranes which may be propagated to all cellular membranes by membrane flow. Changes in serum alkaline phosphatase activity associated with the effect of some steroid hormone derivative represents an example of such association.

A number of compounds may produce primary intrahepatic cholestasis (Table 8). The features of this cholestasis are swelling of the centrolobular hepatocytes and variable degrees of inflammation. Some toxic compounds directly affect the endoplasmic reticulum. Carbon tetrachloride intoxication leads to structural abnormalities of this membrane. In the early stages of intoxication there are changes of rough membranes, dilated cisternae, appearance of dense bodies and liposomes, loss of polysomal aggregates, and degranulation of ribosomes from the membranes. With time the disorientation and fragmentation continues and dense smooth membrane aggregates are formed. At a later phase the presence of aggregates becomes more prominent, the mass is greater, and electron-dense "dirt-like" material appears, followed by increasing fat infiltration, finally leading to death of the cell.[25,93,387,447,510] The extensive damage of the rough endoplasmic reticulum in the liver cell caused by hepatotoxic compounds is seen under the electron microscope (Plate 12).

Biochemical changes are linked with morphological alterations (Table 9). These include a decreased capacity for protein synthesis associated with an effect on protein synthesis due to messenger-ribosome interaction and with increased lability of ribosomes. The result of the toxic interaction is a decreased synthesis of albumin and fibrinogen. Furthermore, a specific interruption of the NADPH-related microsomal electron transport and inhibition of the activity of drug-metabolizing enzymes is apparent, linked to a loss of cytochrome P-450 and the NADPH chain. Several normal enzymes undergo changes, the activity of membrane-bound phosphatases is impaired (some are reduced, some are elevated due to release into free form), and microsomal phospholipid synthesis is decreased.[45,437,509,511]

6. Enzyme Induction

In contrast to the response to hepatotoxic compounds, administration of drugs stimulates drug metabolism. The degree of induction is dependent on the drug and on the dose given.

Table 8
DRUGS CAUSING INTRAHEPATIC CHOLESTASIS

Incidence	Drug	Site of action[a]	Ref.[b]
High <2%	Erythromycin	C	1, 2
	Norethandrolone	C	3, 4
Low ~1%	Anabolic steroids: methyl testosterone	C	5, 6
	Chlorpromazine	C	7
	Contraceptive steroids	C	8
Relatively infrequent	Antibacterials: nitrofurantoin, novobiocin, penicillin, rifampicin, sulfonamides	S, ER	9, 10, 11
	Antirheumatics: gold salts, phenylbutazone	C	12
	Antithyroids: carbimazole, methimazole	C	13, 14
	Anxiolytics: chlordiazepoxide, diazepam		15
	Oral hypoglycemics: tolazamide, tolbutamide	C	
	Phenothiazines: prochlorperazine, promazine		16
	Thioridazine	C	17
	Tricyclic antidepressants	C	18, 19

[a] C, excretory defect; S, uptake defect; ER, bilirubin conjugation defect.

[b]
1. **Slater, T. F. and Delaney, V. B.**, *Toxicol. Appl. Pharmacol.*, 20, 157, 1971.
2. **Dujovne, C. A., Shoeman, B. J., and Lasagna, L.**, *J. Lab. Clin. Med.*, 79, 832, 1972.
3. **Heikel, M. A. and Lathe, G. H.**, *Biochem. J.*, 118, 187, 1970.
4. **Kory, R. C., et al.**, *Am. J. Med.*, 26, 243, 1959.
5. **Wood, J. C.**, *JAMA*, 150, 1484, 1952.
6. **Orlandi, F. and Jesequel, A. M.**, *Rev. Int. Hepat.*, 16, 331, 1966.
7. **Adlercreutz, H. and Tenhunen, R.**, *Am. J. Med.*, 49, 630, 1970.
8. **Bolton, B. H.**, *Am. J. Gastroenterol.*, 48, 497, 1967.
9. **Bridges, R. A., et al.**, *J. Pediatr.*, 50, 579, 1957.
10. **Cohn, H. D.**, *J. Clin. Pharmacol. New Drugs*, 9, 118, 1969.
11. **Dujovne, C. A., et al.**, *N. Engl. J. Med.*, 227, 785, 1967.
12. **Schenker, S., et al.**, *Gastroenterology*, 64, 622, 1973.
13. **Fischer, M. G., et al.**, *J. Am. Med. Assoc.*, 223, 1028, 1973.
14. **Lunzer, M., et al.**, *Gut*, 16, 913, 1975.
15. **Abbruzzese, A. and Swanson, J.**, *N. Engl. J. Med.*, 273, 321, 1965.
16. **Baird, R. W. and Hull, J. G.**, *Ann. Int. Med.*, 53, 194, 1960.
17. **Herron, G. and Boudro, S.**, *Gastroenterology*, 38, 87, 1960.
18. **Karkalas, Y. and Lal, H.**, *Clin. Toxicol.*, 4, 47, 1971.
19. **Morgan, D. H.**, *Br. J. Psychiatry*, 115, 105, 1969.

In the case of high doses, 2 to 3 hr after administration the morphological changes in the endoplasmic reticulum are similar to the early liver damage after experimental toxic injury with carbon tetrachloride or other toxic compounds. The cisternae of the endoplasmic reticulum are distended and lose the ribosomes. The smooth endoplasmic reticulum appears dilated and vesiculated. In contrast to the adverse response to high doses, lower dose levels show a different picture. From the nearly parallel rough membranes a lattice network of smooth tubules grows and sometimes occupies the whole liver cell cytoplasm. These changes can be seen usually 18 to 24 hr after the first dose (Figure 13). The proliferation is quite characteristic of barbiturates or other potent inducers and can be achieved easily in experimental animals (Plate 13). The biochemical changes differ from those due to toxic actions; drugs increase the incorporation of amino acids into the endoplasmic reticulum. Drug-induced protein synthesis consists mainly of enhanced formation of drug-metabolizing enzymes. In contrast to hepatotoxins, following the administration of drugs, albumin or fibrinogen synthesis remains largely unaltered.

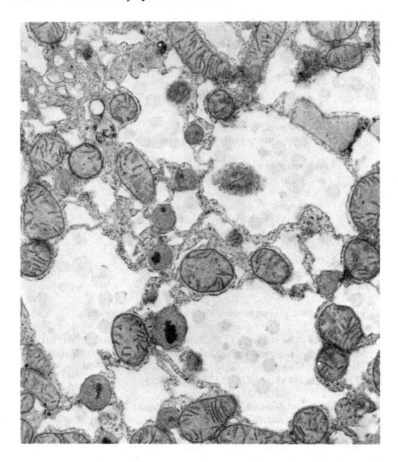

PLATE 12.   Example of endoplasmic reticulum alteration by hepatotoxicants. The rough endoplasmic reticulum cisternae are extremely dilated with lipoprotein accumulation.

Table 9
HEPATIC REACTIONS BY ASSESSING SEVEN PARAMETERS OF ENDOPLASMIC RETICULUM FUNCTION[a]

Morphofunctional parameter	Hepatic reaction[b]					
	Toxicant	Moderate toxicant	Control	Mild inducer	Inducer	Potent inducer
Morphology	Disarrays	Dilatation	Normal	Normal	Proliferation	Large proliferation
Surface density	↓ [c]	NC	NC	NC	↑	↑↑↑
Fatty acid saturation	↑↑↑	↑↑	↑	↑	↓	↓↓
Phospholipid/surface ratio	↓	↓	NC	NC	↑/NC	↑
Methyl transferase	↓↓	↓/NC	NC	↑/NC	↑	↑↑
Hydroxylase	↓↓	↓		↑	↑↑	↑↑↑
Phosphatase	↓	↓/NC	NC	NC	↓	↓↓

[a]   Hepatic reactions characterized as described in Reference 128.
[b]   Archetype compounds used were CCl_4, coumarin, methylcoumarin, and phenobarbital, among others.[167] The morphofunctional parameters help to identify six distinct classes of hepatic reactions and to differentiate between toxicity, normality, and induction.
[c]   Upward arrows indicate a positive parameter change; NC, no change; downward arrows indicate adverse effects or decreased activity. The crossed arrows in the "Phosphatase" row refer to decreased activity due to "dilution effects" reflecting the influence of hepatomegaly.

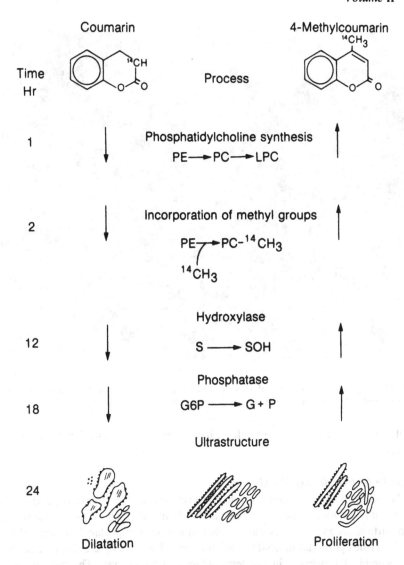

FIGURE 13. Schematic representation of the time course of biochemical and pathologic events occurring in the hepatic endoplasmic reticulum following the administration of coumarin and 4-methylcoumarin. These experimental data represent interesting examples of how the hepatotoxic character of a test compound is converted to an inducer by the introduction of a methyl group. Similar changes have been observed when the action of methyl and nonmethyl derivatives of various drugs on the liver of rats have been investigated.[165,617]

Repeated administration of a drug or several other compounds cause proliferation and increased synthesis of drug-metabolizing enzymes, protein, and phospholipids of the endoplasmic reticulum membranes.[167,395,397] The normal enzymes show no apparent change. On the basis of this inductive action and of toxic hepatic response, foreign compounds can be grouped into two categories (Table 9). Enzyme induction represents an adaptive increase in the amount of specific enzyme systems associated either with an increase in the rate of synthesis or with a decrease in the rate of degradation. Enzyme activity may also be altered without a change in the amount of enzyme by a variety of agents. The course of normal development and differentiation is linked with enzyme induction.

Nutritional control of key metabolic pathways, the response of various organs to a variety of drugs, and the actions of hormones on many cells influence enzyme induction. The

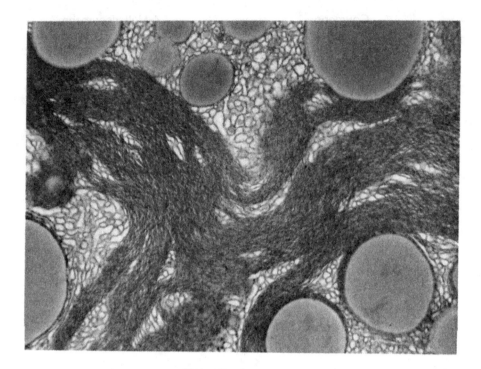

PLATE 13. Experimental induction of smooth endoplasmic reticulum upon long-term treatment. The membranes tend to become arranged in parallel, concentric lamellae.

process of enzyme induction may, therefore, have considerable clinical and pharmacological importance.

a. Enzyme Induction during Development

The adaptation of the newborn baby to the extrauterine life largely depends on the development and maturation of physiological and biochemical functions during the fetal period. Premature birth is accompanied by inadequate maturation of important biochemical functions resulting in serious perinatal morbidity and mortality. Even a single delay in a development of key biochemical processes in the term infant can be serious. The activity of various enzymes appears at a definite time during fetal or postnatal development.[127] Several hormones such as thyroxine, glucocorticoids, and glucagon stimulate many of these enzyme changes and the administration of exogenous hormones can prematurely evoke these enzyme changes.[201,202,432] Nutritional control of key enzymes also has an important role in the development of various pathways. This control frequently operates through induction, or inhibition of synthesis by repression.

Essential enzyme pathways are related to the regulation of the urea cycle and cholesterol biosynthesis, and both processes have considerable clinical importance. The highly toxic end product of protein catabolism, ammonia, is metabolized by enzymes of the Krebs-Henseleit urea cycle. Increased protein catabolism due to starvation or to the administration of corticosteroids or glucagon raises the activity of these enzymes. Subsequently, reduction of protein intake decreases the rate of synthesis and increases the rate of degradation. High-protein diets also cause an adaptive increase of all enzymes involved, particularly the amount of arginase. In contrast, starvation increases arginase activity by lowering the rate of enzyme degradation without altering the rate of synthesis. Alteration in both the rates of synthesis and degradation of arginase are apparently involved in its nutritional regulation.[435,473,474] Inherited deficiencies of many urea cycle enzymes cause intermittent hyperammonemia, vomiting, lethargy, and even coma and death in severely affected patients. If there is a small residual ornithine transcarbamylase activity, the disease condition can be managed successfully with protein restriction. Sometimes the deficient activity linked to structurally altered enzyme molecules can be enhanced by hormone treatment which raises the total catalytic activity.

In premature infants a major clinical problem is the idiopathic respiratory-distress syndrome. This pulmonary disorder is due to the inadequate differentiation of alveolar type II cells and deficiency of surface-active lecithin production. In the fetus of experimental animals the administration of glucocorticoids triggers an accelerated lung maturation, increases the formation of surfactant in the lung, and improves pulmonary function. The increased synthesis of lecithin is connected with the induction of choline phosphotransferase which can be blocked by inhibitors of protein synthesis. The risk of the respiratory distress-syndrome in infants can be assessed before birth by phospholipid analysis of the amniotic fluid. In particular, the lecithin/sphingomyelin ratio and foam stability provide important information on the advancement of fetal phospholipid synthesis. The increased production of phospholipid-synthesizing enzymes by prenatal glucocorticoid treatment of high-risk pregnancies has been shown as an important application of enzyme induction in man.[16,17,20,160,302,508] The administration of glucocorticoids to the mother can normalize lecithin content and decrease the frequency of hyaline-membrane disease in human infants delivered before 32 weeks of gestation.

The effect of glucocorticoids is related to specific receptors existing in human lung from early fetal life. Similar receptors are, however, also present in a variety of fetal tissues. Therefore, several tissues may respond to glucocorticoids and the prenatal administration of these hormones could alter other processes of development, perhaps unfavorably. Therefore, the safe and efficacious application of enzyme induction in therapy in the perinatal period still raises many problems. Thus, it is essential to understand the precise mechanism whereby exogenous agents such as hormones can modulate the synthesis of enzymes during fetal development.[162]

b. Enzyme Induction and Drug Metabolism

Simultaneous administration of two or more drugs may cause interactions that alter the effects of one or both drugs. These interactions are often the result of induced changes in the metabolism of the drugs.[452,487] A wide variety of unrelated compounds can enhance the activity of drug-metabolizing enzymes, and this increased metabolism alters the duration and intensity of action of a variety of pharmacologic agents. Consequently the altered metabolism may be associated with physiological effects dependent on the relative biologic activities of the parent compound and its metabolites.[105,284]

The enzymes involved in the induction process are mixed-function oxygenases or drug-metabolizing enzymes located in the endoplasmic reticulum of the liver and other tissues. The essential effect of these enzymes is to convert lipophilic or fat-soluble compounds to hydrophilic or water-soluble ones, which can be more readily removed from the blood and excreted. The main enzyme actions include aliphatic and aromatic hydroxylation, reduction, dealkylation, and formation of glucuronides. The substrates of these enzymes are not only pharmacologic agents, but a variety of environmental pollutants, food additives, potential carcinogens, and various normal body constituents, such as steroid hormones, thyroxine, fat-soluble vitamins, and bilirubin. Several hundred compounds in man's environment are known to affect the activity of drug-metabolizing enzymes; however, there is no apparent direct relationship between the structure or activity of these chemicals and their inducing ability. Reduced or oxidized derivatives of steroid hormones and increased amounts of steroids produced during pregnancy alter drug metabolism,[162] suggesting that steroids may not only serve as natural substrates of these enzymes but may play an important role in the regulatory mechanism of the induction process.[134,168]

Extensive studies of the induction phenomenon have revealed that the enhancement of the hydroxylating system involves a net synthesis of its various components and the formation and accumulation of new endoplasmic reticulum membranes accompanied by increased hydroxylases, generally of drug-metabolizing enzyme activity. The new endoplasmic reticulum membranes are of the ribosome-free smooth-surfaced type, with a relatively high content of phospholipids.[62,219,397,398,443,444,535]

The drug-induced enzyme synthesis is a *de novo* protein synthesis and involves the formation of a new messenger RNA as indicated by its sensitivity to actinomycin D. Actinomycin D completely blocks the induction of drug-metabolizing enzymes. Puromycin also abolishes both membrane and enzyme synthesis, indicating that the entire induction process, including the formation of new phospholipids, involves the synthesis of new protein molecules.

Turnover studies have revealed, furthermore, that besides an enhancement of the rates of protein and phospholipid synthesis, a decrease also occurs in the rates of breakdown of these compounds as a consequence of the drug treatment, and the increased amounts of membranes and enzymes represent the balance between these two opposite effects.[166]

At the present time, our knowledge is still rather incomplete about the detailed mechanisms of the drug-induced enzyme and membrane synthesis. The earliest event is the binding of the drug to the endoplasmic reticulum which can be detected a few minutes after the injection of the drug, and which reaches a maximum within a short period. The next event that can be observed is an increase in the rate of microsomal phospholipid turnover which begins after about 1/2 to 1 hr following drug administration. The effects of drugs on microsomal phospholipid synthesis are primarily related to changes of phosphatidylcholine production by step-wise methylation from phosphatidylethanolamine and methionine.[2] After about 6 hr, there is the first measurable increase in the content of the microsomal hydroxylating enzyme system. The early increase is found exclusively in the ribosome containing rough-surfaced fractions of the endoplasmic reticulum, whereas the enzyme content in the smooth-surfaced membranes is yet unchanged. Between 8 and 12 hr following drug administration, an increase occurs in the relative content of phospholipids in the microsomes, and a decrease in the rate of breakdown of microsomal phospholipids. At about the same time, an increase in the nuclear RNA polymerase activities can be observed. During the second half of the first 24-hr period following drug administration, the enzyme content begins to increase in the smooth-surfaced microsomes and it is at about 24 hours that the enzyme content of the smooth- and rough-surfaced microsomes becomes equal. After this time repeated administration of drugs brings about continued proliferation of smooth membranes and an increase of hydroxylating enzyme levels. Simultaneously, an increase is found in total liver weight.[150,181,393] The contrasting action of hepatotoxicants on the endoplasmic reticulum may be associated with the production of membrane-bound phospholipids.[2,3] The participation of methyl groups from methyl-containing drugs in membrane phospholipid synthesis has been postulated (Figure 13).

c. Clinical Importance of Enzyme Induction

The effect of inducers in accelerating metabolism may have considerable clinical importance partly by leading to improvements in the therapy of certain diseases or partly by altering effective drug levels or even causing iatrogenic disorders.[103,104,452,478] Phenobarbital lowers serum bilirubin in some patients with intrahepatic cholestasis and in premature or full-term infants with physiologic jaundice or congenital nonhemolytic jaundice. The effect of drug treatment is connected with the induction of microsomal uridine diphosphoglucuronyl transferase. These observations suggest a potential for therapy of patients suffering from certain genetic disorders such as incomplete deficiency of glucuronyl transferase. Barbiturates and other drugs can induce an increased rate of metabolism of anticoagulant drugs which in turn result in decreased anticoagulant effect of a given dose of these agents. Cessation of barbiturate treatment without adjusting the dose of anticoagulants can markedly increase the plasma level of the anticoagulant, causing severe hemorrhagic problems.[30,115,151,491,507]

Chronic consumption of ethanol increases its own metabolism as well as the degradation of a variety of other drugs including barbiturates, diphenylhydantoin, tolbutamide, and coumarin-like anticoagulants. This effect may account for the increased tolerance of alcoholic patients to barbiturates and other sedatives. On the other hand, a large single dose of alcohol which inhibits drug metabolism may be responsible for the elevated sensitivity of intoxicated persons to barbiturates.[301]

Enzyme-induction studies with identical and fraternal twins indicated that this process is dependent on genetic factors.[564] The individual rate of normal, noninduced metabolism, however, is probably associated with environmental changes. Similarly, side effects and drug-induced reaction are determined more by circumstances than genetic conditions. The inductive action of drugs is less efficient or absent in extrahepatic tissues. The placenta also contains drug-metabolizing enzymes which are not inducible by drugs, whereas cigarette smoking produces a marked elevation of several hydroxylating enzymes.[582] The implication of this observation is not fully understood, but it may elicit some impairing action on the placenta manifesting in the significant incidence of low birth weight of babies born from mothers who are heavy smokers. There appears to be a dose-response relationship between birth weight and the number of cigarettes smoked. However, exactly which ingredients in smoke are hazardous to the fetus or mother has not yet been resolved, in spite of the benzo(*a*)pyrene content of smoke condensates.

d. Adverse Effects of Enzyme Induction

Considering the common site of metabolism of drugs and steroids, strong interactions may occur which influence biochemical processes regulating steroid hormone homeostasis. Furthermore, through an interference, drug intake may modify hormone level in the body and alter endocrine actions. Interactions may result in the failure of protection afforded by oral contraceptive steroids, either by changes in their concentrations or by a subsequently altered body response,[50] particularly when the dose of estrogen is reduced to 50 μ.[241,253,362,462]

The interrelationships between drugs and steroid metabolism in the liver is affected by impairment of hepatic functions. Liver damage, as a side effect of drug therapy, is present in substantially greater frequency when the production of steroid hormones is altered. Side effects of various steroid compounds on the liver cell show a variety of clinical and histological patterns, which may be grouped into four categories (Table 10). The risk for developing such toxic responses appears to be greatest in postmenopausal women.[163,177] During pregnancy, endocrine changes are connected with increased estrogen and progesterone levels associated with reduced hepatic drug metabolism.[18,164,375] Subsequently, hepatic susceptibility to drugs is enhanced and hepatic dysfunction is more common in pregnant women than in nonpregnant.

The urinary excretion of 6β-hydroxycortisol is significantly increased and the serum half-life of dexamethasone is reduced in asthmatics treated with barbiturates.[58] Long-term administration of anticonvulsant drugs can accelerate the metabolism of vitamin D and its biologically active metabolites resulting in overt rickets and abnormalities in serum alkaline phosphatase and calcium levels. The dose of vitamin required to treat osteomalacia is higher with simultaneous anticonvulsant therapy than that needed in patients with a simple dietary vitamin deficiency.

Rifampicin brings about an increased activity of drug-metabolizing enzymes in the endoplasmic reticulum of human liver. The clinical effects of simultaneous rifampicin and oral contraceptive treatment include spotting, silent menstruation, amenorrhea, and unwanted pregnancy (Table 11). Diphenylhydantoin, phenylbutazone, phenobarbital, or *N*-phenylbar-

Table 10
VARIOUS TYPES OF STEROID-INDUCED LIVER DISEASES

| | | Manifestation | |
| | | Cholestasis | |
Parameter	Cytotoxicity	Hepatocanalicular	Canalicular
Biochemical			
Alkaline			
Phosphatase	>3[a]	>4	<3
ALT, AST[b]	>7	<7	<7
Histological	Inflammation	Cholestasis	Cholestasis
	Necrosis (±)	Portal inflammation	
Clinical			
Abdominal pain	±	±	–
Anorexia	±	±	–
Hepatic failure	+	–	–
Pruritus	–	+	+
Examples	Methandrostenolone	Ethynyl estradiol	Methyltestosterone
		Fluoxymesterone	
		Mestranol	
		Methandrostenolone	
		Norethindrone	
		Norethynodrel	
Mixed action[c]	Ethynyl estradiol		
	Fluoxymesterone		
	Mestranol		
	Methandrostenolone		
	Norethindrone		
	Norethynodrel		
	Norgestel		
	Oxymetholone		

[a] Multiples of normal serum levels.
[b] ALT, alanine aminotransferase; AST, aspartate aminotransferase.
[c] Hepatic tumors, peliosis hepatis.

Adapted from Feuer, G., Drug control of steroid metabolism by the hepatic endoplasmic reticulum, *Drug Metab. Rev.*, 14, 119, 1983.

Table 11
RIFAMPICIN ACTION ON WOMEN
TAKING ORAL CONTRACEPTIVE STEROIDS

Unwanted pregnancy
Disturbance of menstruation
    Cycle increasing
    Intermediate bleeding
    Amenorrhea
Interference with control of cycle
    Spotting
    Silent menstruation
    Amenorrhea

Adapted from Feuer, G., Drug control of steroid metabolism by the hepatic
endoplasmic reticulum, *Drug Metab. Rev.*, 14, 119, 1983.

bital given to patients increases the urinary elimination of 6β-hydroxycortisol. In epileptics treated with phenobarbital in combination with other antiepileptic drugs, unintended pregnancies occurred despite the use of oral contraceptives.[241,253,268,462]

Interference also occurs with antibiotics such as chloramphenicol, sulfamethoxypyridazine, ampicillin, and tetracycline. Some suppression of steroid hormone level is apparent in almost all women taking oral contraceptives and ampicillin concurrently. Clinical data have also shown that some workers in a pesticide factory exposed to high amounts of DDT excrete more urinary 6β-hydroxycortisol than the control population.[419]

Further interesting aspects of enzyme induction in patients are the faster elimination of another drug given occasionally or significant depletion of essential natural factors in an induced state by a chronically administered drug.[407] In hospitalized mental patients, prolonged treatment with drugs is associated with a significant decrease of folic acid. Folic acid deficiency runs parallel with the duration of enzyme induction. Over a 5-year treatment, enzyme induction is substantially diminished probably due to lack of folic acid, and side effects of the drug given for therapy become marked as a consequence of decreased metabolism rates.

Drug metabolism is responsible for the teratogenicity of thalidomide in man, due to the formation of a unique metabolite which interferes with folic acid utilization[156,605] (Figure 14). Maternal abuse of ethanol during gestation produces a readily identifiable and serious dysmorphic condition called alcohol syndrome. This syndrome has been suggested to represent the most frequent cause of mental deficiency in the Western countries. The chronic consumption of alcohol equivalent to one ounce of absolute ethanol daily by a pregnant woman may produce a significant risk of structural defects and central nervous system dysfunction.[249,250] Beside the effect of the endoplasmic reticulum-bound enzymes on alcohol, ethanol is also metabolized by the cytosolic alcohol dehydrogenase system. The mechanism by which alcohol or its breakdown products might exert their harmful action on the fetus has not been established. Nevertheless, the wide spectrum of alcohol effects on the fetus represents an important example of the role of metabolism in the teratogenic effects of drugs. Moreover, since drugs are rarely consumed one at a time, drug-drug or drug-genotype interactions possibly affect teratogenicity. A great majority of birth defects of unknown etiology might be caused by such interactions.[596,597]

e. Enzyme Induction and Toxic Effects

During metabolism, drugs and other foreign compounds generally lose their pharmacologic or biologic actions and the process really represents deactivation. Metabolic modifications of these compounds lead to inactive products. Thus, the phenomenon of enzyme induction results in beneficial actions with xenobiotics converted to inactive derivatives and eliminated from the body at faster rate. Sometimes, however, metabolism may convert an inactive substance into a pharmacologically active drug or change it from a drug with one type of activity to another drug with different action (Table 12). Reactive metabolites of drugs are important in the initiation of various toxic actions,[438] including methemoglobinemia, blood dyscrasias, cellular necrosis, mutagenesis, carcinogenesis, and perhaps teratogenesis. These observations reflect the importance of metabolic activation of various chemicals by the hepatic endoplasmic reticulum in the pathogenesis of many diseases,[409] even when the origin of some of these diseases has not yet been fully documented.

The exact mechanisms whereby active metabolites elicit toxic reactions are still obscure, and different interactions exist (Figure 15). Namely, in the development of carbon-tetrachloride-induced liver necrosis, the promotion of lipid peroxidation by free radical metabolites represents the activation process. Covalent binding of active metabolites to proteins and/or DNA and RNA can block crucial enzymatic steps in cell metabolism. Reactive metabolites of carcinogens such as aflatoxin or mutagens such as cyclophosphamide might

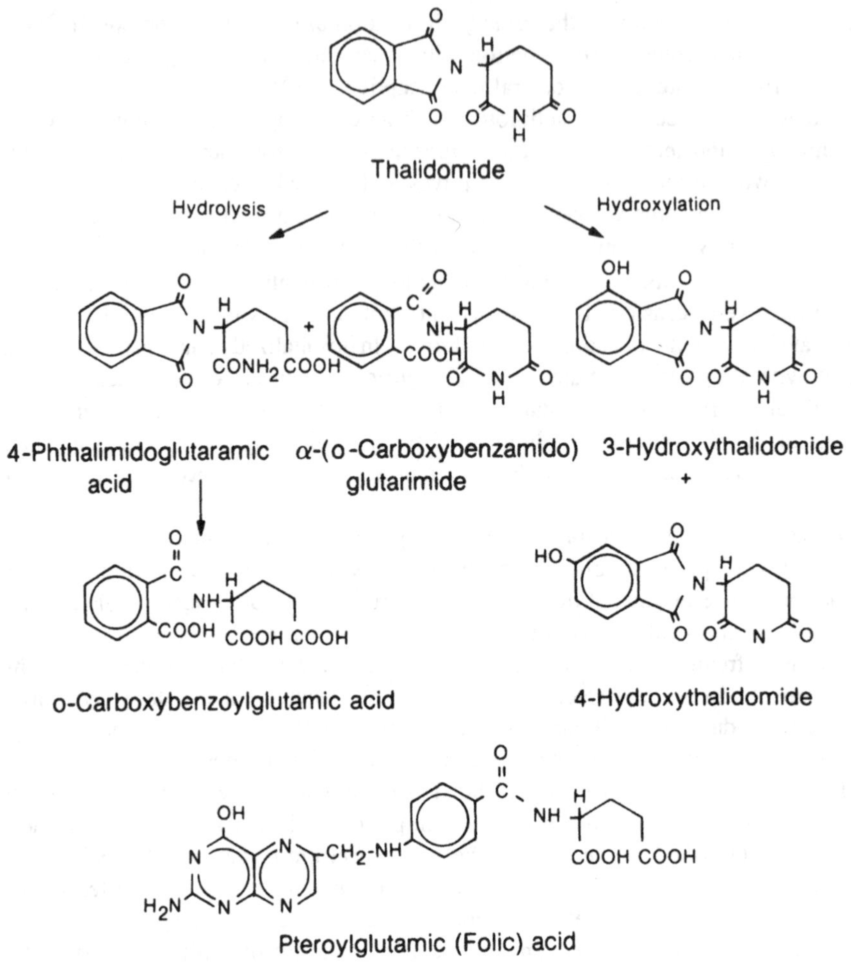

Thalidomide

Hydrolysis   Hydroxylation

4-Phthalimidoglutaramic   α-(o-Carboxybenzamido)   3-Hydroxythalidomide
acid   glutarimide   +

o-Carboxybenzoylglutamic acid   4-Hydroxythalidomide

Pteroylglutamic (Folic) acid

FIGURE 14.   Metabolism of thalidomide. In man, the major metabolite *O*-carboxybenzoylglutamic acid corresponds to part of the folic acid structure. This metabolite is responsible for its teratogenic activity.[156,623]

evoke toxicity by reacting with DNA. Electrophilic metabolites are highly reactive although very unstable. Nevertheless, the covalent binding of electrophilic metabolites has been implied to initiate tissue damage and necrosis, chemical carcinogenesis, mutagenesis, and perhaps teratogenesis. Exposure to certain carcinogenic or toxic chemicals results in binding of these compounds or their reactive metabolites to proteins and nucleic acids; examples include the carcinogen *p*-dimethylazobenzene, benzo(α)pyrene, dimethylbenz(α)anthracene, and aflatoxin, or the hepatotoxicants bromobenzene, furosemide, and acetaminophen. There is a striking correlation between the severity of hepatotoxicity and the extent of covalent binding of these hepatotoxicants to macromolecules. Further, a similar correlation has been established between carcinogenic potential and DNA binding of the aforementioned carcinogens. Degranulation of the hepatic rough endoplasmic reticulum and the conversion of cytochrome P-448 may represent important fundamental changes in the initiation of chemical carcinogenesis.[344,347,348,408,438,596,597,612]

Another aspect of enzyme induction and its relationship with disease concerns the effects of gastrointestinal microflora on foreign compounds and subsequent secondary action on hepatic drug-metabolizing enzymes. In man, cyclamate is metabolized to cyclohexylamine

Table 12
RELATIONSHIP BETWEEN BIOTRANSFORMATION AND BIOLOGICAL/
TOXICOLOGICAL ACTIVITY OF FOREIGN COMPOUNDS

Process	Parent compound	Metabolite	Reaction	Biological/Toxicological
Detoxication	Biliverdin	Bilirubin	Reduction	Inactive → inactive
	Butylated hydroxytoluene	BHT-alcohol BHT-acid	Oxidation	Antioxidant → inactive
	Cholesterol	Bile acid	Hydroxylation	Inactive → inactive
	Phenobarbital	*p*-Hydroxyphenobarbital	Hydroxylation	Hypnotic → inactive
	Procaine	*p*-Aminobenzoic acid Dimethylaminoethanol	Hydrolysis	Analgesic → inactive
Modification	Codeine	Morphine	O-demethylation	Analgesic → narcotic
	3,12-Dimethyl benz (α)anthracene	Hydroxy derivatives	Hydroxylation	Toxic → carcinogenic
	Iproniazid	Isoniazid	N-dealkylation	Antidepressive → antitubercular
	Prominal	Phenobarbital	N-demethylation	Short acting → long acting
Activation	Methanol	Formaldehyde Formic acid	Oxidation	Inactive → toxic
	Parathion	Paraoxon	Oxidation	Inactive → toxic
	Polycyclic hydrocarbons	Epoxy derivatives	Aromatic hydroxylation	Inactive → carcinogen
	Prednisone	Prednisolone	Ketone reduction	Inactive → immuno suppressant
	Prontosil	Sulfanilamide	Azo reduction	Inactive → antibacterial

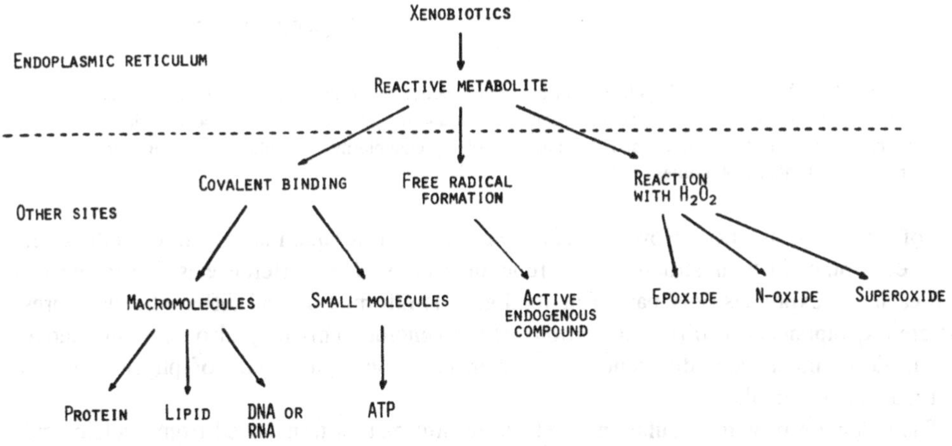

FIGURE 15. Mechanism of interaction between reactive metabolites and tissue components. Many drugs and other foreign compounds are converted to reactive metabolites by the drug-metabolizing enzyme system bound to the hepatic endoplasmic reticulum. The highly reactive electrophile metabolites are then bound to various sites by covalent binding, producing free radicals, or react with H_2O_2 to form further highly toxic metabolites. These events are important in the initiation of toxic side effects, cellular necrosis, mutagenesis, and carcinogenesis.

by the microflora.[138,446] Repeated administration of cyclamate leads to substantially increased desulfonation rates by the inductive action of cyclohexylamine on the endoplasmic reticulum. This accelerated desulfonation process may metabolize cyclohexylamine to further entities (Figure 16).

f. Enzyme Induction in Gene Regulation

The enzyme induction process is also important in the regulatory mechanisms of genetic information.[237] All somatic cells of multicellular differentiated organisms possess similar

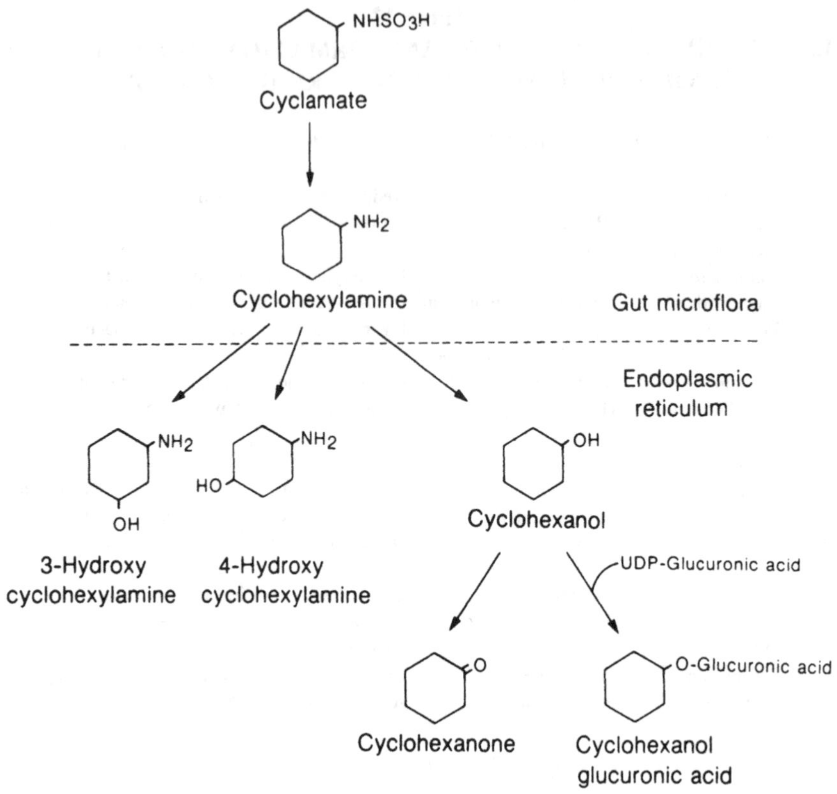

FIGURE 16.   Metabolism of cyclamate in the human body. The primary metabolite of cyclamate is cyclohexylamine produced by the action of the microflora in the gut. Through enterohepatic circulation this metabolite affects the liver, and the induction of the endoplasmic reticulum system initiates the formation of further metabolites.[138,446]

sets of genetic information. However, diverse cell types differ qualitatively and quantitatively from each other in their structural and functional units. This difference is due to the fact that each cell expresses only part of its total genetic information and different cells express different components or different portions of the genome. This may also show in changes during development and differentiation and in response to a variety of physiologic and pharmacologic stimuli.

The induction of gene regulation is related to information transferred from nucleic acids to proteins, and to the organization of the genetic material for macromolecular synthesis.[390] Mammalian cells contain a high amount of tightly wound skeins of DNA, packaged in highly organized chromosomes and associated with various basic and acidic proteins within the nucleus. Other components of gene transcription and translation involved in macromolecule synthesis are in separate sites and messenger RNA is found in polyribosomes. The fine control of protein synthesis is linked with modulation of newly synthetized RNA. In the regulation of RNA sequences many processes are involved, including methylation, phosphorylation, interaction of messenger RNA with protein, cytoplasmic transport of these protein-RNA complexes, and their association with ribosomes and polyribosomal aggregates. The complexity of these processes affects the mechanisms of enzyme induction including initiation, elongation, and termination of peptide chains, posttranslational modification of the protein molecule, and regulation of its degradation. However, relatively little is known about the mechanism of processes regulating the transcription of DNA into messenger RNA, or processes regulating the translation of the message into protein. Certain genes may affect

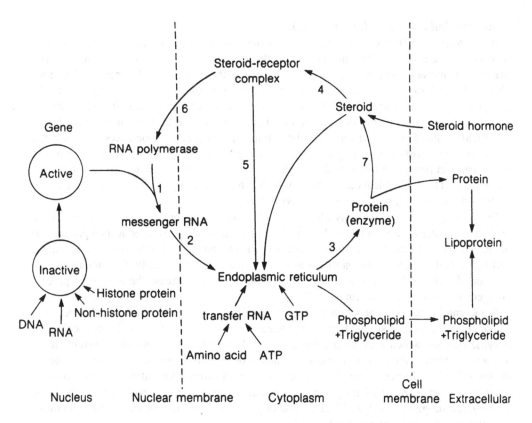

FIGURE 17. Schematic representation of gene-action system and the intracellular effect of steroid hormones. Certain genes are active in the synthesis of messenger RNA while others are inactive. (1) The active DNA is transcribed into messenger RNA in a reaction catalyzed by an enzyme, RNA polymerase. (2) When the messenger RNA is made, it is transferred to the cytoplasm and activates the endoplasmic reticulum. (3) In this ribosome-membrane complex, the messenger RNA directs the assembly of amino acids into specific polypeptide or protein chains. This process requires transfer RNA. The adapter molecules may have a regulatory function in messenger RNA translation. The process of protein synthesis requires a variety of other factors including GTP, transferases, and initiating factors. The latter two may be connected with the role of steroids. (4) Steroid hormones pass through the cell membrane and interact with cytoplasmic receptors. (5) This may affect the endoplasmic reticulum through the synthesis of membrane-bound phospholipid. (6) The hormone-receptor complex subsequently migrates to the nucleus where it activates RNA polymerase, RNA synthesis, and protein synthesis. (7) In turn, the synthesis of enzymes (protein) regulates the composition of intracellular steroids qualitatively and quantitatively.[168,615,618,620]

the activity of others or their sensitivity to environmental agents. These molecules can be endogenous, such as metabolites or hormones, or they may be exogenous, such as chemicals and certain carcinogens.

A simple schematic representation of the action of a gene system is illustrated (Figure 17). The induction of some enzyme activities is related to protein synthesis and linked with the function of inducing steroids. These steroids enter the cells passively and become bound to highly specific cytoplasmic proteins. The action of progesterones, estrogens, and androgens has been described in the induction of several enzymes in the liver, kidney, uterus, and prostrate of experimental animals. The formation of receptor-steroid complexes and many biological effects of steroid hormones appears to involve alterations at the gene level. In particular, nuclear binding is an essential step in the regulation of hormonal induction of enzymes. The understanding of hormone action in enzyme induction is not only relevant in cell differentiation but also in the development of inherited diseases. The regulation of gene expression is therefore critical in these disorders as well as in the understanding of many basic problems related to teratogenesis and neoplasia.

g. Enzyme Induction and Carcinogenesis

It has now been firmly established that various specific types of chemicals or mixtures of chemicals are the determining factors in the incidence of many types of cancer. The majority of human cancers may be caused by naturally occurring substances and various chemicals, and many are due to exposure to industrial substances. Carcinogens can be grouped into several categories: (1) direct-acting carcinogens, (2) procarcinogens, and (3) cocarcinogens. Direct-acting or ultimate carcinogens act on the host organism without any modification and do not require any metabolic transformation either spontaneous or catalyzed by an enzyme. Most direct-acting carcinogens are synthetic compounds, and some represent important occupational hazards with potential risk to people involved in their production or application. These chemicals are generally highly reactive compounds, are usually quickly destroyed, and do not contaminate the environment.

Procarcinogens are active only after metabolic conversion by the host organism. These are relatively stable compounds and exist in the environment. These chemicals represent the greatest potential of cancer risk to man because, when taken up these precursors are activated, and figuratively, these carcinogens are actually synthesized in the body. Cocarcinogens are compounds with little or no carcinogenic potential, but can increase the actions of procarcinogens or direct-acting carcinogens.[33,46]

Some carcinogens are metabolized in the body by a variety of biochemical processes. These reactions include simple hydrolytic breakdown if the chemical is labile, or specialized enzyme systems which may be present in selected tissues. Such metabolic conversion reactions can process the chemical also to less active or less carcinogenic substances by means of detoxification. In many instances these processes of biotransformation can lead to more potent substances, representing bioactivation and other associated with the synthesis of ultimate carcinogenic reactants.[343] The group of procarcinogens require metabolic activation, and without that they are not carcinogenic.

The hepatic endoplasmic reticulum represents the major site where carcinogens are metabolized. The metabolic reactions are connected with the function of the microsomal oxidizing enzyme system which converts carcinogens to a variety of metabolic products.[56] The hepatic drug microsomal metabolizing enzyme system activates carcinogens and this is also the subcellular site for deactivation. In the liver the deactivating and conjugating enzymes are present in rich amounts and the effect of active intermediates is therefore decreased in this tissue. The various reactions catalyzed by the drug-metabolizing enzyme are dependent partly on the chemical structure of the carcinogen; on the age, sex, species, smoking, dietary habits, and microbiological environment; and on the presence of other foreign compounds and endogenous steroids which are also metabolized by the endoplasmic reticulum-bound enzyme system. Induction or inhibition of this enzyme system indeed affects carcinogenicity. It is important that the occurrence of certain substances in food such as flavones, which are potent enzyme inducers,[169,572] or coumarin, which is a hepatotoxicant (to animals), can modify the response to carcinogens.

Most procarcinogens are metabolized to electrophylic intermediates. Some intermediates have a short half-life, whereas some may have a free radical character and could be produced at a site in the cell, such as the endoplasmic reticulum, and transported to the nucleus where it acts on the receptor DNA molecule. It is possible that such potent intermediates are produced in close proximity of the target organelle at a reactive site, or that there is an interaction between the carcinogen, activating enzyme bound to the endoplasmic reticulum, and the nuclear target site. The interface between the nucleus and the endoplasmic reticulum may provide a specific site for such interaction.

Enzymes which metabolize carcinogens are present in smaller amounts, and to varying degrees in most other organs such as kidney, lung, and gastrointestinal tract. The adrenal cortex and brain can also metabolize various carcinogens. The placenta contains a variety of enzymes which can modify drugs and carcinogens and affect transplacental transfer into the fetus.[184,255] Enzyme systems in subcutaneous tissue or skin can activate carcinogens, e.g., the conversion of polycyclic aromatic hydrocarbons to epoxides. These tissues are poor in enzymes which further deactivate these intermediates to water-soluble conjugates; therefore, they are very sensitive to small amounts of carcinogens.

Many carcinogens possess specific affinity for certain target organs. Some aromatic amines cause urinary bladder tumor.[181] There is organ specificity among the carcinogenic nitrosamines related to chemical structure.[139,475] Thus, interactions may modify the site where cancer ensues. Heavy smokers may not develop lung cancer, but such individuals may have urinary bladder or pancreas cancer after a long latency period.[264]

E. Golgi Apparatus

The Golgi apparatus is one of the main sites in the cell where the processes of cell and tissue differentiation take place. Also, this subcellular organelle serves as a major route for intra- and intercellular communications.[384] The Golgi constitutes an important intercommunicating labyrinth of membrane systems interconnected with the endoplasmic reticulum and lysosomes. It is also involved in the formation and packaging of macromolecules for other cells where, by the process of reverse pinocytosis, these substances are transported away from the cell. These cellularly exported materials are secretory proteins, digestive enzymes, or connective tissue matrix.

Many of these macromolecules such as proteins or lipids receive carbohydrate residues in the Golgi apparatus to form glycoproteins and glycolipids or glycolipoproteins. Attachment of sugar moieties to macromolecules by specific glycosyl transferase enzymes determines the character, biological activity, and antigenic properties of the secretory products.[215] The Golgi synthesizes and modifies materials contained in the lumen of the cisternae and vesicles. In addition, vesicles of the Golgi become primary lysosomes which contain hydrolytic enzymes.[383] These primary lysosomes give rise to secondary lysosomes or to digestive vacuoles by fusion with pinocytotic or autophagic vesicles.[98]

Hypertrophy of the Golgi apparatus has been described in human liver in many disease conditions such as dietary changes, chronic alcoholism, drug-related cholestasis, and hepatitis.[360,376] Degeneration or atrophy is found during starvation, or connected with protein or choline deficiency and viral hepatitis.[353] Some of these actions are connected with direct effects on the biosynthesis and transport of glycoproteins and lipoproteins.[332] Lesions of the Golgi also modify the nature and amount of glycoproteins and lipoproteins produced. In many pathologic and genetic disorders, secretion of glycoproteins is impaired.[361,527]

VI. CARCINOGENESIS AND CELLULAR ORGANELLES

A. Structural Changes

Interactions of carcinogens with nucleic acids bring about significant changes in the structure and synthesis of RNA and DNA, modifying gene expression and the regulation of cell development. These changes are associated with chemical modification of nucleic acids and proteins within the nucleus. Ultrastructural studies in experimental animals have shown that interactions occur between a carcinogen or its metabolites and specific cellular organelles. Many of these interactions are related to early, intermediate, or late manifestations including structural changes associated with biochemical abnormalities.[530]

Early changes using selective carcinogens include enlargement of the nuclei and increased nucleoli. Lysosomes are conspicuous and cytoplasmic swelling is followed by the action of some carcinogens. Most carcinogens cause slight to moderate necrosis and variable degrees of inflammation.

Acute changes follow fairly common patterns with the administration of carcinogens. The smooth endoplasmic reticulum shows no change; however, ribosomes are detached from the rough endoplasmic reticulum, resulting in increased free ribosomes in the cytoplasm. The number of lysosomes are slightly increased and mitochondria show variable degrees of swelling. The most uniform changes appear in the nucleus as redistribution and rearrangement of the fibrillar and granular components of the nucleolus. The rearrangement of constituents in the nucleoli may represent different stages of the effects of the carcinogen. These changes have been proposed as the basis for a screening test for chemical carcinogens.[294]

Chronic ultrastructural changes include further alterations in the proportion or configuration of the nucleolar constituents, enhancement of interchromatin or perichromatin granules, or dilation of interchromatinic spaces. Most consistently, the smooth endoplasmic reticulum is increased and rough endoplasmic reticulum is degranulated. Detachment of ribosomes from the rough endoplasmic reticulum and proliferation of smooth endoplasmic reticulum occur with many noncarcinogens. These are considered reversible phenomena and represent either hepatotoxicants or inducer effects.[172,396] In some instances, mitochondria are markedly enlarged, with altered shape and reduced length and number of cristae. Lysosomes show slight to moderate increase, and may only reflect toxic effects.

Many studies revealed changes in nucleoli from tumor cells, specially in the liver.[72] In most cases the nuclei show enlargement due to an increase of the granular component. Perichromatin granules are probably decreased, representing some degree of redistribution between granular and fibrillar nucleolar constituents. Carcinogens inhibit the *de novo* synthesis of nuclear and ribosomal RNA and RNA polymerase activity is decreased. Structural changes induced by carcinogens appear to be nonspecific. Nevertheless, they are related to changes in RNA formation and some other biochemical processes modified in neoplastic tissues. There have been many biochemical changes described in experimental animals.[377] The importance of these changes in neoplastic tissues and the extrapolation of animal data to man have not yet been clearly assessed.

B. Interactions with Nuclear Components

Some toxic compounds or carcinogens react with specific sites of normal components of cell membranes; and proteins, lipids, or nucleic acids may be the target sites. Effects on the target site may alter the viability of the cell and impair its function. The cell nucleus represents specific receptor sites for most chemical carcinogens. Active metabolites of many carcinogens produced by microsomal enzymes[409] interact with nucleophylic macromolecules in the nucleus from cells in a variety of tissues.

The nature of interactions between chemical carcinogens and nucleic acids are varied and can modify the structure and function of these nuclear components. Most carcinogens or their highly reactive metabolites bind covalently to DNA. All four bases of DNA backbone are involved in the binding, with purine nitrogens being the most reactive.[433] Covalent binding occurs with alkylating agents. These include nitrosamines, nitrogen and sulfur mustards, epoxides, ethyleneimines, methanesulfonates, and certain lactones. The major product is 7-methylguanine in most instances. Noncovalent interactions also exist in some cases. These include (1) insertion of the carcinogen between base pairs of the DNA structure, or (2) external binding when the carcinogen is bound on sites not directly involved in base pairing. Some carcinogenic polycyclic aromatic hydrocarbons, aflatoxins, and carcinogenic metals may act upon DNA through noncovalent binding.

Cellular ribonucleic acids play an important role in the transfer of genetic information, and chemical carcinogens also interact with RNA. Alkylating agents act on various sites in RNA with resulting alterations; these are, in decreasing order, N-7 guanine > N-1 adenine > N-3 cytosine > N-7 adenine > N-3 adenine > N-3 guanine.[490] Interactions were first demonstrated with dimethylnitrosamine which methylates liver RNA to a greater extent than those of other organs. Other potent alkylating agents include *N*-methylnitrosamines, methyl methanesulfonate, urethan, ethionine, and aromatic amines and amides.

Alterations of DNA and RNA structure by the carcinogen-nucleic acid interaction lead to abnormalities of protein synthesis and interfere with cell proliferation and DNA as well as RNA synthesis in the nucleus. The interaction of chemical carcinogens with nucleic acids may be of importance in the initiation of carcinogenesis since the binding of carcinogens to DNA or RNA shows good correlation with neoplasm development in many instances. Carcinogens also interact with nuclear proteins. Here the major reactive sites are tyrosine, tryptophan, methionine and, to a lesser extent, histidine; all of these entities are nucleophylic reactants. Conversion of the histidine moiety from nuclear proteins to methylhistidine may lead to mutagenesis. Alterations of lipid methylation processes which are involved in the synthesis of lecithin result in changes in membrane stability and in the function of membrane-bound enzymes. The effects of xenobiotics on lecithin formation are related to the induction process or to hepatotoxicity. Methylation of lipids has not been shown to exert any potential for carcinogenicity.

VII. INTERRELATIONSHIPS BETWEEN CELLULAR ORGANELLES

The close relationships between cellular organelles are based upon common steps in the synthesis of membranes and upon changes resulting from different environmental conditions including diseases or the action of foreign compounds. The assembly of various membranes and vesicles takes place in four phases: (1) synthesis of protein and lipid components, (2) arrangement of proteins and lipids into a membrane, (3) orientation of the formed membrane elements into their structural position which at the same time also represents the functional state, and (4) reorganization of one vesicular membrane to produce the membrane of another organelle. The establishment of the intracellular network is probably the major organizational phenomenon guiding these processes. In this network there are vesicular associations as well as metabolic interactions (Figure 18).

Proteins are produced by ribosomes with the participation of the endoplasmic reticulum. Fatty acids are synthetized and fatty acid chains are elongated by enzymes present on the mitochondrial envelope. The synthesis of unsaturated fatty acid is directed by microsomal enzymes. The incorporation of various fatty acids into phospholipids and their bases by methyl transfer reactions are also rendered by the microsome fraction of the cell. The endoplasmic reticulum is primarily responsible for phospholipid formation but the Golgi apparatus can also be a source. The rate of phospholipid synthesis is related to the amount of available protein; insufficient protein supply causes a reduction or even stops phospholipid production.

The assembly of protein and phospholipids into rough endoplasmic reticulum membranes occurs primarily in the existing reticulum. The smooth endoplasmic reticulum is either directly synthesized from preexisting rough membranes or by the loss of ribosomes from rough membranes. Smooth membranes appear close to the Golgi apparatus, and vesicles budding from existing membranes often extend into the Golgi. These membranes may originate by fusion, but since the two organelles are not strictly homologous, other sources may also contribute to the formation of the Golgi vesicles.

Fusion represents a close association in orientation between cellular organelles. It can take place between Golgi membranes and cell membranes. Golgi vesicles may also be associated with primary lysosomes. The latter membranes can fuse with heterophagosomes and pinocytotic vesicles which are derived from plasma membranes. The production of primary lysosomes from Golgi membranes is initially connected with enzymes bound to the endoplasmic reticulum. The formation of secondary lysosomes and multivesicular bodies is also associated with this fusion process. Under pathologic conditions, toxic compounds or excessive amounts of drugs can break up the endoplasmic reticulum membranes and transform them into secondary lysosomes (Figure 11).

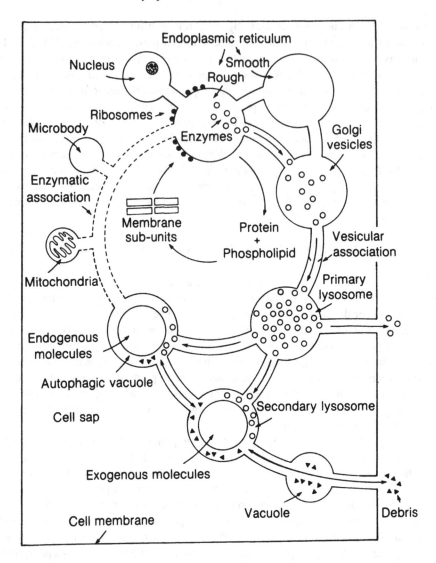

FIGURE 18. Schematic illustration of the intracellular cytocavitary network. Vesicular and enzymatic associations form a close interrelationship between the various subcellular organelles. There is a probable morphological connection between the nucleus, endoplasmic reticulum, Golgi vesicles, and lysosomes. Triggered by the nucleus, the endoplasmic reticulum produces proteins and phospholipids which form membrane subunits for most subcellular organelles. This is also the site of enzyme synthesis which is transferred to the Golgi vesicles and to primary or secondary lysosomes. Mitochondria and microbodies are metabolically integrated into the cytocavitary interactions.

Evidence is available for the vesicular association between subcellular organelles. During cell division, ribosomes attached occasionally to nuclear membranes break up and membranous vesicles are formed which resemble the endoplasmic reticulum. After completion of mitosis the daughter nuclear membranes are derived from the endoplasmic reticulum. Additional morphological observations indicate that membrane fragments derived from the breakdown of plasma membranes are incorporated into Golgi and endoplasmic reticulum membranes.[232]

The outer membranes of mitochondria and peroxisomes probably originate from the endoplasmic reticulum. The inner membranes are formed by the incorporation of new material

Table 13
EFFECT OF HEPATOTOXINS ON VARIOUS CELL ORGANELLES

Organelles	Compound	Ref.[a]	Organelles	Compound	Ref.[a]
Nucleus	Aflatoxin	1	Lysosomes	Aflatoxin	1
	Beryllium	2		Carbon tetrachloride	9
	Dimethylnitrosamine	3		Beryllium	10
	Hydrazine	4		Pyrrolizidine alkaloids	5
	Pyrrolizidine alkaloids	5	Endoplasmic	Allyl formate	6
Mitochondria	Allyl formate	6	reticulum	Carbon tetrachloride	7
	Carbon tetrachloride	7		Dimethylnitrosamine	6
	Dimethylnitrosamine	6		Ethionine	8
	Ethionine	8		Pyrrolizidine alkaloids	5
	Hydrazine	4			
	Pyrrolizidine alkaloids	5			

[a]
1. **Wogan, G. N.,** in *Aflatoxins*, Goldblatt, L. A., Ed., Academic Press, New York, 1969, 151.
2. **Witschi, H. P.,** *Biochem. J.*, 120, 623, 1970.
3. **Emmelot, P. and Benedetti, E. L.,** *J. Biophys. Biochem. Cytol.*, 7, 393, 1960.
4. **Ganote, C. E. and Rosenthal, A. S.,** *Lab. Invest.*, 19, 382, 1968.
5. **McLean, A. E. M.,** *Br. J. Exp. Pathol.*, 51, 317, 1970.
6. **Rouiller, C.,** *The Liver*, Vol. 2, Academic Press, New York, 1964, 335.
7. **Recknagel, R. O.,** *Pharmacol. Rev.*, 19, 145, 1967.
8. **Anthony, D. A., Schaffner, F., Popper, H., and Hutterer, F.,** *Exp. Mol. Pathol.*, 1, 113, 1962.
9. **Williams, D. J. and Rabin, B. R.,** *Nature (London)*, 232, 102, 1971.
10. **Dianzani, M. V.,** in *CIBA Symposium on Lysosomes*, de Reuck, A. V. S. and Cameron, E. P., Eds., Little, Brown, Boston, 1963, 335.
11. **Witschi, H. P. and Aldridge, W. N.,** *Biochem. J.*, 106, 811, 1968.

followed by self-replication. Centrioles are also formed by self-replication. The metabolic association between these subcellular organelles and the nucleus or endoplasmic reticulum makes them integral parts of the cytocavitary network.

The interrelationship between the various cellular organelles is regulated by the availability of phospholipid-protein subunits which are in turn related to the synthesis of new proteins and phospholipids. Thus phospholipid-protein subunits represent the limiting step in the process. Upon death of the cell, primary lysosomes are broken down and their enzyme contents leak out into the cytoplasm. The action of a number of toxic compounds and chemical carcinogens and the effect of many diseases is linked with various cellular organelle interactions and subsequent malfunction (Table 13).

The initial lesion may start on any organelle and this organelle sensitivity is dependent on affinity, differences, on the rate of penetration, and on inhibition of key enzymatic processes related to the structure or metabolism of the drug. The confluence of these conditions will determine the primary lesion brought about by disease. The primary subcellular organelle effect of various substances, including carbon tetrachloride and aflatoxin, begins on the endoplasmic reticulum by ribosome degranulation. The primary subcellular target organelle of several other substances, such as silica or polyene antibiotics, is the lysosome. Lysosomes respond with increased membrane fragility, and the process of heterophagocytosis, a destructive response, is triggered. For example, silica particles taken up by macrophage cells can destroy them. The silica particles taken up by macrophage cells can destroy them. The silica particles are then released and subsequently taken up by other macrophages causing their death in turn. The incorporation of silica particles into the lung of miners is often widespread and results in the destruction of many macrophages, representing an inflammatory process. The replacement of dead cells and collagen deposition can lead to pronounced decrease of pulmonary function and the onset of lung disease, termed silicosis.

Chemical carcinogens also have the ability to interfere with the interrelationships between various cellular organelles. In a similar fashion, various dyes, aflatoxin, and polycyclic hydrocarbons affect the nucleus by binding to DNA and RNA or protein. These chemicals cause degranulation of the rough endoplasmic reticulum membranes. Thus they impair protein synthesis and labilize lysosomal membranes, and through the damaged membranes, nucleases are released into the cytoplasm. The leakage of nucleases into the cytoplasm brings about mutations by secondary reactions, and may initiate the development of processes leading to cancer.

VIII. CELL SHAPE ABNORMALITIES

Abnormalities of cell shape are only related to diseases of the blood cells. The effects of cell shape alterations in other somatic cells are not known, although cancerous cells show bizarre deviations. In these cases the normal geometry of the cells is altered.[5,94,187,285,313,314,359,416,575] This especially occurs in erythrocytes. Stress, due to shearing forces, induces reversible alterations in the shape of normal erythrocytes. However, in a number of disease states, as in hemolytic anemias, the cells may undergo a variety of reversible changes in shape.[189] These changes may be associated with fragmentation of the membrane or with structural variations of the intracellular hemoglobin. Alterations of the erythrocyte shape significantly influence their rheologic properties, the exchange processes between red cell and surrounding plasma, thus affecting survival of the cell.

Under certain conditions normal human erythrocytes may undergo shape changes in a reversible fashion, from the normal biconcave structure to spherical shape.[195,562,576] Slowly, coarse spicules appear on the smooth regular discocyte shape, causing resemblance to a sea urchin. This echinocyte transformation may be triggered by intrinsic or extrinsic factors, including nonpolar anionic amphiphilic compounds such as lysolecithin, bile acids, salicylate, barbiturates, and other drugs.[133] Lysolecithin and the enzyme lecithin-cholesterol acyl transferase are the plasma factors associated with the conformation. The echinocyte form reverts to normal shape when the active agents disappear from the medium. Echinocytes occur in uremia, heart disease, bleeding peptic ulcer, and stomach carcinoma. Parallel with the erythrocyte deformation the cellular ATP content is also reduced.[369]

A. Erythrocyte Disorders

1. Deformation of Erythrocyte Shape

The major role of erythrocytes is to maintain oxygen transport into various tissues. They have a specific disc shape which enables them to go through the microcirculation. Various proteins and glycoproteins participate in the maintenance of the normal shape.[64] These include the spectrin-actin cytoskeletal complex, band-three protein, ankyrin, and sialoglycoproteins.[9,31,51,170,192,212,225,247,251,327,416,517] Some functional proteins such as protein kinases and ATP are required to preserve the proper protein associations in the membrane.[339,371,418,493] In particular, spectrin is involved in the membrane cytoskeletal network responsible for the shape and deformability of the erythrocytes.[373,404]

In severe megaloblastic anemia secondary to vitamin B_{12} or folic acid deficiency, spectrin and band-three protein are deficient and connected with a shortened life span of the erythrocytes.[32] The protein pattern of the erythrocyte membrane returns to normal after replacement therapy with adequate vitamin supplementation.[22,23]

En(a⁻), a rare human erythrocyte phenotype, is characterized by markedly reduced sialic acid content of the red cell membrane, very weak MN bloodgroup activities, and lowered mobility on electrophoresis.[14] It is remarkable that so far no clinically significant abnormalities have been reported in En(a⁻) individuals, since the absent major sialoglycoprotein carries a large proportion of the carbohydrates and sialic acid of the erythrocyte.[118,539]

In patients with congenital dyserythropoietic anemia type II, the red blood cells are abnormal and have a reduced life span. This anemia is characterized by multinuclearity and karyorhexis of the erythroid series in bone marrow. The red blood cells lack protein components which migrate in the band-three region. In addition to the reduction of membrane proteins, membrane lipids of erythrocytes also show abnormal patterns.

2. Hereditary Erythrocyte Disorders

In hereditary spherocytosis small, round erythrocytes (spherocytes) are formed with high average cellular hemoglobin content and a reduced surface area/volume ratio due to deficient surface area. These cells show increased osmotic fragility and exhibit abnormally high sodium ion influx resulting in a high demand for the ATP-dependent sodium pump.[193,196,239,316,371] The defect of sodium permeability may be responsible for producing a cascade of membrane molecular and intracellular events. The cause for the sodium leak in the membrane of spherocytes is unknown; the specific membrane defect may be associated with abnormal phosphorylation.[406,431]

The clinical symptoms of this disease are jaundice, anemia connected with hemolysis due to an increased rate of erythrocyte destruction, and enlarged spleen. The liver is usually normal, but gallstones containing pigments are common, probably derived from the precipitation of elevated amounts of bilirubin produced. The disease manifests itself in early infancy and causes a diminished life span when the spleen is present. After splenectomy the survival rate is almost normal. The defects of the red cell formation are intrinsic, with rapid degenerative changes and significant destruction occurring only in the spleen. Although the main characteristic feature of the disease is the formation of spherocytes, similar cells are formed in acquired hemolytic anemia brought about by drugs.

In the case of normal erythrocytes, cations from the plasma cross erythrocyte membranes at a relatively slow rate, in contrast to the anions which are freely permeable. Osmotic balance is maintained in the normal red cell: sodium ions enter by passive processes, but potassium is transported into the cell by active cation transport. This regulation is achieved by the membrane-bound ATP, the sodium pump. In the erythrocytes, ATP is produced and this energy is almost entirely derived from the metabolism of glucose by glycolysis. In spherocytosis the biochemical abnormalities include ATP breakdown, abnormal permeability to sodium, and disruption of cation gradients through the membrane.[43] The loss of sodium is compensated by an activation of ATPase activity, which in turn increases the transport of sodium from the cell into the serum. The rate of glycolysis is also enhanced to replace ATP molecules. Glycolysis is not complete in spherocytosis and there is an overall deficiency in glucose consumption whereby the cells are deficient in their ability to generate high-energy-containing phosphate compounds such as ATP and creatine phosphate. Rapid utilization of energy sources is necessary for the maintenance of the cell structure. However, due to the incomplete generation of energy-containing compounds, increased susceptibility to stasis occurs. Thus, the cell membrane is defective. The loss of lipid from the stroma is due to degeneration of the cell wall and increased osmotic fragility.[236]

In hereditary stomatocytosis, stomatocytes exhibit decreased osmotic fragility, which may be connected with an increased membrane phosphatidylcholine content.[340] In this disease, a permeability defect also exists, although the pattern is somewhat complex. In one type the stomatocytic cells have low potassium content, resulting in cell swelling and increased osmotic fragility. In the second type, cells have high sodium and potassium content, resulting in cellular dehydration and decreased osmotic fragility. Protein phosphorylation is affected in stomatocytes connected with abnormal cross-linking. Some conditions cause stomatocyte transformation, provoking reversible membrane loss. Subsequently, the cell initially becomes uniconcave, and then spherical cells result, following endovesicle formation. Agents responsible for this shape change include cationic amphiphilic compounds, nonpenetrating anions, and some drugs, such as chlorpromazine.[24]

Hereditary elliptocytosis is associated with the formation of oval or elliptical red cells. There is no serious anemia in most patients; in some cases, however, the circulating erythrocytes have short survival. This condition is due to membrane permeability defects and the cells exhibit greater thermal sensitivity.

3. Hemoglobinopathies Affecting Membrane Structure

The primary defect in hemoglobinopathies is in the abnormal genetic coding of hemoglobin synthesis. In some instances pathological manifestations of these conditions include modifications in the structure and function of the erythrocyte membranes.[186,187,273,584] Sickle cell anemia is due to a single amino acid mutation in which a glutamate of the normal hemoglobin β-chain is replaced by a valine residue. The result of this amino acid substitution is that the physical properties of the hemoglobin molecule are altered and under the conditions of reduced oxygenation, large molecular aggregates are formed.[220,316] The end result of this change is sickling of the red cells. Most sickle cells can be reversed by increasing oxygen tension, but some remain permanently altered. Repeated changes lead to irreversibly sickled erythrocytes.[297,405]

Biochemical changes in sickle cell membranes include higher levels of phospholipids and cholesterol. The cross-linking of spectrin-actin cytoskeletal network proteins is altered in sickled cells, connected with abnormal protein phosphorylation and increased accumulation of Ca^{2+} inside the cell.[44,141,306,373,405]

Stomatocytic and echinocytic transformations become to a greater extent irreversible in pathologic erythrocytes such as the abnormal sickle cell. Erythrocytes from patients suffering from thalassemia major show poikilocyte transformation. These cells usually contain significant amounts of Heinz bodies.

4. Enzyme Deficiencies in Erythrocytes

Defects of various cytoplasmic enzymes in erythrocytes produce hemolytic anemias.[36,175] The most common enzymopathies, glucose 6-phosphate dehydrogenase and pyruvate kinase deficiencies, rarely affect the shape of red cells, and thus these are generally recognized as nonspherocytic anemias. Glucose 6-phosphate dehydrogenase deficiency has been identified by the effect of certain drugs such as primaquine or fava beans causing an oxidative stress. The resulting change is the production of hydrogen peroxide which oxidizes glutathione and also leads to the formation of mixed disulfides of glutathione and hemoglobin. The loss of free sulfhydryl groups leads to changes of the normal erythrocyte shape and function.[170] In the case of pyruvate kinase deficiency the specific reason for hemolysis is probably related to the reduction in ATP synthesis. This molecule affects the mobility of intrinsic membrane proteins and modifies the membrane skeleton.[472]

B. Muscular Distrophy

Congenital muscular distrophy has been shown to be connected with a generalized membrane defect. Changes in membrane properties occur in muscle and in other cells. Erythrocytes from patients suffering from various types of muscular dystrophy show intense surface deformation.[309,330]

The two major types of muscular dystrophy, Duchenne dystrophy and myotonic muscular dystrophy, differ in their pattern of genetic inheritance.[42,84,303,477] A small proportion of abnormal erythrocytes have been found in the Duchenne form of dystrophy in female carriers. These are echinocytic in appearance. Erythrocytes from myotonic muscular dystrophy patients are similar to stomatocytes. Although membrane phospholipid and cholesterol content are within normal range, the abnormal membrane structure may be related to fatty acid changes, particularly since palmitoleic acid has been found to be reduced significantly.[464,466] Several membrane-bound enzymes are altered, including protein phosphorylation reaction and Ca^{2+}-stimulated-Mg^{2+}-dependent ATPase activity.[465,566]

IX. RADIATION EFFECTS ON BIOMEMBRANES

The sources of ionizing radiations to which man may be exposed can be grouped into four categories: (1) natural sources, (2) medical sources (diagnostic and therapeutic X-ray examinations and radiopharmaceuticals), (3) nuclear reactions (nuclear power generating stations and nuclear weapons), and (4) occupational sources (X-ray machines and industrial instruments emitting ionizing irradiations). Exposure to the natural sources cannot be avoided for the most part, but the degree of exposure to man-made sources is subject to change, depending on the use and demand of such sources.

Cellular damage can be produced by directly ionizing radiations such as α particles, β particles, and protons, or indirectly by γ- and X-rays which cause ejections of fast electrons from target atoms, or by neutrons which generate recoil protons and other nuclei. In all cases the resulting charged particles excite or ionize other molecules in a so-called primary event.[292] The energy of the absorbed radiation is deposited in randomly distributed packages which may be separated by several thousand nanometers for γ- and X-rays or as little as a few nanometers for densely ionizing radiation such as α particles. In the primary event, about 50 to 100 eV (1200 to 200/cal/mol) are released or transferred. Although this energy is large, few of these releases ($\sim 10^6$ cells) are required to cause cell death.

Most absorbed radiation energy induces molecules into excited electronic states. These molecules then return to ground state and dissipate the excess energy as heat in a relatively harmless way. Cellular damage brought about by radiation is primarily derived from ionization. This ionization may cause critical local changes, some of which remain permanent, while others may reverse by rapid biochemical processes or slower metabolic reactions. When radiation causes biological damage, this is usually manifest in chromosomes after the impact of the ionization.

Ionization processes occur more frequently in condensed systems such as macromolecules and membranes. Direct hits by radiation cause inactivation of some membrane enzymes,[269] but some free radicals may be also involved in the damage by permeating these structures.[523] In most living matter, water is the dominant component and its oxidizing radicals can alter membrane proteins, lipids, and carbohydrates.

Many factors modify the response to a given dose of radiation. These include extrinsic factors such as quality of radiation, dose rate, geometry of the exposure, and the portion of the body exposed. Intrinsic factors are sex, age, race, metabolic status, and oxygen tension. There are wide differences in the sensitivity of human tissues to radiation-induced cancer (Table 14).

A. Effects of Exposure

Early and intermediate effects of ionizing radiation include somatic effects. These occur early but only after exposures to a relatively high dose (>50 rads), resulting primarily from cell death. The consequences are diverse, such as acute radiation sickness and pulmonary fibrosis. The radiosensitivity of cells is related to the frequency of undergoing mitosis. Cells with frequent mitosis are the most sensitive, whereas cells with no mitosis are the most radioresistant.[463] Most radiosensitive cells are the hematopoietic stem cells, dividing glandular cells of the intestine and stomach, granulosa cells of ovarian follicles, lymphocytes, and germinal cells of the epidermis. Most radioresistant cells are those cells which normally do not divide and are well differentiated and specialized in function. These include neurons, erythrocytes, neutrophils, some muscle cells, epithelial cells of the sebaceous glands, and superficial cells of the alimentary tract. There are some relatively radioresistant cells, such as the cells of the pituitary, thyroid, parathyroid, and adrenal glands, liver parenchymal cells, duct cells of kidney, pancreas, and salivary glands. These cells have long lives and divide at low rates except under conditions of stimulation.

Table 14
RELATIVE SENSITIVITY OF HUMAN TISSUES TO
RADIATION-INDUCED CANCER

Sensitivity	Tissue	Radiation dose (rads)	Ref.[a]
High	Thyroid gland	>30	1, 2
	Myelopoietic tissue	50—100	3—5
	Acute leukemia		
	Myeloid leukemia		
Moderate	Breast	<100	5—7
	Salivary gland	<300	8, 9
	Lung	400	10—12
Low	Bone	<1000	13—15
	Skin	<1000	16, 17
	Stomach	<1000	16, 18
Relatively resistant	Larynx, pharynx, esophagus, pancreas, other neoplasms of reticular tissues		19—21

[a]

1. **Pifer, J. W., Hempelmann, L. H., Dodge, H. J., and Hodges, F. J.,** *Am. J. Roentgenol. Radium Ther. Nucl. Med.*, 103, 13, 1968.
2. **Conrad, R. A., Dobyns, B. M., and Suttow, W. W.,** *J. Am. Med. Assoc.*, 214, 316, 1968.
3. **Doll, R. and Smith, P. G.,** *Br. J. Radiol.*, 41, 362, 1968.
4. **Ishimaru, T., Hoshino, T., Ichimaru, M., Okada, H., Tomisayu, T., Tsuchimoto, T., and Yamamoto, T.,** *Radiat. Res.*, 45, 216, 1971.
5. **Jablon, S. and Kato, H.,** *Lancet*, 2, 1000, 1970.
6. **Wanebo, C. K., Johnson, K. G., Sato, K., and Thorslund, T. W.,** *N. Engl. J. Med.*, 279, 667, 1968.
7. **Mettler, F. A., Hempelmann, L. H., Dutton, A. M., Pifer, J. W., Toyooka, E. T., and Ames, W. R.,** *J. Natl. Cancer Inst.*, 43, 803, 1969.
8. **Hazen, R. W., Pifer, J. W., Tayooka, T., Livengood, J., and Hempelmann, L. H.,** *Cancer Res.*, 26, 305, 1966.
9. **Hempelmann, L. H., Pifer, J. W., Burke, G. W., Terry, R., and Ames, W. R.,** *J. Natl. Cancer Inst.*, 38, 317, 1967.
10. **Lundin, F. E., Jr., Wagoner, J. K., and Archer, V. E.,** *Radon Daughter Exposure and Respiratory Cancer: Quantitative and Temporal Aspects*, NIOSH-NIEHS Joint Monograph No. 1, Springfield, Va., 1971, 204.
11. **Steinitz, R.,** *Am. Rev. Resp. Dis.*, 92, 758, 1965.
12. **Wanebo, C. K., Johnson, K. G., Sato, K., and Thorslund, T. W.,** *Am. Rev. Resp. Dis.*, 98, 778, 1968.
13. **Evans, R. D., Keane, A. T., and Shanahan, M. M.,** in *Radiobiology of Plutonium*, Stover, B. J. and Jee, W. S. S., Eds., J. W. Press, Salt Lake City, 1972, 431.
14. **Spiess, H. and Mays, C. W.,** *Health Phys.*, 19, 713, 1970.
15. **Yamamoto, T. and Wakabayashi, T.,** *Atomic Bomb Casualty Commission Technical Reports*, Hiroshima, 1968, 26.
16. **Court Brown, W. M. and Doll, R.,** *Br. Med. J.*, 2, 1327, 1965.
17. **Johnson, M. L. T., Land, C. E., Gregory, P. B., Taura, T., and Milton, R. C.,** *Atomic Bomb Casualty Commission Technical Reports*, Hiroshima, 1969, 20.
18. **Yamamoto, T., Kato, H., Ishida, K., Tahara, E., and McGregor, D. H.,** *Gann*, 61, 473, 1970.
19. **Beebe, G. W., Kato, K., and Land, C. E.,** *Radiat. Res.*, 48, 613, 1971.
20. The Effects of Populations of Exposure to Low Levels of Ionizing Radiation, National Academy of Sciences-National Research Council, Washington, D.C., 1972.
21. Committee on the Biological Effects of Ionizing Radiations, The Effects on Populations of Exposure to Low Levels of Ionizing Radiation: 1980 (BEIR Report), National Academy Press, Washington, D.C., 1980.

In cases of acute exposure and depending on the size and the distribution of the absorbed dose, the clinical manifestations can be grouped into three categories: (1) hematopoietic depression, (2) gastrointestinal tract denudation, and (3) central nervous system disruption. These are related to acute whole-body irradiation doses.[289] At higher doses, the symptoms may effect more forms. Acute exposure of the whole body to penetrating ionizing radiations causes a combination of gastrointestinal and neuromuscular symptoms within 1 or 2 hr. Anorexia, vomiting, diarrhea, and nausea are the most common symptoms. The time of onset and the severity of the symptoms show great individual variations and are related to the total absorbed dose. At doses approaching 5000 rads or more, death is caused by neurologic and cardiovascular degeneration within short periods up to 48 hr following exposure. Even very low doses of irradiation can produce deleterious effects in man. Experimental data and human observations have clearly established that late somatic effects of ionizing radiation are the result of carcinogenesis or mutagenesis. The induction of leukemia in experimental animals and man after exposure to external ionizing radiation is well documented.[558] High-level radiation is potentially carcinogenic to all tissues under the proper set of circumstances.[559] Following exposures of the whole body or of significant portions of the bone marrow to ionizing radiation, leukemia develops after a short latency period, whereas other malignant neoplasms appear sometime later. Numerous reports show increased incidence of malignant and benign thyroid nodules following radiation exposure.[372] Skeletal tumors have also been reported at the site of therapeutic irradiation in many instances. Osteosarcomas were found in patients of various ages who received doses greater than 3000 rads. The average latency period until the onset of neoplasms was about 9 years. In children receiving about 500 rads irradiation, the tumors were mainly benign osteochondromas.[155] An increased incidence of mammary carcinomas was found in women who received (1) radiation therapy for acute postpartum mastitis, (2) multiple fluoroscopic examinations while treated for tuberculosis with artificial pneumothorax, and (3) female atom-bomb survivors. The dose to the mammary gland was about 200 rads for the postpartum mastitis group, greater than 1000 rads for those receiving fluoroscopic examination, and greater than 90 rem for those exposed to the atom bomb.[38] Evaluation of the atom-bomb survivors in Japan also indicated that excess incidence of myelogenous leukemia, lymphomas, cancer of the breast, esophagus, stomach, lung, thyroid, and urinary organs have been induced by radiation exposure.[29]

Ionizing radiation can produce chromosome aberrations in both germ cells and somatic cells. These aberrations are connected with chromosome breakage. Many of the broken chromosomes rejoin, leaving no visible effect of the damage in the cell. Others may fail to rejoin, or abnormal configurations are produced by deletions, duplications, or translocations. Many chromsome aberrations produce cell death.[497] In the case of germ cells, some of the chromosome aberrations are passed to the offspring. The estimated risk to man for these abnormalities from chronic radiation exposure is between 20 and 200 rem.[38]

B. Molecular Mechanism of Radiation Damage

1. Protein Damage

Protein constitutes more than 60% of most biomembranes which can be the sites of radiation-induced reactions. Many radiation damage which affects membrane function mostly arises from injury to membrane proteins. Substrates and coenzymes generally exert a protective action to enzymes; however, due to hydration they are inactivated via water radiolysis.[47] Enzymes with essential radicals and labile cofactors are particularly radiosensitive, and SH-reagents can provide some protection against damage.[525] Ionizing irradiation can disrupt S–S bonds, leading to reaction with other proteins or with unsaturated fatty acids.[279,368]

2. Lipid Damage

Whole-body irradiation of experimental animals causes severe lipemia, particularly with elevated unsaturated fatty acids in plasma.[142] Membrane phospholipids are oxidized by free radicals arising from water radiolysis.[142,593] The presence of oxygen increases lipid oxidation in various cellular organelles, and the increasing order of sensitivity is nuclei < mitochondria < lysosomes < microsomes. Malonaldehyde, the major end product of lipid peroxidation, is mainly formed from damage of unsaturated membrane phospholipids which subsequently influence the conformation of membrane proteins.

3. Sugar Radiolysis

Irradiation of certain carbohydrate molecules such as D-glucose, D-fructose, D-galactose, D-mannose, D-fucose, and L-rhamnose in oxygen-free solution can produce highly cytotoxic γ,β-unsaturated carbonyls.[479]

4. Effects on Membranes

It has not yet been settled as to what extent radiation damage to biomembranes accounts for the effects of radiotherapy, although it is well established that after low doses ionizing radiation alters these structures. Some cells such as erythrocytes and lymphoid cells show serious membrane changes. After high X-ray doses erythrocytes show swelling.[367,489] Lymphoid cells are quickly and very extensively damaged by even very low doses of ionizing radiation. A few hours after 200 to 400 rad whole-body irradiation of mice, their thymocytes show breakdown of nuclear envelopes, fragmentation of the endoplasmic reticulum, and detachment of ribosomes. Mitochondria also show swelling and clumping of the cristae, and outer membranes are damaged.[49] Lysosomes are easily perturbed by ionizing radiations. Autolysis of these membranes is followed by loss of hydrolases.[540]

5. Damage of Membrane Function

Irradiation of erythrocytes and other cells impairs the normal electrolyte gradients across their membranes eventually causing lysis of the cell.[367,489] In erythrocytes the K^+ loss and Na^+ gain are directly proportional to the radiation exposure. The radiation damage to membrane permeability is primarily derived from impaired active transport. Probably membrane SH groups are involved, since exposure of human erythrocytes to SH inhibitors prior to irradiation blocks the inhibition of both nonspecific and Na^+/K^+-sensitive ATPases. In contrast, preincubation with disulfide group-reducing compounds inhibits both ion-specific and nonspecific ATPases of the membrane, probably by splitting critical S–S bridges.[365] High radiation also causes damage of ion permeability in the membrane of subcellular organelles such as nuclei and mitochondria, even though there is no loss in ATP production by mitochondria.[238]

Thymocytes are extremely radiosensitive, probably through membrane-related mechanisms.[366] Lysosomes and other cytoplasmic membranes show leakage of various enzymes upon radiation. β-Glucuronidase and certain other hydrolases are released from these particles,[594] and linked to the oxidation of protein SH groups. In microsomes the damage is connected with marked lipid peroxidation, associated with a loss of the detoxifying capability.[595]

A further important aspect of the biological action of ionizing radiations is the balance between the beneficial effects of radiotherapy and side effects on the immunological system of the patient. Radiation exerts an impact on the immune response.[536] This effect is a membrane-related phenomenon, probably due to a damage on processing of antigens by macrophages,[350] although some other experimental evidence showed that only the lymphocytes are involved in the immune response to be radiosensitive and not the macrophages.[456] It is also unsettled as to what extent radiation damage to biomembranes accounts for the effects of radiotherapy causing radiation toxicity. Certainly some cells (lymphoid cells, in particular) and even erythrocytes show serious membrane impairment under certain circumstances.[367,489,570]

REFERENCES

1. **Abraham, E. P. and Duthie, E. S.,** Effect of pH of the medium on activity of streptomycin and penicillin and other chemotherapeutic substances, *Lancet,* 1, 455, 1946.
2. **Acheampong-Mensah, D. and Feuer, G.,** Effect of phenobarbital on methyl transfer between methylated drugs and hepatic microsomal phospholipids, *Toxicol. Appl. Pharmacol.,* 32, 577, 1975.
3. **Acheampong-Mensah, D. and Feuer, G.,** Relation between the hepatic action of drugs and the synthesis of subcellular structures, *Int. J. Clin. Pharmacol. Ther. Toxicol.,* 9, 49, 1974.
4. **Alfrey, A. C. and Solomons, C. C.,** Bone pyrophosphate in uremia and its association with extraosseous calcification, *J. Clin. Invest.,* 57, 700, 1976.
5. **Allen, D. W., Cadman, S., McCann, S. R., and Finkel, B.,** Increased membrane binding of erythrocyte catalase in hereditary spherocytosis and in metabolically stressed normal cells, *Blood,* 49, 113, 1977.
6. **Allera, A., Rao, G. S., and Breuer, H.,** Specific interaction of corticosteroids with components of the cell membrane which are involved in the translocation of the hormone into the intravesicular space of purified rat liver plasma membrane vesicles, *J. Steroid Biochem.,* 12, 259, 1980.
7. **Alvares, A. P., Kappas, A., Eiseman, J. L., Anderson, K. E., Pantuck, C. B., Pantuck, E. J., Hsiao, K. C., Garland, W. A., and Conney, A. H.,** Intraindividual variation in drug disposition, *Clin. Pharmacol. Ther.,* 26, 407, 1979.
8. **Anderson, D. R., Davis, J. L., and Carraway, K. L.,** Calcium-promoted changes of the human erythrocyte membrane. Involvement of spectrin, transglutaminase, and a membrane-bound protease, *J. Biol. Chem.,* 252, 6617, 1977.
9. **Anderson, J. M. and Tyler, J. M.,** State of spectrin phosphorylation does not affect erythrocyte shape or spectrin binding to erythrocyte membranes, *J. Biol. Chem.,* 255, 1259, 1980.
10. **Anderson, N. S., III and Fanestil, D. D.,** Biology of mineralocorticoid receptors, in *Receptors and Hormone Action,* Vol. 2, O'Malley, B. W. and Birnbaumer, L., Eds., Academic Press, New York, 1977, 323.
11. **Anderson, S., Bankier, A. T., Barrell, B. G., de Bruijn, M. H. L., Coulson, A. R., Drouin, J., Eperon, I. C., Nierlich, D. P., Roe, B. A., Sanger, F., Schreier, P. H., Smith, A. J. H., Staden, R., and Young, I. G.,** Sequence and organization of the human mitochondrial genome, *Nature (London),* 290, 457, 1981.
12. **Ansell, G. B. and Hawthorne, J. N.,** *Phospholipids: Chemistry, Metabolism and Function,* Elsevier, Amsterdam, 1964.
13. **Archer, J. A., Gorden, P., Gavin, J. R., III, Lesniak, M. A., and Roth, J.,** Insulin receptors in human circulating lymphocytes: application to the study of insulin resistance in man, *J. Clin. Endocrinol. Metab. Chem.,* 36, 627, 1973.
14. **Anstee, D. J., Barker, D. M., Judson, P. A., and Tanner, M. J. A.,** Inherited sialoglycoprotein deficiencies in human erythrocytes of type En(a⁻), *Br. J. Haematol.,* 35, 309, 1977.
15. **Ariyoshi, T. and Takobatake, E.,** Effect of diphenylhydantoin on the drug metabolism and the fatty acid composition of phospholipids in hepatic microsomes, *Chem. Pharm. Bull.,* 20, 180, 1972.
16. **Avery, M. E.,** Pharmacological approaches to the acceleration of fetal lung maturation, *Br. Med. Bull.,* 31, 13, 1975.
17. **Avery, M. E.,** Prenatal diagnosis and prevention of hyaline-membrane disease, *N. Engl. J. Med.,* 292, 157, 1975.
18. **Back, D. J., Breckenridge, A. M., Crawford, F. E., MacIver, M., Orme, M. L. E., and Rowe, P. H.,** Interindividual variation and drug interactions with hormonal steroid contraceptives, *Drugs,* 21, 46, 1981.
19. **Bacq, Z. M.,** *Fundamentals of Biochemical Pharmacology,* Pergamon Press, Oxford, 1971.
20. **Ballard, P. L. and Ballard, R. A.,** Cytoplasmic receptor for glucocorticoids in lung of the human fetus and neonate, *J. Clin. Invest.,* 53, 477, 1974.

21. **Ballard, P. L. and Tomkins, G. M.**, Glucocorticoid-induced alteration of the surface membrane of cultured hepatoma cells, *J. Cell Biol.*, 47, 222, 1970.
22. **Ballas, S. K.**, Abnormal erythrocyte membrane protein pattern in severe megaloblastic anemia, *J. Clin. Invest.*, 61, 1097, 1978.
23. **Ballas, S. K., Saidi, P., and Constantino, M.**, Reduced erythrocytic deformability in megaloblastic anemia, *Am. J. Clin. Pathol.*, 66, 953, 1976.
24. **Balzer, H., Makinose, M., and Hasselbach, W.**, The inhibition of the sarcoplasmic calcium pump by prenylamine, reserpine, chlorpromazine and imipramine, *Naunyn Schmiedebergs Arch. Pharmakol. Exp. Pathol.*, 260, 444, 1968.
25. **Bassi, M.**, Electron microscopy of rat liver after carbon tetrachloride poisoning, *Exp. Cell Res.*, 20, 313, 1960.
26. **Baulieu, E. E., Godeau, F., Schoroderet, M., and Schorderet-Slatkine, S.**, Steroid-induced meiotic division in *Xenopus laevis* oocytes: surface and calcium, *Nature (London)*, 275, 593, 1978.
27. **Baxter, J. D., Eberhardt, N. L., Apriletti, J. W., Johnson, L. K., Ivarie, R. D., Schachter, B. S., Morris, J. A., Seeburg, P. H., Goodman, H. M., Latham, K. R., Polansky, J. R., and Martial, J. A.**, Thyroid hormone receptors and responses, *Recent Prog. Horm. Res.*, 35, 97, 1979.
28. **Baxter, J. D. and Funder, J. W.**, Hormone receptors, *N. Engl. J. Med.*, 301, 1149, 1979.
29. **Beebe, G. W., Kato, H., and Land, C. E.**, Studies of the mortality of A-bomb survivors. VI. Mortality and radiation dose, 1950—1974, *Radiat. Res.*, 75, 138, 1978.
30. **Behrman, R. E. and Fisher, D. E.**, Phenobarbital for neonatal jaundice, *J. Pediatr.*, 76, 945, 1970.
31. **Bennett, V. and Stenbuck, P. J.**, Human erythrocyte ankyrin. Purification and properties, *J. Biol. Chem.*, 255, 2540, 1980.
32. **Bennett, V. and Stenbuck, P. J.**, Association between ankyrin and the cytoplasmic domain of band 3 isolated from the human erythrocyte membrane, *J. Biol. Chem.*, 255, 6424, 1980.
33. **Berenblum, I.**, A re-evaluation of the concept of cocarcinogenesis, *Prog. Exp. Tumor Res.*, 11, 21, 1969.
34. **Bergel'son, L. D., Dyatlovitskaya, E. V., Torkhovskaya, T. I., Sorokina, J. B., and Gor'kova, N. P.**, Differentiation on phospholipid composition in subcellular particles of cancer cell, *FEBS Lett.*, 2, 87, 1968.
35. **Bergel'son, L. D., Dyatlovitskaya, E. V., Sorokina, I. B., and Gor'kova, N. P.**, Phospholipid composition of mitochondria and microsomes from regenerating rat liver and hepatomas of different growth rate, *Biochim. Biophys. Acta*, 360, 361, 1974.
36. **Beutler, E., Guinto, E., and Johnson, C.**, Human red cell protein kinase in normal subjects and patients with hereditary spherocytosis, sickle cell disease, and autoimmune hemolytic anemia, *Blood*, 48, 887, 1976.
37. **Bieberdorf, F. A., Gorden, P., and Fordtran, J. S.**, Pathogenesis of congenital alkalosis with diarrhea. Implications for the physiology of normal ileal electrolyte absorption and secretion, *J. Clin. Invest.*, 51, 1958, 1972.
38. Biological Effects of Ionizing Radiation (BEIR) Advisory Committee, *Report: The Effects on Populations of Exposure to Low Levels of Ionizing Radiation*, National Academy of Sciences-National Research Council, Washington, D.C., 1972.
39. **Birchmeier, W. and Singer, S. J.**, On the mechanism of ATP-induced shape changes in human erythrocyte membranes. II. The role of ATP, *J. Cell Biol.*, 73, 647, 1977.
40. **Black, M. and Billing, B. H.**, Hepatic bilirubin UDP-glucuronyl transferase activity in liver disease and Gilbert's syndrome, *N. Engl. J. Med.*, 280, 1266, 1969.
41. **Blyth, C. A., Freedman, R. B., and Rabin, B. R.**, Sex specific binding of steroid hormones to microsomal membranes of rat liver, *Nature (London) New Biol.*, 230, 137, 1971.
42. **Bodensteiner, J. B. and Engel, A. G.**, Intracellular calcium accumulation in Duchenne dystrophy and other myopathies: study of 567,000 muscle fibers in 114 biopsies, *Neurology*, 28, 439, 1978.
43. **Boivin, P. and Galand, C.**, Erythrocyte membrane phosphorylation in hereditary spherocytosis, *Biomed. Express (Paris)*, 27, 34, 1977.
44. **Bookchin, R. M. and Lew, V. L.**, Progressive inhibition of the Ca pump and Ca:Ca exchange in sickle red cells, *Nature (London)*, 284, 561, 1980.
45. **Börnig, H., Richter, G., and Frunder, H.**, Der Stoffwechsel geschädigter Gewebe. XII. Mitochondrien and Nucleinsäuren in der Zell Fraktionen der Leber CCl$_4$-geschädigter Mäuse, *Z. Physiol. Chem.*, 322, 213, 1960.
46. **Boutwell, R. K.**, The function and mechanism of promoters of carcinogenesis, *Crit. Rev. Toxicol.*, 2, 419, 1974.
47. **Braams, R.**, The effects of ionizing radiations on proteins, in *Radiation Research*, Silini, G., Ed., North-Holland, Amsterdam, 1967, 371.
48. **Brady, R. O.**, Disorders of lipid metabolism, in *Current Trends in the Biochemistry of Lipids*, Ganguly, J. and Smellie, R. M. S., Eds., Academic Press, London, 1972, 113.

49. **Braun, H.**, Beiträge zur Histologie und Zytologie des Bestrahlten Thymus. IV. Die Wirkung Von Teilkörperbestrahlungen, *Strahlentherapie*, 133, 411, 1967.

50. **Breckenridge, A. M., Back, D. J., Cross, K., Crawford, F., MacIver, M., L'E Orme, M., Rowe, P. H., and Smith, E.**, Influence of environmental chemicals on drug therapy in humans: studies with contraceptive steroids, in *Environmental Chemicals, Enzyme Function and Human Disease* (Ciba Found. Symp. No. 76), Amsterdam, Excerpta Medica, 1980, 289.

51. **Brenner, S. L. and Korn, E. D.**, Spectrin-actin interaction. Phosphorylated and dephosphorylated spectrin tetramer cross-link F-actin, *J. Biol. Chem.*, 254, 8620, 1979.

52. **Bretscher, M. S.**, Assymetrical lipid bilayer structure for biological membranes, *Nature (London) New Biol.*, 236, 11, 1972.

53. **Bretscher, M. S.**, Membrane structure: some general principles, *Science*, 181, 622, 1973.

54. **Brewster, M. E., Ihm, J., Brainard, J. R., and Harmony, J. A. K.**, Transfer of phosphatidylcholine facilitated by a component of human plasma, *Biochim. Biophys. Acta*, 529, 147, 1978.

55. **Brodie, B. B., Gillette, J. R., and La Du, B. N.**, Enzymatic metabolism of drugs and other foreign compounds, *Annu. Rev. Biochem.*, 27, 427, 1958.

56. **Brodie, B. B. and Gillette, J. R., Eds.**, *Concepts in Biochemical Pharmacology*, Part II, Springer-Verlag, Berlin, 1971.

57. **Bronstein, H. D., Haeffner, L. J., and Kowlessar, O. D.**, Enzymatic digestion of gliadin: the effect of the resultant peptides in adult celiac disease, *Clin. Chim. Acta*, 14, 141, 1966.

58. **Brooks, S. M., Werk, E. E., Ackerman, S. J., Sullivan, I., and Thrasher, R.**, Adverse effects of phenobarbital on corticosteroid metabolism in patients with bronchial asthma, *N. Engl. J. Med.*, 286, 1125, 1972.

59. **Brown, J. H., Fabre, L. F., Jr., Farrell, G. L., and Adams, E. D.**, Hyperlysinuria with hyperammonemia, a new metabolic disorder, *Am. J. Dis. Child.*, 124, 127, 1972.

60. **Brown, M. S., Dana, S. E., and Goldstein, J. L.**, Regulation of 3-hydroxy-3-methylglutaryl coenzyme A reductase activity in cultured human fibroblasts: comparison of cells from a normal subject and from a patient with homozygous familial hypercholesterolemia, *J. Biol. Chem.*, 249, 789, 1974.

61. **Brown, M. S. and Goldstein, J. L.**, Receptor-mediated endocytosis: insights from the lipoprotein receptor system, *Proc. Natl. Acad. Sci. U.S.A.*, 76, 3330, 1979.

62. **Brown, R. R., Miller, J. A., and Miller, E. C.**, The metabolism of methylated aminoazo dyes; dietary factors enhancing demethylation in vitro, *J. Biol. Chem.*, 209, 211, 1954.

63. **Brown, R. and Darnelli, J. F., Eds.**, *Active transport and secretion* (Soc. Exp. Biol. Symp. No. 8), Cambridge, Cambridge University, Cambridge, 1954.

64. **Bruder, G., Bretscher, A., Franke, W. W., and Jarasch, E. D.**, Plasma membranes from intestinal microvilli and erythrocytes contain cytochromes b_5 and P-450, *Biochim. Biophys. Acta*, 600, 739, 1980.

65. **Brunner, J., Hauser, H., Braun, H., Wilson, K. J., Wacker, H., O'Neill, B., and Semenza, G.**, The mode of association of the enzyme complex sucrase-isomaltase with the intestinal brush border membrane, *J. Biol. Chem.*, 254, 1821, 1979.

66. **Buffa, P., Guarriera-Bobyleva, V., Muscatello, U., and Pasquali-Ronchetti, I.**, Conformational changes of mitochondria associated with uncoupling of oxidative phosphorylation in vivo and in vitro, *Nature (London)*, 226, 272, 1970.

67. **Buja, L. M., Hagler, H. K., Burton, K. P., and Willeson, J. T.**, Subcellular localization and significance of pathological calcium accumulation in myocardium, in *Pathobiology of Cell Membranes*, Vol. 3, Trump, B. F. and Arstila, A. U., Eds., Academic Press, New York, 1983, 87.

68. **Bultmann, B., Melzner, I., Haferkamp, O., and Gruler, H.**, Selective inhibition of human neutrophilic chemotaxis by echo virus, type 9. Virus-induced changes in membrane fluidity, *Pathol. Res. Pract.*, 175, 308, 1982.

69. **Burger, M. M.**, Proteolytic enzymes initiating cell division and escape from contact inhibition of growth, *Nature (London)*, 227, 170, 1970.

70. **Burnett, C. H., Dent, C. E., Harper, C., and Warland, B. J.**, Vitamin D-resistant rickets. Analysis of twenty-four pedigrees with hereditary and sporadic cases, *Am. J. Med.*, 36, 222, 1964.

71. **Busch, H., Ballal, N. R., Olson, M. O. J., and Yeoman, L. C.**, Chromatin and its nonhistone proteins, *Methods Cancer Res.*, 11, 43, 1975.

72. **Busch, H. and Smetana, K.**, *The Nucleolus*, Academic Press, New York, 1970.

73. **Cagianut, B., Rhyner, K., Furrer, W., and Schnebli, H. P.**, Thiosulphate-sulphur transferase (Rhodanese) deficiency in Leber's hereditary optic atrophy, *Lancet*, 2, 981, 1981.

74. **Callahan, J. W. and Lowden, J. A., Eds.**, *Lysosomes and Lysosomal Storage Diseases*, Raven Press, New York, 1981.

75. **Campbell, P. N.**, *The Interaction of Drugs and Subcellular Components In Animal Cells*, Churchill, London, 1968.

76. **Campbell, P. N., Cooper, C., and Hicks, M.,** Studies on the role of the morphological constituents of the microsome fraction from rat liver in protein synthesis, *Biochem. J.,* 92, 225, 1964.

77. **Campbell, P. N., Greengard, O., and Kernot, B. A.,** Studies on the synthesis of serum albumin by the isolated microsome fraction from rat liver, *Biochem. J.,* 74, 107, 1960.

78. **Campbell, P. N., Lowe, E., and Serck-Hanssen, G.,** Protein synthesis by microsomal particles from regenerating rat liver, *Biochem. J.,* 103, 280, 1967.

79. **Cannon, W. B. and Rosenblueth, A.,** *The Supersensitivity of Denervated Structures: A Law of Denervation,* New York, Macmillan, 1949, 1.

80. **Capaldi, R. A. and Vanderkooi, G.,** The low polarity of many membrane proteins, *Proc. Natl. Acad. Sci. U.S.A.,* 69, 930, 1972.

81. **Carlson, S. A. and Gelehrter, T. D.,** Hormonal regulation of membrane phenotype, *J. Supramol. Struct.,* 6, 325, 1977.

82. **Carnegie, P. R.,** Amino acid sequence of the encephalitogenic basic protein from human myelin, *Biochem. J.,* 123, 57, 1971.

83. **Caro, L. G. and Palade, G. E.,** Protein synthesis, storage and discharge in the pancreatic exocrine cell. An autoradiographic study, *J. Cell Biol.,* 20, 473, 1964.

84. **Carpenter, S. and Karpati, G.,** Duchenne muscular dystrophy. Plasma membrane loss initiates muscle cell necrosis unless it is repaired, *Brain,* 102, 147, 1979.

85. **Carraway, K. L., Triplett, R. B., and Anderson, D. R.,** Calcium-promoted aggregation of erythrocyte membrane proteins, *Biochim. Biophys. Acta,* 379, 571, 1975.

86. **Cater, B. R., Chapman, D., Hawes, S. M., and Saville, J.,** Lipid phase transitions and drug interactions, *Biochim. Biophys. Acta,* 363, 54, 1974.

87. **Chan, L. and O'Malley, B. W.,** Mechanism of action of the sex steroid hormones, *N. Engl. J. Med.,* 294, 1322; 1372; 1430; 1976.

88. **Chan, L. and Tindall, D. J.,** Steroid hormone action, in *Pediatric Endocrinology,* Collu, R., Ducharme, J. R., and Guyda, H., Eds., Raven Press, New York, 1981, 63.

89. **Chaplin, M. D. and Mannering, G. J.,** Role of phospholipids in the hepatic microsomal drug-metabolizing system, *Mol. Pharmacol.,* 6, 631, 1970.

90. **Chapman, D.,** Some recent studies of lipids, lipid-cholesterol, and membrane systems, in *Biological Membranes,* Vol. 2, Chapman, D. and Wallach, D. F. H., Eds., Academic Press, New York, 1973, 91.

91. **Chesson, A. L., Jr., Schochet, S. S., Jr., and Peters, B. H.,** Biphasic periodic paralysis, *Arch. Neurol. (Chicago),* 36, 700, 1979.

92. **Christensen, H. N.,** *Biological Transport,* W. A. Benjamin, Reading, Mass., 1975.

93. **Christie, G. S. and Judah, J. D.,** Mechanism of action of carbon tetrachloride on liver cells, *Proc. R. Soc. London,* B142, 241, 1954.

94. **Chui, D. H. K. and Clarke, B. J.,** Abnormal erythroid progenitor cells in human preleukemia, *Blood,* 60, 362, 1982.

95. **Clark, J. H. and Peck, E. J., Jr.,** Nuclear retention of receptor-oestrogen complex and nuclear acceptor sites, *Nature (London),* 260, 635, 1976.

96. **Claude, A.,** Fractionation of mammalian liver cells by differential centrifugation. I. Problems, methods, and preparation of extract, *J. Exp. Med.,* 84, 51, 1946.

97. **Clemetson, K. J., Capitanio, A., Pareti, F. I., McGregor, J. L., and Luscher, E. F.,** Additional platelet membrane glycoprotein abnormalities in Glanzmann's thrombasthenia: a comparison with normals by high-resolution two-dimensional polyacrylamide gel electrophoresis, *Thromb. Res.,* 18, 797, 1980.

98. **Cohn, Z. A. and Fedorko, M. E.,** The formation and fate of lysosomes, in *Lysosomes in Biology and Pathology,* Vol. 1, Dingle, J. T. and Fell, H. B., Eds., London, North-Holland, 1969, 43.

99. **Colbeau, A. and Maroux, S.,** Integration of alkaline phosphatase in the intestinal brush border membrane, *Biochim. Biophys. Acta,* 511, 39, 1978.

100. **Colbeau, A., Nachbaur, J., and Vignais, P. M.,** Enzymic characterization and lipid composition of rat liver subcellular membranes, *Biochim. Biophys. Acta,* 249, 462, 1971.

101. **Cohn, Z. A. and Parks, E.,** The regulation of pinocytosis in mouse macrophages. II. Factors inducing vesicle formation, *J. Exp. Med.,* 125, 213, 1967.

102. **Conney, A. H. and Burns, J. J.,** Stimulatory effect of foreign compounds on ascorbic acid biosynthesis and on drug-metabolizing enzymes, *Nature (London),* 184, 363, 1959.

103. **Conney, A. H. and Burns, J. J.,** Metabolic interactions among environmental chemicals and drugs, *Science,* 178, 576, 1972.

104. **Conney, A. H.,** Drug metabolism and therapeutics, *N. Engl. J. Med.,* 280, 653, 1969.

105. **Conney, A. H.,** Pharmacological implications of microsomal enzyme induction, *Pharmacol. Rev.,* 19, 317, 1967.

106. **Cooper, R. A.,** Influence of increased membrane cholesterol on membrane fluidity and cell function in human red blood cells, *J. Supramol. Struct.,* 8, 413, 1978.

107. **Cooper, S. D. and Feuer, G.,** Relation between drug-metabolizing activity and phospholipids in hepatic microsomes. I. Effects of phenobarbital, carbon tetrachloride and actinomycin D, *Can. J. Physiol. Pharmacol.,* 50, 568, 1972.

108. **Cornu, G., Michaux, J. L., Sokal, G., and Trouet, A.,** Daunorubicin-DNA: further clinical trials in acute non-lymphoblastic leukemia, *Eur. J. Cancer,* 10, 695, 1974.

109. **Couchman, J. R. and Rees, D. A.,** Organelle-cytoskeleton relationships in fibroblasts: mitochondria, Golgi apparatus and endoplasmic reticulum in phases of movement and growth, *Eur. J. Cell Biol.,* 27, 47, 1982.

110. **Crane, R. K.,** Enzymes and malabsorption: a concept of brush border membrane disease, *Gastroenterology,* 50, 254, 1966.

111. **Crick, F.,** General model for the chromosomes of higher organisms, *Nature (London),* 234, 25, 1971.

112. **Critchley, D. R. and Vicker, M. G.,** Glycolipids as membrane receptors important in growth regulation and cell-cell interactions, in *Dynamic Aspects of Cell Surface Organization,* Poste, G. and Nicolson, G. L., Eds., Elsevier, Amsterdam, 1977, 307.

113. **Cull-Candy, S. G., Lundh, H., and Thesleff, S.,** Effects of botulinum toxin on neuromuscular transmission in the rat, *J. Physiol.,* 260, 177, 1976.

114. **Cull-Candy, S. G., Miledi, R., Trautmann, A., and Uchitel, O. D.,** On the release of transmitter at normal, myasthenia gravis and myasthenic syndrome affected human end-plates, *J. Physiol.,* 299, 621, 1980.

115. **Cunningham, M. D., Mace, J. W., and Peters, E. R.,** Clinical experience with phenobarbitone in icterus neonatorum, *Lancet,* 1, 550, 1969.

116. **Cusworth, D. C. and Dent, C. E.,** Renal clearances of amino acids in normal adults and in patients with amino-aciduria, *Biochem. J.,* 74, 550, 1960.

117. **Dahlqvist, A.,** Specificity of the human intestinal disaccharidases and implications for hereditary disaccharide intolerance, *J. Clin. Invest.,* 41, 463, 1962.

118. **Dahr, W., Uhlenbruck, G., Leikola, J., Wagstaff, W., and Landfried, K.,** Studies on the membrane glycoprotein defect of En(a−) erythrocytes. I. Biochemical aspects, *J. Immunogenet.,* 3, 329, 1976.

119. **Dallman, P. R. and Goodman, J. R.,** The effects of iron deficiency on the hepatocyte: a biochemical and ultrastructural study, *J. Cell Biol.,* 48, 79, 1971.

120. **Dallner, G. and Ericsson, J. L. E.,** Molecular structure and biological implication of the liver endoplasmic reticulum, *Prog. Liver Dis.,* 5, 35, 1976.

121. **Dallner, G. and Nilsson, R.,** Mechanism of the cation effect in subfractionation of microsomes, *J. Cell Biol.,* 31, 181, 1966.

122. **Dallner, G.,** Studies on the structural and enzymic organization of the membranous elements of liver microsomes, *Acta Pathol. Microbiol. Scand. Suppl.,* 166, 1, 1963.

123. **De Duve, C.,** The lysosome in retrospect, in *Lysosomes in Biology and Pathology,* Vol. 1, Dingle, J. T. and Fell, H. B., Eds., Amsterdam, North Holland, 1969, 3.

124. **De Duve, C. and Wattiaux, R.,** Functions of lysosomes, *Annu. Rev. Physiol.,* 28, 435, 1966.

125. **De Duve, C., de Barsy, T., Poole, B., Trouet, A., Tulkens, P., and Van Hoof, F.,** Commentary. Lysosomotropic agents, *Biochem. Pharmacol.,* 23, 2495, 1974.

126. **de la Iglesia, F. A., Feuer, G., Takada, A., and Matsuda, Y.,** Morphologic studies on secondary phospholipidosis in human liver, *Lab. Invest.,* 30, 539, 1974.

127. **de la Iglesia, F. A., Sturgess, J. M., McGuire, E. J., and Feuer, G.,** Quantitative microscopic evaluation of the endoplasmic reticulum in developing human liver, *Am. J. Pathol.,* 82, 61, 1976.

128. **de la Iglesia, F. A., Sturgess, J. M., and Feuer, G.,** New approaches for the assessment of hepatotoxicity by means of quantitative functional-morphological interrelationships, in *Toxicology of the Liver (Target Organ Toxicology Series),* Plaa, G. L. and Hewitt, W. R., Eds., Raven Press, New York, 1982, 47.

129. **De Lellis, R. A. and Fishman, W. H.,** The dual localization of beta-glucuronidase in rat thyroid, *Histochemie,* 13, 4, 1968.

130. **Dean, M. E. and Stock, B. H.,** Hepatic microsomal metabolism of drugs during pregnancy in the rat, *Drug Metab. Dispos.,* 3, 325, 1975.

131. **Dent, C. E. and Rose, G. A.,** Aminoacid metabolism in cystinuria, *Q. J. Med.,* 20, 205, 1951.

132. **DePierre, J. W. and Dallner, G.,** Structural aspects of the membrane of the endoplasmic reticulum, *Biochim. Biophys. Acta,* 415, 411, 1975.

133. **Deuticke, B.,** Transformation and restoration of biconcave shape of human erythrocytes induced by amphiphilic agents and changes of ionic environment, *Biochim. Biophys. Acta,* 163, 494, 1968.

134. **Dhami, M. S. I., de la Iglesia, F. A., and Feuer, G.,** Effects of progesterone metabolites on fatty acids of the hepatic endoplasmic reticulum membranes, *Toxicology,* 14, 99, 1979.

135. **Dingle, J. T. and Fell, H. B., Eds.,** *Lysosomes in Biology and Pathology,* North Holland, Amsterdam, 1969.

136. **Dise, C. A., Goodman, D. B. P., and Rasmussen, H.,** Definition of the pathway for membrane phospholipid fatty acid turnover in human erythrocytes, *J. Lipid Res.,* 21, 292, 1980.

137. **Drangova, R. and Feuer, G.,** Progesterone binding by the hepatic endoplasmic reticulum of the female rat, *J. Steroid Biochem.,* 13, 629, 1980.

138. **Drasar, B. S., Renwick, A. G., and Williams, R. T.,** The role of the gut flora in the metabolism of cyclamate, *Biochem. J.,* 129, 881, 1972.

139. **Druckrey, H.,** Specific carcinogenic and teratogenic effects of "indirect" alkylating methyl and ethyl compounds, and their dependency on stages of oncogenic developments, *Xenobiotica,* 3, 271, 1973.

140. **Drummond, K. N., Michael, A. F., Ulstrom, R. A., and Good, R. A.,** The blue diaper syndrome: familial hypercalcemia with nephrocalcinosis and indicanuria. A new familial disease with definition of the metabolic abnormality, *Am. J. Med.,* 37, 928, 1964.

141. **Dzandu, J. K. and Johnson, R. M.,** Membrane protein phosphorylation in intact normal and sickle cell erythrocytes, *J. Biol. Chem.,* 255, 6382, 1980.

142. **Eberhagen, D. and Horn, U.,** Die Lipoidveränderungen in verschiedenen Rattenorganen nach Ganz-Körperbestrahlung mit letalen Roentgendosen, *Strahlentherapie,* 135, 364, 1968.

143. **Efron, M. L. and Ampola, M. G.,** The aminoacidurias, *Pediatr. Clin. North Am.,* 14, 881, 1967.

144. **Egger, J. and Wilson, J.,** Mitochondrial inheritance in a mitochondrially mediated disease, *N. Engl. J. Med.,* 309, 142, 1983.

145. **Einstein, E. R., Robertson, D. M., DiCaprio, J. M., and Moore, W.,** The isolation from bovine spinal cord of a homogeneous protein with encephalitogenic activity, *J. Neurochem.,* 9, 353, 1962.

146. **Elsas, L. J., Hillman, R. E., Patterson, J. H., and Rosenberg, L. E.,** Renal and intestinal hexose transport in familial glucose-galactose malabsorption, *J. Clin. Invest.,* 49, 576, 1970.

147. **Emmelot, P. and van Hoeven, R. P.,** Phospholipid unsaturation and plasma membrane organization, *Chem. Phys. Lipids,* 14, 236, 1975.

148. **Emmelot, P.,** The organization of the plasma membrane of mammalian cells: structure in relation to function, in *Mammalian Cell Membranes,* Vol. 2, Jamieson, G. A. and Robinson, D. M., Eds., Butterworths, London, 1977, 1.

149. **Engel, A. G. and Schotland, D. L.,** Myotonic disorders, in *Scientific Approaches to Clinical Neurology,* Goldenshon, E. S. and Appel, S. H., Eds., Lea & Febiger, Philadelphia, 1977, 1656.

150. **Ernster, L. and Orrenius, S.,** Substrate-induced synthesis of the hydroxylating enzyme system of liver microsomes, *Fed. Proc.,* 24, 1190, 1965.

151. **Ertel, I. J. and Newton, W. A., Jr.,** Therapy in congenital hyperbilirubinemia: phenobarbital and diethylnicotinamide, *Pediatrics,* 44, 43, 1969.

152. **Evans, I. H., Diala, E. S., Earl, A., and Wilkie, D.,** Mitochondrial control of cell surface characteristics in *Saccharomyces cerevisiae,* *Biochim. Biophys. Acta,* 602, 201, 1980.

153. **Evans, W. H.,** Hepatic plasma-membrane modifications in disease, *Clin. Sci.,* 58, 439, 1980.

154. **Eylar, E. H., Brostoff, S., Hashim, G., Caecam, J., and Burnett, P.,** Basic A_1 protein of the myelin membrane: the complete amino acid sequence, *J. Biol. Chem.,* 246, 5770, 1971.

155. **Fabrikant, J. I.,** *Radiobiology,* Year Book Medical Publishers, Chicago, 1972.

156. **Fabro, S., Shull, G., and Dixon, R.,** Further studies on the mechanism of teratogenic action of thalidomide, *Pharmacologist,* 18, 231, 1976.

157. **Fallon, H. J., Gertman, P. M., and Kemp, E. L.,** The effects of ethanol ingestion and choline deficiency on hepatic lecithin biosynthesis in the rat, *Biochim. Biophys. Acta,* 187, 94, 1969.

158. **Fambrough, D. M., Drachman, D. B., and Satyamurti, S.,** Neuromuscular junction in myasthenia gravis: decreased acetylcholine receptors, *Science,* 182, 293, 1973.

159. **Farber, J. L., Chien, K. R., and Mittnacht, S., Jr.,** The pathogenesis of irreversible cell injury in ischemia, *Am. J. Pathol.,* 102, 271, 1981.

160. **Farrell, P. M. and Zachman, R. D.,** Induction of choline phosphotransferase and lecithin synthesis in the fetal lung by corticosteroids, *Science,* 179, 297, 1973.

161. **Feinberg, H.,** Platelet membrane proteins, in *Membrane Abnormalities and Disease,* Vol. 1, Tao, M., Ed., CRC Press, Boca Raton, Fla., 1982, 91.

162. **Feuer, G.,** Action of pregnancy and various progesterones on hepatic microsomal activities, *Drug. Metab. Rev.,* 9, 143, 1979.

163. **Feuer, G.,** Drug interaction with female sex steroids: effects on hepatic metabolism, in *The Role of Drugs and Electrolytes in Hormonogenesis,* Fotherby, K. and Pal, S. B., Eds., Walter de Gruyter, Berlin, 1984, 193.

164. **Feuer, G. and Kardish, R.,** Hormonal regulation of drug metabolism during pregnancy, *Int. J. Clin. Pharmacol. Biopharm.,* 11, 366, 1975.

165. **Feuer, G., Belina, H., Farkas, R., and de la Iglesia, F. A.**, The importance of phospholipid methylation in liver microsomes in the sequential reactions of the induction process by 4-methylcoumarin, *Proc. Eur. Soc. Study Drug Toxic.* (Int. Congress Ser. No. 376), Excerpta Medica, Amsterdam, 17, 323, 1976.

166. **Feuer, G., Cooper, S. D., de la Iglesia, F. A., and Lumb, G.**, Microsomal phospholipids and drug action. Quantitative biochemical and electron microscopic studies, *Int. Z. Klin. Pharmakol. Ther. Toxikol.*, 5, 389, 1972.

167. **Feuer, G., de la Iglesia, F. A., and Cooper, S.**, Role of hepatic endoplasmic reticulum enzyme markers in the preliminary safety evaluations of drugs and other foreign compounds, in *Proc. Eur. Soc. Study Drug Toxic.* (Int. Cong. Ser. No. 311), Excerpta Medica, Amsterdam, 15, 142, 1974.

168. **Feuer, G., Kardish, R., and Farkas, R.**, Differential action of progesterones on hepatic microsomal activities in the rat, *Biochem. Pharmacol.*, 26, 1495, 1977.

169. **Feuer, G., Kellen, J. A., and Kovacs, K.**, Suppression of 7,12-dimethyl-benz(α)anthracene-induced breast carcinoma by coumarin in the rat, *Oncology*, 33, 35, 1976.

170. **Fischer, T. M., Haest, C. W. M., Stoehr, M., Kamp, D., and Deuticke, B.**, Selective alteration of erythrocyte deformability by SH-reagents. Evidence for an involvement of spectrin in membrane shear elasticity, *Biochim. Biophys. Acta*, 510, 270, 1978.

171. **Fishman, P. H. and Brady, R. O.**, Biosynthesis and function of gangliosides, *Science*, 194, 906, 1976.

172. **Fouts, J. R. and Rogers, L. A.**, Morphological changes in liver accompanying stimulation of microsomal drug metabolising enzyme activity by phenobarbital, chlordane, benzopyrene, or methylcholanthrene in rats, *J. Pharmacol. Exp. Ther.*, 147, 112, 1965.

173. **Freiburghaus, A. U., Dubs, R., Hadorn, B., Gaze, H., Hauri, H. P., and Gitzelmann, R.**, The brush border membrane in hereditary sucrase-isomaltase deficiency: abnormal protein pattern and presence of immunoreactive enzyme, *Eur. J. Clin. Invest.*, 7, 455, 1977.

174. **Frenkel, G. D. and Harrington, L.**, Inhibition of mitochondrial nucleic acid synthesis by methyl mercury, *Biochem. Pharmacol.*, 32, 1454, 1983.

175. **Friedmann, H. and Rapoport, S. W.**, Enzymes of the red cell; a critical catalogue, in *Cellular and Molecular Biology of Erythrocytes*, Yoshikawa, H. and Rapoport, S. M., Eds., University Park Press, Baltimore, 1974, 181.

176. **Frontali, N., Ceccarelli, B., Gorio, A., Mauro, A., Siekevitz, M. P., Tzeng, M., and Hurlbut, W. P.**, Purification from black widow spider venom of a protein factor causing the depletion of synaptic vesicles at neuromuscular junctions, *J. Cell Biol.*, 68, 462, 1976.

177. **Furhoff, A. K.**, Adverse reactions with methyldopa — a decade's reports, *Acta Med. Scand.*, 203, 425, 1978.

178. **Gahmberg, C. G.**, Cell surface problems: changes during cell growth and malignant transformation, in *Dynamic Aspects of Cell Surface Organization*, Poste, G. and Nicolson, G. L., Eds., Elsevier, Amsterdam, 1977, 371.

179. **Geiduschek, E. P. and Haselkorn, R.**, Messenger RNA, *Annu. Rev. Biochem.*, 38, 647, 1969.

180. **Geiduschek, J. B. and Singer, S. J.**, Molecular changes in the membranes of mouse erythroid cells accompanying differentiation, *Cell*, 16, 149, 1979.

181. **Gelboin, H. V., Wortham, J. S., and Wilson, R. G.**, 3-Methylcholanthrene and phenobarbital stimulation of rat liver RNA polymerase, *Nature (London)*, 214, 281, 1967.

182. **Getz, G. S., Bartley, W., Stirpe, F., Notton, B. M., and Renshaw, A.**, The lipid composition of rat-liver mitochondria, fluffy layer and microsomes, *Biochem. J.*, 83, 181, 1962.

183. **Gillette, J. R., Davis, D. C., and Sasame, H. A.**, Cytochrome P-450 and its role in drug metabolism, *Annu. Rev. Pharmacol.*, 12, 57, 1972.

184. **Gillette, J. R., Menard, R. H., and Stripp, B.**, Active products of fetal drug metabolism, *Clin. Pharmacol. Ther.*, 14, 680, 1973.

185. **Gjone, E., Torsvik, H., and Norum, K. R.**, Familial plasma cholesterol ester deficiency. A study of the erythrocytes, *Scand. J. Clin. Lab. Invest.*, 21, 327, 1968.

186. **Glader, B. E. and Nathan, D. G.**, Cation permeability alterations during sickling: relationship to cation composition and cellular hydration of irreversibly sickled cells, *Blood*, 51, 983, 1978.

187. **Glader, B. E. and Sullivan, D. W.**, Erythrocyte disorders leading to potassium loss and cellular dehydration, in *Normal and Abnormal Red Cell Membranes* (Prog. Clin. Biol. Res., 30), Lux, S. E., Marchesi, V. T., and Fox, C. F., Eds., Alan R. Liss, New York, 1979, 503.

188. **Glaumann, H. and Dallner, G.**, Lipid composition and turnover of rough and smooth microsomal membranes in rat liver, *J. Lipid Res.*, 9, 720, 1968.

189. **Goebel, K. M., Goebel, F. D., Schubotz, R., and Schneider, J.**, Red cell metabolic and membrane features in haemolytic anaemia of alcoholic liver disease (Zieve's syndrome), *Br. J. Haematol.*, 35, 573, 1977.

190. **Goldstein, J. L. and Brown, M. S.**, The low-density lipoprotein pathway and its relation to atherosclerosis, *Annu. Rev. Biochem.*, 46, 897, 1977.

191. **Goldstein, L.,** Movement of molecules between nucleus and cytoplasm, in *The Cell Nucleus,* Vol. 1., Busch, H., Ed., Academic Press, New York, 1974, 387.

192. **Golovtchenko-Matsumoto, A. M. and Osawa, T.,** Heterogeneity of band 3, the major intrinsic protein of human erythrocyte membranes. Studies by crossed immunoelectrophoresis and crossed immuno-affinoelectrophoresis, *J. Biochem. (Tokyo),* 87, 847, 1980.

193. **Gomperts, E. D., Metz, J., and Zail, S. S.,** A red cell membrane protein abnormality in hereditary spherocytosis, *Br. J. Haematol.,* 23, 363, 1972.

194. **Goodman, S. I., McIntyre, C. A., Jr., and O'Brien, D.,** Impaired intestinal transport of proline in a patient with familial iminoaciduria, *J. Pediatr.,* 71, 246, 1967.

195. **Goodman, S. R. and Shiffer, K.,** The spectrin membrane skeleton of normal and abnormal human erythrocytes: a review, *Am. J. Physiol.,* 244, C121, 1983.

196. **Goodman, S. R., Shiffer, K. A., Casoria, L. A., and Eyster, M. E.,** Identification of the molecular defect in the erythrocyte membrane skeleton of some kindreds with hereditary spherocytosis, *Blood,* 60, 772, 1982.

197. **Gordesky, S. E. and Marinetti, G. V.,** The asymmetric arrangement of phospholipids in the human erythrocyte membrane, *Biochem. Biophys. Res. Commun.,* 50, 1027, 1973.

198. **Goyer, R. A. and Rhyne, B. C.,** Toxic changes in mitochondrial membranes and mitochondrial function, in *Pathobiology of Cell Membranes,* Vol 1, Trump, B. F. and Arstila, A. U., Eds., Academic Press, New York, 1975, 383.

199. **Gray, G. M., Conklin, K. A., and Townley, R. R. W.,** Sucrase-isomaltase deficiency. Absence of an inactive enzyme variant, *N. Engl. J. Med.,* 294, 750, 1976.

200. **Gray, G. M.,** Intestinal disaccharidase deficiencies and glucose-galactose malabsorption, in *The Metabolic Basis of Inherited Disease,* 4th ed., Stanbury, J. B., Wyngaarden, J. B., and Fredrickson, D. S., Eds., McGraw-Hill, New York, 1978, 1526.

201. **Greengard, O.,** Effects of hormones on development of fetal enzymes, *Clin. Pharmacol. Ther.,* 14, 721, 1973.

202. **Greengard, O.,** Enzymic differentiation in mammalian liver, *Science,* 163, 891, 1969.

203. **Griffith, W. H. and Wade, N. J.,** Relation of methionine, cystine and choline to renal lesions occurring on low-choline diets, *Proc. Soc. Exp. Biol. Med.,* 41, 333, 1939.

204. **Guidotti, G.,** Membrane proteins: structure and arrangement in the membrane, in *Physiology of Membrane Disorders,* Andreoli, T. E., Hoffman, J. F., and Fanestil, D. D., Eds., Plenum Press, New York, 1978, 49.

205. **Gurdon, J. B. and Woodland, H. R.,** The cytoplasmic control of nuclear activity in animal development, *Biol. Rev. Cambridge Philos. Soc.,* 43, 233, 1968.

206. **Hagler, H. K., Lopez, L. E., Murphy, M. E., Greico, C. A., and Buja, L. M.,** Quantitative X-ray microanalysis of mitochondrial calcification in damaged myocardium, *Lab. Invest.,* 45, 241, 1981.

207. **Haimes, H. B., Stockert, R. J., Morell, A. G., and Novikoff, A. B.,** Carbohydrate-specified endocytosis: localization of ligand in the lysosomal compartment, *Proc. Natl. Acad. Sci. U.S.A.,* 78, 6936, 1981.

208. **Haines, D. S. M.,** The effects of choline deficiency and choline re-feeding upon the metabolism of plasma and liver lipids, *Can. J. Biochem.,* 44, 45, 1966.

209. **Hakomori, S.,** Glycolipids of tumor cell membrane, *Adv. Cancer Res.,* 18, 265, 1973.

210. **Hakomori, S.,** Structures and organization of cell surface glycolipids. Dependency on cell growth and malignant transformation, *Biochim. Biophys. Acta,* 417, 55, 1975.

211. **Holmberg, C., Perheentupa, J., and Launiala, K.,** Colonic electrolyte transport in health and in congenital chloride diarrhea, *J. Clin. Invest.,* 56, 302, 1975.

212. **Hardy, B., Bensch, K. G., and Schrier, S. L.,** Spectrin rearrangement early in erythrocyte ghost endocytosis, *J. Cell Biol.,* 82, 654, 1979.

213. **Harmon, H. J., Hall, J. D., and Crane, F. L.,** Structure of mitochondrial cristae membranes, *Biochim. Biophys. Acta,* 344, 119, 1974.

214. **Harris, H.,** *Nucleus and Cytoplasm,* Clarendon, Oxford, 1968.

215. **Harris, P. J. and Northcote, D. H.,** Polysaccharide formation in plant Golgi bodies, *Biochim. Biophys. Acta,* 237, 56, 1971.

216. **Hartroft, W. S.,** Pathogenesis of renal lesions in weanling and young adult rats fed choline-deficient diets, *Br. J. Exp. Pathol.,* 29, 483, 1948.

217. **Haslam, R. J. and Lynham, J. A.,** Relationship between phosphorylation of blood platelet proteins and secretion of platelet granule constituents. II. Effects of different inhibitors, *Thromb. Res.,* 12, 619, 1968.

218. **Hawkins, H. K., Ericsson, J. L. E., Biberfeld, P., and Trump, B. F.,** Lysosome and phagosome stability in lethal cell injury. Morphologic tracer studies in cell injury due to inhibition of energy metabolism, immune cytolysis, and photosensitization, *Am. J. Pathol.,* 68, 255, 1972.

219. **Hayaishi, O.,** *Oxygenases,* Academic Press, New York, 1962.

220. **Hebbel, R. P., Yamada, O., Moldow, C. F., Jacob, H. S., White, J. G., and Eaton, J. W.,** Abnormal adherence of sickle erythrocytes to cultured vascular endothelium. Possible mechanism for microvascular occlusion in sickle cell disease, *J. Clin. Invest.,* 65, 154, 1980.

221. **Hellier, M. D., Perrett, D., and Holdsworth, C. D.,** Dipeptide absorption in cystinuria, *Br. Med. J.,* 4, 782, 1970.

222. **Hers, H. G.,** Inborn lysosomal diseases, *Gastroenterology,* 48, 625, 1965.

223. **Higashi, T. and Peters, T., Jr.,** Studies on rat liver catalase. II. Incorporation of ^{14}C-leucine into catalase of liver cell fractions in vivo, *J. Biol. Chem.,* 238, 3952, 1963.

224. **Hinton, R. H. and Reid, E.,** Enzyme distribution in mammalian membranes, in *Mammalian Cell Membranes,* Vol. 1, Jamieson, G. A. and Robinson, D. M., Eds., Butterworth, London, 1976, 161.

225. **Hirata, F. and Axelrod, J.,** Enzymatic synthesis and rapid translocation of phosphatidylcholine by two methyltransferases in erythrocyte membranes, *Proc. Natl. Acad. Sci. U.S.A.,* 75, 2348, 1978.

226. **Hirsch, P. C. and Szego, C. M.,** Estradiol receptor functions of soluble proteins from target-specific lysosomes, *J. Steroid Biochem.,* 5, 533, 1974.

227. **Hoffmann, N., Thees, M., and Kinne, R.,** Phosphate transport by isolated renal brush border vesicles, *Pfluegers Arch. Gesamte Physiol. Menschen Tiere,* 362, 147, 1976.

228. **Hokin, L. E. and Hokin, M. R.,** Biological transport, *Annu. Rev. Biochem.,* 32, 553, 1963.

229. **Holtzman, J. L. and Gillette, J. R.,** The effect of phenobarbital on the turnover of microsomal phospholipid in male and female rats, *J. Biol. Chem.,* 243, 3020, 1968.

230. **Holzel, A.,** Defects of sugar absorption: sugar malabsorption and sugar intolerance in childhood, *Proc. R. Soc. Med.,* 61, 1095, 1968.

231. **Hopkins, G. J. and West, C. E.,** Diet-induced changes in the fatty acid composition of mouse hepatocyte plasma membranes, *Lipids,* 12, 327, 1977.

232. **Howell, K. E. and Palade, G. E.,** Hepatic Golgi fractions resolved into membrane and content subfractions, *J. Cell Biol.,* 92, 822, 1982.

233. **Hurley, L. S., Theriault, L. L., and Dreosti, I. E.,** Liver mitochondria from manganese-deficient and pallid mice: function and ultrastructure, *Science,* 170, 1316, 1970.

234. **Hynes, R. O.,** Cell surface proteins and malignant transformation, *Biochim. Biophys. Acta,* 458, 73, 1976.

235. **Hynes, R. O., Destree, A. T., and Mautner, V.,** Spatial organization at the cell surface, *Prog. Clin. Biol. Res.,* 9, 189, 1976.

236. **Iida, H., Hasegawa, I., and Nozawa, Y.,** Biochemical studies on abnormal erythrocyte membranes. Protein abnormality of erythrocyte membrane in biliary obstruction, *Biochim. Biophys. Acta,* 443, 394, 1976.

237. **Ilyin, Y. V. and Georgiev, G. P.,** The main types of organization of genetic material in eukaryotes, *Crit. Rev. Biochem.,* 12, 237, 1982.

238. **Jackson, K. L. and Christensen, G. M.,** Sodium and potassium binding in X-irradiated nuclei, *Radiat. Res.,* 27, 434, 1966.

239. **Jacob, H. S.,** The defective red blood cell in hereditary spherocytosis, *Annu. Rev. Med.,* 20, 41, 1969.

240. **Jain, M. K., Wu, N. Y. M., and Wray, L. V.,** Drug-induced phase change in bilayer as possible mode of action of membrane expanding drugs, *Nature (London),* 255, 494, 1975.

241. **Janz, D. and Schmidt, D.,** Anti-epileptic drugs and failure of oral contraceptives, *Lancet,* 1, 1113, 1974.

242. **Jennings, R. B. and Ganote, C. E.,** Mitochondrial structure and function in acute myocardial ischemic injury, *Circ. Res.,* 38(Suppl. 1), 80, 1976.

243. **Jensen, E. V. and DeSombre, E. R.,** Mechanism of action of the female sex hormones, *Annu. Rev. Biochem.,* 41, 203, 1972.

244. **Jezequel, A. M., Koch, M., and Orlandi, F.,** A morphometric study of the endoplasmic reticulum in human hepatocytes. Correlation between morphological and biochemical data in subjects under treatment with certain drugs, *Gut,* 15, 737, 1974.

245. **Jezequel, A. M.,** Ultrastructural changes induced by drugs in the liver, in *The Hepatobiliary System. Fundamental and Pathological Mechanisms,* Taylor, W., Ed., Plenum Press, New York, 1976, 179.

246. **Ji, T. H.,** The application of chemical crosslinking for studies on cell membranes and the identification of surface reporters, *Biochim. Biophys. Acta,* 559, 39, 1979.

247. **Johnson, R. M., Taylor, G., and Meyer, D. B.,** Shape and volume changes in erythrocyte ghosts and spectrin-actin networks, *J. Cell Biol.,* 86, 371, 1980.

248. **Johnson, R. T.,** Selective vulnerability of neural cells to viral infections, *Brain,* 103, 447, 1980.

249. **Jones, K. L. and Smith, D. W.,** Recognition of the fetal alcohol syndrome in early infancy, *Lancet,* 2, 999, 1973.

250. **Jones, K. L. and Smith, D. W.,** The fetal alcohol syndrome, *Teratology,* 12, 1, 1975.

251. **Jones, M. N. and Nickson, J. K.,** Identifying the monosaccharide transport protein in the human erythrocyte membrane, *FEBS Lett.,* 115, 1, 1980.

252. **Jones, M. S. and Jones, O. T. G.,** Evidence for the location of ferrochelatase on the inner membrane of rat liver mitochondria, *Biochem. Biophys. Res. Commun.,* 31, 977, 1968.

253. Joshi, J. V., Joshi, U. M., Sankolli, G. M., Gupta, K., Rao, A. P., Hazari, K., Sheth, U. K., and Saxena, B. N., A study of interaction of a low-dose combination oral contraceptive with anti-tubercular drugs, *Contraception*, 21, 617, 1980.

254. Jost, J. P. and Rickenberg, H. V., Cyclic AMP, *Annu. Rev. Biochem.*, 40, 741, 1971.

255. Juchau, M. R., Zachariah, P. K., Colson, J., Symms, K. G., Krasner, J., and Yaffe, S. J., Studies on human placental carbon monoxide-binding cytochromes, *Drug Metab. Dispos.*, 2, 79, 1974.

256. Iseri, O. A. and Gottlieb, L. S., Alcoholic hyalin and megamitochondria as separate and distinct entities in liver disease associated with alcoholism, *Gastroenterology*, 60, 1027, 1971.

257. Kahlenberg, A., Walker, C., and Rohrlick, R., Evidence for an asymmetric distribution of phospholipids in the human erythrocyte membrane, *Can. J. Biochem.*, 52, 803, 1974.

258. Kahonen, M. T. and Ylikahri, R. H., Effect of clofibrate and gemfibrozil on the activities of mitochondrial carnitine acyltransferases in rat liver, dose-response relations, *Atherosclerosis*, 32, 47, 1979.

259. Kao, I., Drachman, D. B., and Price, D. L., Botulinum toxin: mechanism of presynaptic blockade, *Science*, 193, 1256, 1976.

260. Kardish, R. and Feuer, G., Relationships between maternal progesterones and the delayed drug metabolism in the neonate, *Biol. Neonate*, 20, 58, 1972.

261. Kay, M. M. B., Role of physiologic autoantibody in the removal of senescent human red cells, *J. Supramol. Struct.*, 9, 555, 1978.

262. Kebabian, J. W., Zatz, M., Romero, J. A., and Axelrod, J., Rapid changes in rat pineal β-adrenergic receptor: Alterations in ℓ [³H] alprenolol binding and adenylate cyclase, *Proc. Natl. Acad. Sci. U.S.A.*, 72, 3735, 1975.

263. Keenan, T. W. and Moore, D. J., Phospholipid class and fatty acid composition of Golgi apparatus isolated from rat liver and comparison with other cell fractions, *Biochemistry*, 9, 19, 1970.

264. Kellermann, G., Shaw, C. R., and Luyten-Kellerman, M., Aryl hydrocarbon hydroxylase inducibility and bronchogenic carcinoma, *N. Engl. J. Med.*, 289, 934, 1973.

265. Kekomaki, M., Visakorpi, J. K., Perheentupa, J., and Saxen, L., Familial protein intolerance with deficient transport of basic amino acids, *Acta Paediatr. Scand.*, 56, 617, 1967.

266. Kempson, S., Marinetti, G. V., and Shaw, A., Hormone action at the membrane level. VII. Stimulation of dihydroalprenolol binding to beta-adrenergic receptors in isolated rat heart ventricle slices by triiodothyronine and thyroxine, *Biochim. Biophys. Acta*, 540, 320, 1978.

267. Kenny, A. J. and Booth, A. G., Microvilli: their ultrastructure enzymology and molecular organization, *Essays Biochem.*, 14, 1, 1978.

268. Kenyon, I. E., Unplanned pregnancy in an epileptic, *Br. Med. J.*, 1, 686, 1972.

269. Kay, R. E. and Bean, R. C., Effects of radiation on artificial lipid membranes, *Adv. Biol. Med. Phys.*, 13, 235, 1970.

270. Kerenyi, N. A., Morava-Protzner, I., Feuer, G., and Sotonyi, P., The pineal gland and breast cancer — the presence and potential of melatonin receptor, *J. Steroid Biochem.*, 20, 1454, 1984.

271. Kerry, K. R. and Townley, R. R. W., Genetic aspects of intestinal sucrase-isomaltase deficiency, *Aust. Pediatr. J.*, 1, 223, 1965.

272. Keuning, J. J., Pena, A. S., van Hooff, J. P., van Leeuwen, A., and van Rood, J. J., HLA-DW3 associated with coeliac disease, *Lancet*, 1, 506, 1976.

273. Kim, H. C., Friedman, S., Asakura, T., and Schwartz, E., Inclusions in red blood cells containing Hb S or Hb C, *Br. J. Haematol.*, 44, 547, 1980.

274. Kimberg, D. V. and Loeb, J. N., Effects of cortisone administration on rat liver mitochondria: support for the concept of mitochondrial fusion, *J. Cell Biol.*, 55, 635, 1972.

275. Kirkpatrick, F. H., New models of cellular control: membrane cytoskeletons, membrane curvature potential, and possible interactions, *BioSystems*, 11, 93, 1979.

276. Klein, W. L. and Wolf, M., Regulation of cell surface receptors, in *Membrane Abnormalities and Disease*, Vol. 2, Tao, M., Ed., CRC Press, Boca Raton, Fla., 1982, 97.

277. Kleinzeller, A., The role of potassium and calcium in the regulation of metabolism in kidney cortex slices, in *Membrane Transport and Metabolism*, Kleinzeller, A. and Kotyk, A., Eds., Academic Press, London, 1961, 527.

278. Knopp, J., Stolc, V., and Tong, W., Evidence for the induction of iodide transport in bovine thyroid cells treated with thyroid-stimulating hormone or dibutyryl cyclic adenosine 3′,5′-monophosphate, *J. Biol. Chem.*, 245, 4403, 1970.

279. Kollmann, G. and Shapiro, B., The mechanism of action of AET. VI. The protection of proteins against ionizing radiation by GED, *Radiat. Res.*, 27, 474, 1966.

280. Kosek, J. C., Mazze, R. I., and Cousins, M. J., Nephrotoxicity of gentamicin, *Lab. Invest.*, 30, 48, 1974.

281. Krane, S. M., Renal glycosuria, in *The Metabolic Basis of Inherited Disease*, 4th ed., Stanbury, J. B., Wyngaarden, J. B., and Fredrickson, D. S., Eds., McGraw-Hill, New York, 1978, 1607.

282. **Kratzing, C. C. and Wetzig, G. A.,** Hypertension produced by choline deficiency in rats, *Pathology*, 2, 125, 1970.

283. **Krejs, G. J. and Fordtran, J. S.,** Physiology and pathophysiology of iron and water movement in the human intestine, in *Gastrointestinal Disease*, 2nd ed., Sleisenger, M. H. and Fordtran, J. S., Eds., W. B. Saunders, Philadelphia, 1978, 297.

284. **Kuntzman, R. and Southren, A. L.,** The effects of CNS active drugs on the metabolism of steroids in man, *Adv. Biochem. Psychopharmacol.*, 1, 205, 1969.

285. **La Celle, P. L., Weed, R. I., and Santillo, P. A.,** Pathophysiologic significance of abnormalities of red cell shape, in *Membranes and Disease*, Bolis, L., Hoffman, J. F., and Leaf, A., Eds., Raven Press, New York, 1976, 1.

286. **Land, J. M., Hockaday, J. M., Hughes, J. T., and Ross, B. D.,** Childhood mitochondrial myopathy with ophthalmoplegia, *J. Neurol. Sci.*, 51, 371, 1981.

287. **Lange, Y. and D'Alessandro, J. S.,** The exchangeability of human erythrocyte membrane cholesterol, *J. Supramol. Struct.*, 8, 391, 1978.

288. **Langer, G. A.,** Events at the cardiac sarcolemma: localization and movement of contractile-dependent calcium, *Fed. Proc.*, 35, 1274, 1976.

289. **Langham, W. H., Brooks, P. M., and Grahn, D.,** Biological effects of ionizing radiation, *Aerosp. Med.*, 36, 1, 1965.

290. **LaNoue, K. F. and Schoolwerth, A. C.,** Metabolite transport in mitochondria, *Annu. Rev. Biochem.*, 48, 871, 1979.

291. **Lausch, R. N. and Rapp, F.,** Surface antigens on human herpesvirus-infected and -transformed cells, in *Virus-Transformed Cell Membranes*, Nicolau, C., Ed., Academic Press, London, 1978, 373.

292. **Lea, D. E.,** *Actions of Radiations on Living Cells*, 2nd ed., Cambridge University Press, Cambridge, 1955.

293. **Lee, C. P., Schatz, G., and Dallner, G.,** Eds., *Mitochondria and Microsomes*, Addison-Wesley, Reading, Mass., 1981.

294. **Lefevre, P. A.,** The degranulation test. Six tests for carcinogenicity, *Br. J. Cancer*, 37, 937, 1978.

295. **Lehninger, A. L., Cara Poli, E., and Rossi, C. S.,** Energy-linked ion movements in mitochondrial systems, *Adv. Enzymol.*, 29, 259, 1967.

296. **Lenaz, G.,** Lipid-protein interactions in the structure of biological membranes, *Subcell. Biochem.*, 3, 167, 1974.

297. **Lessin, L. S., Kurantsin-Mills, J., Wallas, C., and Weems, H.,** Membrane alterations in irreversibly sickled cells: hemoglobin-membrane interaction, *J. Supramol. Struct.*, 9, 537, 1978.

298. **Levey, G. S., Ed.,** *Hormone-Receptor Interaction:Molecular Aspects*, Marcel Dekker, New York, 1976.

299. **Lie, S. O.,** Induction of "storage disease" in normal human fibroblasts, *Birth Defects*, 10, 203, 1974.

300. **Lie, S. O., Lie, K. K., and Langslet, A.,** in *Adriamycin Review*, EORTC Int. Symp., Staquet, M., Ed., European Press Medikon, Ghent, Belgium, 1975, 226.

301. **Lieber, C. S., Teschke, R., Hasumura, Y., and DeCarli, L. M.,** Differences in hepatic and metabolic changes after acute and chronic alcohol consumption, *Fed. Proc.*, 34, 2060, 1975.

302. **Liggins, G. C. and Howie, R. N.,** A controlled trial of antepartum glucocorticoid treatment for prevention of the respiratory distress syndrome in premature infants, *Pediatrics*, 50, 515, 1972.

303. **Lipicky, R. J.,** Studies in human myotonic dystrophy, in *Pathogenesis of Human Muscular Dystrophies*, Rowland, L. P., Ed., Excerpta Medica, Amsterdam, 1977, 729.

304. **Lombardi, B. and Recknagel, R. O.,** Interference with secretion of triglycerides by the liver as a common factor in toxic liver injury; with some observations on choline deficiency fatty liver, *Am. J. Pathol.*, 40, 571, 1962.

305. **Lombardi, B.,** Effect of choline deficiency on rat hepatocytes, *Fed. Proc.*, 30, 139, 1971.

306. **Lorand, L., Siefring, G. E., Jr., and Lowe-Krentz, L.,** Enzymatic basis of membrane stiffening in human erythrocytes, *Semin. Hematol.*, 16, 65, 1979.

307. **Lu, A. Y. H. and West, S. B.,** Multiplicity of mammalian microsomal cytochromes P-450, *Pharmacol. Rev.*, 31, 277, 1980.

308. **Lucas, C. C. and Ridout, J. H.,** Fatty livers and lipotropic phenomena, *Prog. Chem. Fats Other Lipids*, 10(1), 1, 1967.

309. **Lucy, J. A.,** Is there a membrane defect in muscle and other cells?, *Br. Med. Bull.*, 36, 187, 1980.

310. **Lucy, J. A.,** Ultrastructure of membranes: micellar organization, *Br. Med. Bull.*, 24, 127, 1968.

311. **Ludwig, J.,** Drug effects on the liver. A tabular compilation of drugs and drug-related hepatic diseases, *Dig. Dis. Sci. (New York)*, 24, 785, 1979.

312. **Luna, E. J., Kidd, G. H., and Branton, D.,** Identification by peptide analysis of the spectrin-binding protein in human erythrocytes, *J. Biol. Chem.*, 254, 2526, 1979.

313. **Lux, S. E., John, K. M., and Karnovsky, M. J.,** Irreversible deformation of the spectrin-actin lattice in irreversibly sickled cells, *J. Clin. Invest.*, 58, 955, 1976.

314. **Lux, S. E., John, K. M., and Ukena, T. E.,** Diminished spectrin extraction from ATP depleted human erythrocytes. Evidence relating spectrin to changes in erythrocyte shape and deformability, *J. Clin. Invest.,* 61, 815, 1978.

315. **Lux, S. E.,** Dissecting the red cell membrane skeleton, *Nature (London),* 281, 426, 1979.

316. **Lux, S. E.,** Spectrin-actin membrane skeleton of normal and abnormal red blood cells, *Semin. Hematol.,* 16, 21, 1979.

317. **MacLennan, D. H. and Wong, P. T. S.,** Isolation of a calcium-sequestering protein from sarcoplasmic reticulum, *Proc. Natl. Acad. Sci. U.S.A.,* 68, 1231, 1971.

318. **MacLennan, D. H.,** Resolution of the calcium transport system of sarcoplasmic reticulum, *Can. J. Biochem.,* 53, 251, 1975.

319. **MacLennan, D. H., Seeman, P., Iles, G. H., and Yip, C. C.,** Membrane formation by the adenosine triphosphatase of sarcoplasmic reticulum, *J. Biol. Chem.,* 246, 2702, 1971.

320. **MacLennan, D. H., Yip, C. C., Iles, G. H., and Seeman, P.,** Isolation of sarcoplasmic reticulum proteins, *Cold Spring Harbor Symp. Quant. Biol.,* 37, 469, 1972.

321. **Maddy, A. H.,** The chemical organization of the plasma membrane of animal cells, *Int. Rev. Cytol.,* 20, 1, 1966.

322. **Maestracci, D., Preiser, H., Hedges, T., Schmitz, J., and Crane, R. K.,** Enzymes of the human intestinal brush border membrane: identification after gel electrophoretic separation, *Biochim. Biophys. Acta,* 382, 147, 1975.

323. **Maestracci, D., Schmitz, J., Preiser, H., and Crane, R. K.,** Proteins and glycoproteins of the human intestinal brush border membrane, *Biochim. Biophys. Acta,* 323, 113, 1973.

324. **Bittar, E. E.,** *Cell Biology in Medicine,* John Wiley & Sons, New York, 1973.

325. **Mansbach, C. M., II, Wilkins, R. M., Dobbins, W. O., and Tyor, M. P.,** Intestinal mucosal function and structure in the steatorrhea of Zollinger-Ellison syndrome, *Arch. Intern. Med.,* 121, 487, 1968.

326. **Marchesi, V. T.,** Functional proteins of the human red blood cell membrane, *Semin. Hematol.,* 16, 3, 1979.

327. **Marchesi, V. T.,** Spectrin: present status of a putative cytoskeletal protein of the red cell membrane, *J. Membr. Biol.,* 51, 101, 1979.

328. **Marx, J. L.,** Biochemistry of cancer: focus on the cell surface, *Science,* 183, 1279, 1974.

329. **Mason, H. S., North, J. C., and Vanneste, M.,** Microsomal mixed-function oxidations: the metabolism of xenobiotics, *Fed. Proc.,* 24, 1172, 1965.

330. **Matheson, D. W. and Howland, J. L.,** Erythrocyte deformation in human muscular dystrophy, *Science,* 184, 165, 1974.

331. **Matlib, M. A. and Schwartz, A.,** Selective effects of diltiazem, a benzothiazepine calcium channel blocker, and diazepam, and other benzodiazepines on the Na^+/Ca^{2+} exchange carrier system of heart and brain mitochondria, *Life Sci.,* 32, 2837, 1983.

332. **Matsuura, S. and Tashiro, Y.,** Immunoelectron-microscopic studies of endoplasmic reticulum-Golgi relationships in the intracellular transport process of lipoprotein particles in rat hepatocytes, *J. Cell Sci.,* 39, 273, 1979.

333. **McEwen, B. S.,** Gonadal steroids: humoral modulators of nerve-cell function, *Mol. Cell. Endocrinol.,* 18, 151, 1980.

334. **McGuigan, J. E.,** The Zollinger-Ellison syndrome, in *Gastrointestinal Disease,* 2nd ed., Sleisenger, M. H. and Fordtran, J. S., Eds., W. B. Saunders, Philadelphia, 1978, 860.

335. **McNutt, N. S., Culp, L. A., and Black, P. H.,** Contact-inhibited revertant cell lines isolated from SV40-transformed cells. IV. Microfilament distribution and cell shape in untransformed, transformed, and revertant Balb/c 3T3 cells, *J. Cell Biol.,* 56, 412, 1973.

336. **Meeuwisse, G. W.,** Glucose-galactose malabsorption: studies on renal glucosuria, *Helv. Paediatr. Acta,* 25, 13, 1970.

337. **Meister, A. and Tate, S. S.,** Glutathione and related γ-glutamyl compounds, biosynthesis and utilization, *Annu. Rev. Biochem.,* 45, 549, 1976.

338. **Meister, A.,** On the enzymology of amino acid transport, *Science,* 180, 33, 1973.

339. **Mentzer, W. C., Jr. and Lubin, B. H.,** The effect of crosslinking reagents on red-cell shape, *Semin. Hematol.,* 16, 115, 1979.

340. **Mentzer, W. C., Jr., Smith, W. B., Goldstone, J., and Shohet, S. B.,** Hereditary stomatocytosis. Membrane and metabolism studies, *Blood,* 46, 659, 1975.

341. **Merlie, J. P., Heinemann, S., Einarson, B., and Lindstrom, J. M.,** Degradation of acetylcholine receptor in diaphragms of rats with experimental autoimmune myasthenia gravis, *J. Biol. Chem.,* 254, 6328, 1979.

342. **Metge, W. R., Owen, C. A., Jr., Foulk, W. T., and Hoffman, H. N.,** Bilirubin glucuronyl transferase activity in liver disease, *J. Lab. Clin. Med.,* 64, 89, 1964.

343. **Miller, E. C. and Miller, J. A.,** The mutagenicity of chemical carcinogens: correlations, problems, and interpretations, in *Chemical Mutagens:* Principles and Methods for their Detection, Hollaender, A., Ed., Plenum Press, New York, 1971, 83.

344. **Miller, J. A.**, Carcinogenesis by chemicals: an overview. G.H.A. Clowes Memorial Lecture, *Cancer Res.*, 30, 559, 1970.

345. **Milne, M. D., Asatoor, A. M., Edwards, K. D. G., and Loughridge, L. W.**, The intestinal absorption defect in cystinuria, *Gut*, 2, 323, 1961.

346. **Milne, M. D., Crawford, M. A., Girao, C. B., and Loughridge, L. W.**, The metabolic disorder in Hartnup disease, *Q. J. Med.*, 29, 407, 1960.

347. **Mitchell, J. R. and Jollows, D. J.**, Metabolic activation of drugs to toxic substances, *Gastroenterology*, 68, 392, 1975.

348. **Mitchell, J. R., Jollow, D. J., Potter, W. Z., Davis, D. C., Gillette, J. R., and Brodie, B. B.**, Acetaminophen-induced hepatic necrosis. I. Role of drug metabolism, *J. Pharmacol. Exp. Ther.*, 187, 185, 1973.

349. **Mitchell, P.**, Translocations through natural membranes, *Adv. Enzymol.*, 29, 33, 1967.

350. **Mitchison, N. A.**, The immunogenic capacity of antigen taken up by peritoneal exudate cells, *Immunology*, 16, 1, 1969.

351. **Molday, R. S. and Molday, L. L.**, Identification and characterization of multiple forms of rhodopsin and minor proteins in frog and bovine rod outer segment disc. Electrophoresis, lectin labeling, and proteolysis studies, *J. Biol. Chem.*, 254, 4653, 1979.

352. **Mookerjea, S.**, Action of choline in lipoprotein metabolism, *Fed. Proc.*, 30, 143, 1971.

353. **Mookerjea, S., Jeng, D., and Black, J.**, Studies on the synthesis of plasma glycolipoprotein and hepatic sub-cellular glycoprotein in early choline deficiency, *Can. J. Biochem.*, 45, 825, 1967.

354. **Moore, W. T. and Smith, L. H., Jr.**, Experience with a calcium infusion test in parathyroid disease, *Metabolism*, 12, 447, 1963.

355. **Mooseker, M. S. and Tilney, L. G.**, Organization of an actin filament-membrane complex: filament polarity and membrane attachment in the microvilli of intestinal epithelial cells, *J. Cell Biol.*, 67, 725, 1975.

356. **Mooseker, M. S.**, Actin filament-membrane attachment in the microvilli of intestinal epithelial cells, *Cell Motility* (Cold Spring Harbor Conferences on Cell Proliferation, Vol. 3B), 1976, 631.

357. **Morell, A. G., Gregoriadis, G., Scheinberg, I. H., Hickman, J., and Ashwell, G.**, The role of sialic acid in determining the survival of glycoproteins in the circulation, *J. Biol. Chem.*, 246, 1461, 1971.

358. **Morgan-Hughes, J. A., Hayes, D. J., Clark, J. B., Landon, D. N., Swash, M., Stark, R. J., and Rudge, P.**, Mitochondrial encephalomyopathies: biochemical studies in two cases revealing defects in the respiratory chain, *Brain*, 105, 553, 1982.

359. **Morrison, M., Mueller, T. J., and Edwards, H. H.**, Protein architecture of the erythrocyte membrane, *Prog. Clin. Biol. Res.*, 51, 17, 1981.

360. **Moscarello, M. A., Abe, H., and Sturgess, J. M.**, Morphological and biochemical approaches to characterize the structure and function of the Golgi complex, *Mod. Probl. Paediatr.*, 19, 88, 1977.

361. **Moscarello, M. A., Sutherland, L., and Jackson, S. H.**, Stimulation of incorporation of glucosamine-1-^{14}C into rat plasma glycoprotein after renal damage, *Can. J. Biochem.*, 45, 136, 1967.

362. **Mumford, J. P.**, Drugs affecting oral contraceptives, *Br. Med. J.*, 2, 333, 1974.

363. **Murphy, J. R.**, Erythrocyte metabolism. VI. Cell shape and the location of cholesterol in the erythrocyte membrane, *J. Lab. Clin. Med.*, 65, 756, 1965.

364. **Muting, D.**, Detoxication capacity of the diseased liver, *Dtsch. Med. Wochenschr.*, 88, 130, 1963.

365. **Myers, D. K. and Church, M. L.**, Inhibition of stromal enzymes by x-radiation, *Nature (London)*, 213, 636, 1967.

366. **Myers, D. K. and Skov, K.**, Nucleic acid synthesis in x-irradiated thymocytes, *Can. J. Biochem.*, 44, 839, 1966.

367. **Myers, D. K.**, Some aspects of radiation effects on cell membranes, *Adv. Biol. Med. Phys.*, 13, 219, 1970.

368. **Myers, D. K., Tribe, T. A., and Mortimer, R.**, On the radiation-induced reaction of iodoacetamide with albumin and with the erythrocyte membrane, *Radiat. Res.*, 40, 580, 1969.

369. **Nakao, M., Nakao, T., and Yamazoe, S.**, Adenosine triphosphate and maintenance of shape of the human red cells, *Nature (London)*, 187, 945, 1960.

370. **Nakashima, K. and Beutler, E.**, Erythrocyte cellular and membrane deformability in hereditary spherocytosis, *Blood*, 53, 481, 1979.

371. **Nakashima, K. and Beutler, E.**, Effect of anti-spectrin antibody and ATP on deformability of resealed erythrocyte membranes, *Proc. Natl. Acad. Sci. U.S.A.*, 75, 3823, 1978.

372. **National Council on Radiation Protection and Measurements**, *Protection of the Thyroid Gland in the Event of Releases of Radioiodine* (NCRP Report No. 55), National Council on Radiation Protection and Measurements, Washington, 1977.

373. **Natta, C. L. and Kremzner, L. T.**, Polyamines and membrane proteins in sickle cell disease, *Blood Cells*, 8, 273, 1982.

374. **Nayler, W. G.,** The role of calcium in the ischemic myocardium, *Am. J. Pathol.,* 102, 262, 1981.
375. **Neale, M. G. and Parke, D. V.,** Effects of pregnancy on the metabolism of drugs in the rat and rabbit, *Biochem. Pharmacol.,* 22, 1451, 1973.
376. **Neuhaus, O. W., Balegno, H., and Milauskas, A. T.,** Biochemical significance of serum glycoproteins. II. Effects of dietary protein on changes in rat serum following injury, *Exp. Mol. Pathol.,* 2, 183, 1963.
377. **Newberne, P. M.,** Assessment of the hepatocarcinogenic potential of chemicals: response of the liver, in *Toxicology of the Liver (Target Organ Toxicology Series),* Plaa, G. L. and Hewitt, W. R., Eds., Raven Press, New York, 1982, 243.
378. **Newberne, P. M. and Young, V. R.,** Effect of diets marginal in methionine and choline with and without vitamin B12 on rat liver and kidney, *J. Nutr.,* 89, 69, 1966.
379. **Nicholls, D.,** Some recent advances in mitochondrial calcium transport, *Trends Biochem. Sci.,* 6, 36, 1981.
380. **Nicolson, G. L.,** Transmembrane control of the receptors on normal and tumor cells. I. Cytoplasmic influence over cell surface components, *Biochim. Biophys. Acta,* 457, 57, 1976.
381. **Noel, G., Trouet, A., Zenebergh, A., and Tulkens, P.,** in *Adriamycin Review, EORTC Int. Symp.,* Staquet, M., Ed., European Press Medikon, Ghent, Belgium, 1975, 99.
382. **Novikoff, A. B. and Novikoff, P. M.,** Cell organelles and production of secretory products: hormone action upon cell organelles, *Biol. Cell.,* 36, 101, 1979.
383. **Novikoff, A. B., Essner, E., and Quintana, N.,** Golgi apparatus and lysosomes, *Fed. Proc.,* 23, 1010, 1964.
384. **Northcote, D. H.,** The Golgi apparatus, *Endeavour,* 30, 26, 1971.
385. **Nurden, A. T. and Caen, J. P.,** Specific roles for platelets surface glycoproteins in platelet function, *Nature (London),* 255, 720, 1975.
386. **Nurden, A. T. and Caen, J. P.,** The different glycoprotein abnormalities in thrombasthenic and Bernard-Soulier platelets, *Semin. Hematol.,* 16, 234, 1979.
387. **Oberling, C. and Rouiller, C.,** Effects of acute carbon tetrachloride poisoning on the liver of rats; electron microscopic study, *Ann. Anat. Pathol.,* 1, 401, 1956.
388. **O'Brien, J. S. and Sampson, E. L.,** Lipid composition of the normal human brain: gray matter, white matter, and myelin, *J. Lipid Res.,* 6, 537, 1965.
389. **Okumura, K., Hayakawa, K., and Tada, T.,** Cell to cell interaction controlled by immunoglobulin genes; role of THY-1$^-$, LYT-1$^+$, IG$^+$(B′) cell in allotype-restricted antibody production, *J. Exp. Med.,* 156, 443, 1982.
390. **Olson, M. O. J. and Busch, H.,** Nuclear proteins, in *The Cell Nucleus,* Vol. 3., Busch, H., Ed., Academic Press, New York, 1974, 211.
391. **O'Malley, B. W. and Means, A. R.,** Female steroid hormones and target cell nuclei, *Science,* 183, 610, 1974.
392. **Omura, T.,** Cytochrome P-450 linked mixed function oxidase turnover of microsomal components and effects of inducers on the turnover phospholipids, proteins, and specific enzymes, *Pharmacol. Ther.,* 8, 489, 1980.
393. **Omura, T., Siekevitz, P., and Palade, G. E.,** Turnover of constituents of the endoplasmic reticulum membranes of rat hepatocytes, *J. Biol. Chem.,* 242, 2389, 1967.
394. **Orlandi, F.,** Electron-microscopic observations on human liver during cholestasis, *Acta Hepato-Splenol.,* 9, 155, 1962.
395. **Orrenius, S. and Ericsson, J. L. E.,** Enzyme-membrane relationship in phenobarbital induction of synthesis of drug-metabolizing enzyme system and proliferation of endoplasmic membranes, *J. Cell Biol.,* 28, 181, 1966.
396. **Orrenius, S. and Ericsson, J. L. E.,** On the relationship of liver glucose-6-phosphatase to the proliferation of endoplasmic reticulum in phenobarbital induction, *J. Cell Biol.,* 31, 243, 1966.
397. **Orrenius, S., Ericsson, J. L. E., and Ernster, L.,** Phenobarbital-induced synthesis of the microsomal drug metabolizing enzyme system and its relationship to the proliferation of endoplasmic membranes. A morphological and biochemical study, *J. Cell Biol.,* 25, 627, 1965.
398. **Orrenius, S.,** On the mechanism of drug hydroxylation in rat liver microsomes, *J. Cell Biol.,* 26, 713, 1965.
399. **Owens, J. W., Mueller, T. J., and Morrison, M.,** A minor sialoglycoprotein of the human erythrocyte membrane, *Arch. Biochem. Biophys.,* 204, 247, 1980.
400. **Oyanagi, K., Miura, R., and Yamanouchi, T.,** Congenital lysinuria: a new inherited transport disorder of dibasic amino acids, *J. Pediatr.,* 77, 259, 1970.
401. **Packer, D., Deamer, D. W., and Heath, R. L.,** Regulation and deterioration of structure in membranes, *Adv. Gerontol. Res.,* 2, 77, 1967.
402. **Palade, G. E.,** Electron microscopy of cytoplasmic structures, in *Enzymes: Units of Biological Structure and Function,* Gaebler, O. H., Ed., Academic Press, New York, 1956, 185.
403. **Palade, G.,** Intracellular aspects of the process of protein synthesis, *Science,* 189, 347, 1975.

404. **Palek, J. and Liu, S. C.**, Dependence of spectrin organization in red blood cell membranes on cell metabolism: implications for control of red cell shape, deformability, and surface area, *Semin. Hematol.*, 16, 75, 1979.

405. **Palek, J. and Liu, S. C.**, Membrane protein organization in ATP-depleted and irreversibly sickled red cells, *J. Supramol. Struct.*, 10, 79, 1979.

406. **Palek, J., Liu, S. C., and Snyder, L. M.**, Metabolic dependence of protein arrangement in human erythrocyte membranes. I. Analysis of spectrin-rich complexes in ATP-depleted red cells, *Blood*, 51, 385, 1978.

407. **Parke, D. V.**, Phase I metabolic reactions in man, in *Drug Metabolism in Man*, Gorrod, J. W. and Beckett, A. H., Eds., Taylor and Francis, London, 1978, 61.

408. **Parke, D. V.**, Biochemical aspects, in *Principles of Surgical Oncology*, Raven, R. W., Ed., Plenum Press, New York, 1977, 113.

409. **Parke, D. V.**, The role of the endoplasmic reticulum in carcinogenesis, in *Regulatory Aspects of Carcinogenesis and Food Additives: The Delaney Clause*, Coulston, F., Ed., Academic Press, New York, 1979, 173.

410. **Parsons, D. F.**, Recent advances correlating structure and function in mitochondria, *Int. Rev. Exp. Pathol.*, 4, 1, 1965.

411. **Perheentupa, J. and Visakorpi, J. K.**, Protein intolerance with deficient transport of basic aminoacids. Another inborn error of metabolism, *Lancet*, 2, 813, 1965.

412. **Peterson, M. L. and Herberg, R.**, Intestinal sucrase deficiency, *Trans. Assoc. Am. Physicians*, 80, 275, 1967.

413. **Petitou, M., Tuy, F., Rosenfeld, C., Mishal, Z., Paintrand, M., Jasnin, C., Mathe, G., and Inbar, M.**, Decreased microviscosity of membrane lipids in leukemic cells: two possible mechanisms, *Proc. Natl. Acad. Sci. U.S.A.*, 75, 2306, 1978.

414. **Phillips, M. J. and Steiner, J. W.**, Electron microscopical studies on the liver cells in hyperplastic nodules of human cirrhosis, *Rev. Int. Hepatol.*, 16, 307, 1966.

415. **Pietras, R. J. and Szego, C. M.**, Partial purification and characterization of oestrogen receptors in subfractions of hepatocyte plasma membranes, *Biochem. J.*, 191, 743, 1980.

416. **Pinder, J. C., Ungewickell, E., Bray, D., and Gratzer, W. B.**, The spectrin-actin complex and erythrocyte shape, *J. Supramol. Struct.*, 8, 439, 1978.

417. **Platt, B. S., Heard, C. R. C., and Stewart, R. J. C.**, The effects of protein-calorie deficiency of the gastrointestinal tract, in *The Role of the Gastrointestinal Tract in Protein Metabolism*, Munro, H., Ed., F. A. Davis, Philadelphia, 1964, 227.

418. **Plut, D. A., Hosey, M. M., and Tao, M.**, Evidence for the participation of cytosolic protein kinases in membrane phosphorylation in intact erythrocytes, *Eur. J. Biochem.*, 82, 333, 1978.

419. **Poland, A., Smith, D., Kuntzman, R., Jacobson, M., and Conney, A. H.**, Effect of intensive occupational exposure to DDT on phenylbutazone and cortisol metabolism in human subjects, *Clin. Pharmacol. Ther.*, 11, 724, 1970.

420. **Pollakis, G., Goormaghtigh, E., and Ruysschaert, J. M.**, Role of the quinone structure in the mitochondrial damage induced by antitumor anthracyclines, comparison of adriamycin and 5-iminodaunorubicin, *FEBS Lett.*, 155, 267, 1983.

421. **Porter, K. R.**, The ground substance; observations from electron microscopy, in *The Cell*, Brachet, J. and Mirsky, A. E., Eds., Vol. 2, Academic Press, New York, 1961, 621.

422. **Posner, B. I., Josefsberg, Z., and Bergeron, J. J. M.**, Intracellular polypeptide hormone receptors; characterization of insulin binding sites in golgi fractions from the liver of female rats, *J. Biol. Chem.*, 253, 4067, 1978.

423. **Poznansky, M. J. and Lange, Y.**, Transbilayer movement of cholesterol in phospholipid vesicles under equilibrium and non-equilibrium conditions, *Biochim. Biophys. Acta*, 506, 256, 1977.

424. **Preiser, H., Menard, D., Crane, R. K., and Cerda, J. J.**, Deletion of enzyme protein from the brush border membrane in sucrase-isomaltase deficiency, *Biochim. Biophys. Acta*, 363, 279, 1974.

425. **Price, D. L., Griffin, J., Young, A., Peck, K., and Stocks, A.**, Tetanus toxin: direct evidence for retrograde intraaxonal transport, *Science*, 188, 945, 1975.

426. **Price, D. L., Stocks, A., Griffin, J. W., Young, A., and Peck, K.**, Glycine-specific synapses in rat spinal cord; identification by electron microscope autoradiography, *J. Cell Biol.*, 68, 389, 1976.

427. **Prince, D. A.**, Neurophysiology of epilepsy, *Annu. Rev. Neurosci.*, 1, 395, 1978.

428. **Quastel, J. H.**, Transport at cell membranes and regulation of cell metabolism, in *Membrane Transport and Metabolism*, Kleinzeller, A. and Kotyk, A., Eds., Academic Press, London, 1961, 512.

429. **Quinn, P. J. and Chapman, D.**, The dynamics of membrane structure, *Crit. Rev. Biochem.*, 8, 1, 1980.

430. **Quinn, P. J.**, *The Molecular Biology of Cell Membranes*, University Park Press, Baltimore, 1976.

431. **Quist, E. E. and Reece, K. L.**, The role of diphosphatidylinositol in erythrocyte membrane shape regulation, *Biochem. Biophys. Res. Commun.*, 95, 1023, 1980.

432. **Räihä, N. C. R. and Schwartz, A. L.**, Enzyme induction in human fetal liver in organ culture, *Enzyme*, 15, 330, 1973.

433. **Rajalakshmi, S., Rao, P. M., and Sarma, D. S. R.**, Chemical carcinogenesis: interactions of carcinogens with nucleic acids, in *Cancer, A Comprehensive Treatise*, 2nd ed., Vol. 1, Becker, F. F., Ed., Plenum Press, New York, 1982, 335.

434. **Rappaport, A. M.**, Physioanatomical basis of toxic liver injury, in *Toxic Injury of the Liver*, Farber, E. and Fisher, M. M., Eds., Marcel Dekker, New York, 1979, 1.

435. **Ratner, S.**, Enzymes of arginine and urea synthesis, *Adv. Enzymol. Relat. Areas Mol. Biol.*, 39, 1, 1973.

436. **Rawlins, F. A., Villegas, G. M., and Uzman, B. G.**, Myelin, in *Mammalian Cell Membranes*, Vol. 2, Jamieson, G. A. and Robinson, D. M., Eds., Butterworths, Boston, 1977, 266.

437. **Recknagel, R. O. and Lombardi, B.**, Studies of biochemical changes in subcellular particles of rat liver and their relationship to a new hypothesis regarding the pathogenesis of carbon tetrachloride fat accumulation, *J. Biol. Chem.*, 236, 564, 1961.

438. **Recknagel, R. O., Glende, E. A., Jr., and Hruszkewycz, A. M.**, New data supporting an obligatory role for lipid peroxidation in carbon tetrachloride-induced loss of aminopyrine demethylase, cytochrome P450, and glucose-6-phosphatase, in *Biological Reactive Intermediates*, Jollow, D. J., Kocsis, J. J., Snyder, R., Eds., Plenum Press, New York, 1977, 417.

439. **Redman, C. M.**, Studies on the transfer of incomplete polypeptide chains across rat liver microsomal membranes in vitro, *J. Biol. Chem.*, 242, 761, 1967.

440. **Reid, E.**, Membrane systems, in *Enzyme Cytology*, Roodyn, D. B., Ed., Academic Press, New York, 1967, 321.

441. **Reidenberg, M. M.**, Obesity and fasting: effects on drug metabolism and drug action in man, *Clin. Pharmacol. Ther.*, 22, 729, 1977.

442. **Reinherz, E. L., Weiner, H. L., Hauser, S. L., Cohen, J. A., and Schlossman, S. F.**, Loss of suppressor T cells in active multiple sclerosis. Analysis with monoclonal antibodies, *N. Engl. J. Med.*, 303, 125, 1980.

443. **Remmer, H. and Merker, H. J.**, Drug-induced changes in the liver endoplasmic reticulum: association with drug-metabolizing enzymes, *Science*, 142, 1657, 1963.

444. **Remmer, H. and Merker, H. J.**, Enzymic induction and increase of endoplasmic reticulum in liver cells during phenobarbital (Luminal) therapy, *Klin. Wochenschr.*, 41, 276, 1963.

445. **Remmer, H.**, The accelerated decomposition of drugs in the liver microsomes under the effect of Luminal, *Naunyn-Schmiedebergs Arch. Exp. Pathol. Pharmakol.*, 235, 279, 1959.

446. **Renwick, A. G. and Williams, R. T.**, The fate of cyclamate in man and other species, *Biochem. J.*, 129, 869, 1972.

447. **Reynolds, E. S.**, Liver parenchymal cell injury. I. Initial alterations of the cell following poisoning with carbon tetrachloride, *J. Cell Biol.*, 19, 139, 1963.

448. **Cavill, I., Ricketts, C., Napier, J. A. F., and Jacobs, A.**, Ferrokinetics and erythropoiesis in man: red-cell production and destruction in normal and anaemic subjects, *Br. J. Haematol.*, 35, 33, 1977.

449. **Rintoul, D. A., Chou, S. M., and Silbert, D. F.**, Physical characterization of sterol-depleted LM-cell plasma membranes, *J. Biol. Chem.*, 254, 10070, 1979.

450. **Robel, P., Blondeau, J. P., and Baulieu, E. E.**, Androgen receptors in rat ventral prostate microsomes, *Biochim. Biophys. Acta*, 373, 1, 1974.

451. **Robert, A.**, Current history of cytoprotection, *Prostaglandins*, 21(Suppl.), 89, 1981.

452. **Roberts, R. K., Grice, J., Wood, L., Petroff, V., and McGuffie, C.**, Cimetidine impairs the elimination of theophylline and antipyrine, *Gastroenterology*, 81, 19, 1981.

453. **Robertson, A. F. and Lands, W. E. M.**, Metabolism of phospholipids in normal and spherocytic human erythrocytes, *J. Lipid Res.*, 5, 88, 1964.

454. **Rodbell, M., Lin, M. C., Salomon, Y., Londos, C., Harwood, J. P., Martin, B. R., Rendell, M., and Berman, M.**, Role of adenine and guanine nucleotides in the activity and response of adenylate cyclase systems to hormones: evidence for multisite transition states, *Adv. Cyclic Nucleotide Res.*, 5, 3, 1975.

455. **Rohr, H. P. and Riede, U. N.**, Experimental metabolic disorders and the subcellular reaction pattern, *Curr. Topics Pathol.*, 58, 1, 1973.

456. **Roseman, J.**, X-ray resistant cell required for the induction of *in vitro* antibody formation, *Science*, 165, 1125, 1969.

457. **Rosenberg, L. E., Albrecht, I., and Segal, S.**, Lysine transport in human kidney: evidence for two systems, *Science*, 155, 1426, 1967.

458. **Rosenberg, S. A. and Guidotti, G.**, The proteins of the erythrocyte membrane: structure and arrangement in the membrane, in *Red Cell Membrane: Structure and Function*, Jamieson, G. A. and Greenwalt, T. J., Eds., Lippincott, New York, 1969, 93.

459. **Rosenfeld, B.**, Effects of a single choline-deficient meal on the phospholipids of hepatic particulate lipid of the rat, *Can. J. Physiol. Pharmacol.*, 46, 495, 1968.

460. **Rouser, G., Nelson, G. J., Fleischer, S., and Simon, G.,** Lipid composition of animal cell membranes, organelles, and organs, in *Biological Membranes, Physical Fact and Function,* Chapman, D., Ed., Academic Press, London, 1968, 5.

461. **Rubenstein, B. and Rubenstein, D.,** The effect of carbon tetrachloride on hepatic lipid metabolism, *Can. J. Biochem.,* 42, 1263, 1964.

462. **Rubin, D. F.,** Antibiotics and oral contraceptives, *Arch. Dermatol.,* 117, 189, 1981.

463. **Rubin, P. and Casarett, G. W.,** *Clinical Radiation Pathology,* W. B. Saunders, Philadelphia, 1968, 33.

464. **Ruitenbeek, W. and Scholte, H. R.,** Fatty acid activation and transfer in blood cells of patients with muscular dystrophy, *J. Neurol. Sci.,* 41, 191, 1979.

465. **Ruitenbeek, W.,** Membrane-bound enzymes of erythrocytes in human muscular dystrophy: (sodium-potassium) ion-stimulated ATPase, calcium ion-stimulated ATPase, potassium ion, and calcium ion-stimulated p-nitrophenylphosphatase, *J. Neurol. Sci.,* 41, 71, 1979.

466. **Ruitenbeek, W.,** The fatty acid composition of various lipid fractions isolated from erythrocytes and blood plasma of patients with Duchenne and congenital myotonic muscular dystrophy, *Clin. Chim. Acta,* 89, 99, 1978.

467. **Ryser, H. J. P.,** Uptake of protein by mammalian cells. An underdeveloped area, *Science,* 159, 390, 1968.

468. **Saks, V. A., Chernousova, G. B., Voronkov, I. I., Smirnov, V. N., and Chazov, E. I.,** Study of energy transport mechanism in myocardial cells, *Circ. Res.,* 34/35(Suppl. III), 138, 1974.

469. **Sandstead, H. H., Michelakis, A. M., and Temple, T. E.,** Lead intoxication, its effect on the renin-aldosterone response to sodium deprivation, *Arch. Environ. Health,* 20, 356, 1970.

470. **Sato, S. B., Yanagida, M., Maruyama, K., and Ohnishi, S.,** Seeding role of spectrin in polymerization of skeletal muscle actin, *Biochim. Biophys. Acta,* 578, 436, 1979.

471. **Savouret, J. F., Loosfelt, H., Atger, M., and Milgrom, E.,** Differential hormonal control of a messenger RNA in two tissues: uteroglobin mRNA in the lung and the endometrium, *J. Biol. Chem.,* 255, 4131, 1980.

472. **Schindler, M., Koppel, D. E., and Sheetz, M. P.,** Modulation of membrane protein lateral mobility by polyphosphates and polyamines, *Proc. Natl. Acad. Sci. U.S.A.,* 77, 1457, 1980.

473. **Schimke, R. T.,** Enzymes of arginine metabolism in cell culture: studies on enzyme induction and repression, *Natl. Cancer Inst. Monogr.,* 13, 197, 1964.

474. **Schimke, R. T.,** The importance of both synthesis and degradation in the control of arginase levels in rat liver, *J. Biol. Chem.,* 239, 3808, 1964.

475. **Schmähl, D.,** *Entstehung, Wachstum, und Chemotherapie maligner Tumoren,* Editio Cantor, Aulendorf, Germany, 1970.

476. **Schneider, J. A. and Schulman, J. D.,** Cystinosis and the Fanconi syndrome, in *Physiology of Membrane Disorders,* Andreoli, T. E., Hoffman, J. F., and Fanestil, D. D., Eds., Plenum Press, New York, 1978, 1019.

477. **Schotland, D. L., Bonilla, E., and van Meter, M.,** Duchenne dystrophy: alteration in muscle plasma membrane structure, *Science,* 196, 1005, 1977.

478. **Schriefers, H., Ghraf, R., and Pohl, F.,** Zur Frage des Verhaltens der UDP glucuronyl — transferase — Aktivität unter Nahrungsentzug und in Alloxandiabetes, *Hoppe-Seyler's Z. Physiol. Chem.,* 344, 25, 1966.

479. **Schubert, J. and Sanders, E. B.,** Cytotoxic radiolysis products of irradiated α,β-unsaturated carbonyl sugars as the carbohydrates, *Nature (London) New Biol.,* 233, 199, 1971.

480. **Schulman, J. D., Goodman, S. I., Mace, J. W., Patrick, A. D., Tietze, F., and Butler, E. J.,** Glutathionuria: inborn error of metabolism due to tissue deficiency of gamma glutamyl transpeptidase, *Biochem. Biophys. Res. Commun.,* 65, 68, 1975.

481. **Schwartz, A.,** Active transport in mammalian myocardium, in *The Mammalian Myocardium,* Langer, G. A. and Brady, A. J., Eds., John Wiley & Sons, New York, 1974, 81.

482. **Scriver, C. R.,** Familial iminoglycinuria, in *The Metabolic Basis of Inherited Disease,* 4th ed., Stanbury, J. B., Wyngaarden, J. B., and Fredrickson, D. S., Eds., McGraw-Hill, New York, 1978, 1593.

483. **Seldin, D. W. and Wilson, J. D.,** Renal tubular acidosis, in *The Metabolic Basis of Inherited Disease,* 4th ed., Stanbury, J. B., Wyngaarden, J. B., and Fredrickson, D. S., Eds., McGraw-Hill, New York, 1978, 1618.

484. **Seljelid, R.,** Thyroid lysosomes in health and disease, in *Pathobiology of Cell Membranes,* Vol. 1, Trump, B. F. and Arstila, A. U., Eds., Academic Press, New York, 1975, 325.

485. **Semenza, G.,** Mode of insertion of the sucrase-isomaltase complex in the intestinal brush border membrane: implications for the biosynthesis of this stalked intrinsic membrane protein, *Ciba Symp.,* 70, 133, 1979.

486. **Semenza, G.,** Small intestinal disaccharidases: their properties and role as sugar translocators across natural and artificial membranes, in *The Enzymes of Biological Membranes,* Vol. 3, Martonosi, A., Ed., Plenum Press, New York, 1976, 349.

487. **Serlin, M. J., Sibeon, R. G., Mossman, S., Breckenridge, A. M., Williams, J. R. B., Atwood, J. L., and Willoughby, J. M. T.,** Cimetidine: interaction with oral anticoagulants in man, *Lancet*, 2, 318, 1979.

488. **Serur, J. R., Galyean, J. R., Urschel, C. W., and Sonnenblick, E. H.,** Experimental myocardial ischemia: dynamic alterations in ventricular contractility and relaxation with dissociation of speed and force in isovolumic dog heart, *Circ. Res.*, 39, 602, 1976.

489. **Shapiro, B. and Kollmann, G.,** The nature of the membrane injury in irradiated human erythrocytes, *Radiat. Res.*, 34, 335, 1968.

490. **Shapiro, R.,** Reactions with purines and pyrimidines, *Ann. N. Y. Acad. Sci.*, 163, 624, 1969.

491. **Sharp, H. L. and Mirkin, B. L.,** Effect of phenobarbital on hyperbilirubinemia, bile acid metabolism, and microsomal enzyme activity in chronic intrahepatic cholestasis of childhood, *J. Pediatr.*, 81, 116, 1972.

492. **Sheetz, M. P. and Singer, S. J.,** On the mechanism of ATP-induced shape changes in human erythrocyte membranes. I. The role of the spectrin complex, *J. Cell Biol.*, 73, 638, 1977.

493. **Sheetz, M. P., Sawyer, D., and Jackowski, S.,** The ATP-dependent red cell membrane shape change: a molecular explanation, in *The Red Cell*, (Prog. Clin. Biol. Res. No. 21), Brewer. G. J., Ed., New York, Alan R. Liss, New York, 1978, 431.

494. **Shiga, T., and Maeda, N.,** Influence of membrane fluidity on erythrocyte functions, *Biorheology*, 17, 485, 1980.

495. **Shih, V. E., Bixby, E. M., Alpers, D. H., Bartsocas, C. S., and Thier, S. O.,** Studies of intestinal transport defect in Hartnup disease, *Gastroenterology*, 61, 445, 1971.

496. **Shinitzky, M. and Henkart, P.,** Fluidity of cell membranes — current concepts and trends, *Int. Rev. Cytol.*, 60, 121, 1979.

497. **Shishiba, Y., Solomon, D. H., and Davidson, W. D.,** Comparison of the effect of thyrotropin and long-acting thyroid stimulator on glucose oxidation and endocytosis in canine thyroid slices, *Endocrinology*, 86, 183, 1970.

498. **Shohet, S. B.,** Mechanisms of red cell membrane lipid renewal, in *Membranes and Disease*, Bolis, L., Hoffman, J. F., and Leaf, A., Eds., Raven Press, New York, 1976, 61.

499. **Shohet, S. B.,** Release of phospholipid fatty acid from human erythrocytes, *J. Clin. Invest.*, 49, 1668, 1970.

500. **Siekevitz, P.,** Biological membranes: the dynamics of their organization, *Annu. Rev. Physiol.*, 34, 117, 1972.

501. **Silbert, D. F.,** Genetic modification of membrane lipid, *Annu. Rev. Biochem.*, 44, 315, 1975.

502. **Simpson, J. A.,** Myasthenia gravis: a new hypothesis, *Scott. Med. J.*, 5, 419, 1960.

503. **Sinensky, M., Minneman, K. P., and Molinoff, P. B.,** Increased membrane acyl chain ordering activates adenylate cyclase, *J. Biol. Chem.*, 254, 9135, 1979.

504. **Singer, S. J. and Nicolson, G. L.,** The fluid mosaic model of the structure of cell membranes, *Science*, 175, 720, 1972.

505. **Singer, S. J.,** The molecular organization of membranes, *Annu. Rev. Biochem.*, 43, 805, 1974.

506. **Slater, T. F. and Greenbaum, A. L.,** Changes in lysosomal enzymes in acute experimental liver injury, *Biochem. J.*, 96, 484, 1965.

507. **Smith, D. M. and Druse, M. J.,** Effects of maternal protein deficiency on synaptic plasma membranes in offspring, *Dev. Neurosci. (Basel)*, 5, 403, 1982.

508. **Smith, B. T. and Torday, J. S.,** Factors affecting lecithin synthesis by fetal lung cells in culture, *Pediatr. Res.*, 8, 848, 1974.

509. **Smuckler, E. A. and Hultin, T.,** Effects of SKF 525-A and adrenalectomy on the amino acid incorporation by rat liver microsomes from normal and CCl_4-treated rats, *Exp. Mol. Pathol.*, 5, 504, 1966.

510. **Smuckler, E. A., Iseri, O. A., and Benditt, E. P.,** An intracellular defect in protein synthesis induced by carbon tetrachloride, *J. Exp. Med.*, 116, 55, 1962.

511. **Smuckler, E. A., Iseri, O. A., and Benditt, E. P.,** Studies on carbon tetrachloride intoxication. I. The effect of carbon tetrachloride on incorporation of labelled amino acids into plasma proteins, *Biochem. Biophys. Res. Commun.*, 5, 270, 1961.

512. **Stanbury, J. B., Wyngaarden, J. B., and Fredrickson, D. S., Eds.,** *The Metabolic Basis of Inherited Disease*, 4th ed., McGraw-Hill, New York, 1978.

513. **Staros, J. V., Haley, B. E., and Richards, F. M.,** Human erythrocytes and resealed ghosts: a comparison of membrane topology, *J. Biol. Chem.*, 249, 5004, 1974.

514. **Staubli, W. and Hess, R.,** Lipoprotein formation in the liver cell. Ultrastructural and functional aspects relevant to hypolipidemic action, in *Handbook of Experimental Pathology*, Vol. 41, Hypolipidemic Agents, Kritchevky, D., Ed., Springer-Verlag, Berlin, 1975, 229.

515. **Steck, T. L. and Dawson, G.,** Topographical distribution of complex carbohydrates in the erythrocyte membrane, *J. Biol. Chem.*, 249, 2135, 1974.

516. **Steck, T. L.,** The band 3 protein of the human red cell membrane: a review, *J. Supramol. Struct.*, 8, 311, 1978.

517. **Steck, T. L.,** The organization of proteins in human erythrocyte membranes, in *Membrane Research*, Fox, C. F., Ed., Academic Press, New York, 1972, 71.

518. **Steck, T. L.,** The organization of proteins in the human red blood cell membrane. A review, *J. Cell Biol.*, 62, 1, 1974.

519. **Steiner, S. and Steiner, M. R.,** Glycolipids in virus-transformed cells, in *Virus-Transformed Cell Membranes*, Nicolau, C., Ed., Academic Press, London, 1978, 91.

520. **Steinmetz, P. R.,** Cellular defects in urinary acidification and renal tubular acidosis, in *Physiology of Membrane Disorders*, Andreoli, T. E., Hoffman, J. F., and Fanestil, D. D., Eds., Plenum Press, New York, 1978, 987.

521. **Stickler, G. B.,** External calcium and phosphorous balances in vitamin D resistant rickets, *J. Pediatr.*, 63, 942, 1963.

522. **Stier, A.,** Commentary. Lipid structure and drug metabolizing enzymes, *Biochem. Pharmacol.*, 25, 109, 1976.

523. **Stocken, L. A.,** Some observations of the biochemical effects of x-radiation, *Radiat. Res. Suppl.*, 1, 53, 1959.

524. **Stoeckenius, W. and Engelman, D. M.,** Current models for the structure of biological membranes, *J. Cell Biol.*, 42, 613, 1969.

525. **Streffer, C.,** *Strahlen-Biochemie*, Springer-Verlag, Berlin, 1969.

526. **Stoll, R., Kinne, R., and Murer, H.,** Effect of dietary phosphate intake on phosphate transport by isolated rat renal brush-border vesicles, *Biochem. J.*, 180, 465, 1979.

527. **Sturgess, J. M. and Moscarello, M. A.,** Alterations in the Golgi complex and glycoprotein biosynthesis in normal and diseased tissues, in *Patho. Biol. Annu.*, 6, 1, 1976.

528. **Sturgess, J. M., Moscarello, M., and Schachter, H.,** The structure and biosynthesis of membrane glycoproteins, *Curr. Top. Membr. Transp.*, 11, 15, 1978.

529. **Svedenhag, J., Henriksson, J., and Sylvén, C.,** Dissociation of training effects on skeletal muscle mitochondrial enzymes and myoglobin in man, *Acta Physiol. Scand.*, 117, 213, 1983.

530. **Svoboda, D. J. and Reddy, J. K.,** Some effects of carcinogens on cell organelles, in *Cancer, A Comprehensive Treatise*, 2nd ed., Vol. 1, Becker, F. F., Ed., Plenum Press, New York, 1982, 411.

531. **Sweeley, C. C. and Dawson, G.,** Lipids of the erythrocyte, in *Red Cell Membrane: Structure and Function*, Jamieson, G. A. and Greenwalt, T. J., Eds., Lippincott, Philadelphia, 1969, 172.

532. **Sweeley, C. C. and Siddiqui, B.,** Chemistry of mammalian glycolipids, in *The Glycoconjugates*, Vol. 1, Horowitz, M. I. and Pigman, W., Eds., Academic Press, New York, 1977, 459.

533. **Syrjänen, S. M. and Syrjänen, K. J.,** Plasma cells and their immunoglobulins in the normal and delayed healing of the extraction wound in man, *Br. J. Oral Surg.*, 18, 100, 1980.

534. **Taketa, K.,** Enzymic studies of glucuronide formation in impaired liver. IV. Liver glucuronyl transferase activity and uridine diphosphate glucuronic acid content in viral hepatitis patients, *Acta Med. Okayama*, 16, 115, 1962.

535. **Talalay, P.,** Enzymatic mechanisms in steroid biochemistry, *Annu. Rev. Biochem.*, 34, 347, 1965.

536. **Taliaferro, W. H., Taliaferro, L. G., and Jaroslow, B. N.,** *Radiation and Immune Mechanisms*, Academic Press, New York, 1964.

537. **Tanabe, T., Pricer, W. E., Jr., and Ashwell, G.,** Subcellular membrane topology and turnover of a rat hepatic binding protein specific for asialoglycoproteins, *J. Biol. Chem.*, 254, 1038, 1979.

538. **Tandler, B., Erlandson, R. A., Smith, A. L., and Wynder, E. L.,** Riboflavin and mouse hepatic cell structure and function. II. Division of mitochondria during recovery from simple deficiency, *J. Cell Biol.*, 41, 477, 1969.

539. **Tanner, M. J. A. and Anstee, D. J.,** The membrane change in En(a−) human erythrocytes. Absence of the major erythrocyte sialoglycoprotein, *Biochem. J.*, 153, 271, 1976.

540. **Tappel, A. L., Sawant, P. L., and Shibko, S.,** Lysosomes: distribution in animals, hydrolytic capacity, and other properties, in *Ciba Foundation Symposium on Lysosomes*, de Reuck, A. V. S. and Cameron, M. P., Eds., Little-Brown, Boston, 1963, 78.

541. **Tebecis, A. K.,** *Transmitters and Identified Neurons in the Mammalian Central Nervous System*, Scientechnica, Bristol, England, 1974.

542. **Teijema, H. L., Van Gelderen, H. H., Giesberts, M. A. H., and Laurent de Angulo, M. S. L.,** Dicarboxylic aminoaciduria: an inborn error of glutamate and aspartate transport with metabolic implications, in combination with a hyperprolinemia, *Metabolism*, 23, 115, 1974.

543. **Thayer, W. S. and Rubin, E.,** Effects of chronic ethanol intoxication on oxidative phosphorylation in rat liver submitochondrial particles, *J. Biol. Chem.*, 254, 7717, 1979.

544. **Thomas, P. K.,** The peripheral nervous system as a target for toxic substances, in *Experimental and Clinical Neurotoxicology*, Spender, P. S. and Schaumburg, H. H., Eds., Williams & Wilkins, Baltimore, 1980, 35.

545. **Thompson, T. E. and Huang, C.**, Dynamics of lipids in biomembranes, in *Physiology of Membrane Disorders*, Andreoli, T. E., Hoffman, J. F., and Fanestil, D. D., Eds., Plenum Press, New York, 1978, 27.

546. **Thompson, T. E.**, Transmembrane compositional asymmetry of lipids in bilayers and biomembranes, in *Molecular Specilization and Symmetry in Membrane Function*, Solomon, A. K. and Karnovsky, M., Eds., Harvard University Press, Cambridge, 1978, 78.

547. **Trier, J. S.**, Celiac sprue disease, in *Gastrointestinal Disease*, 2nd ed., Sleisenger, M. H. and Fordtran, J. S., Eds., W.B. Saunders, Philadelphia, 1978, 1029.

548. **Triplett, R. B., Wingate, J. M., and Carraway, K. L.**, Calcium effects on erythrocyte membrane proteins, *Biochem. Biophys. Res. Commun.*, 49, 1014, 1972.

549. **Trouet, A., Deprez-De Campaneere, D., De Smedt-Malengreaux, M., and Ataassi, G.**, Experimental leukemia chemotherapy with a lysosomotropic adriamycin-DNA complex, *Eur. J. Cancer*, 10, 405, 1974.

550. **Trouet, A., Deprez-de Companeere, D., and de Duve, C.**, Chemotherapy through lysosomes with a DNA-daunorubicin complex, *Nature (London) New Biol.*, 239, 110, 1972.

551. **Trump, B. F. and Arstila, A. U.**, Cell membranes and disease processes, in *Pathobiology of Cell Membranes*, Vol. 1, Trump, B. F. and Arstila, A. U., Eds., Academic Press, New York, 1975, 1.

552. **Trump, B. F.**, Effects of hyperosmotic mannitol in reducing ischemic cell swelling and minimizing myocardial necrosis, *Circulation*, 53(Suppl. I), I-52, 1976.

553. **Trump, B. F. and Ginn, F. L.**, The pathogenesis of subcellular reaction to lethal injury, in *Methods and Achievements in Experimental Pathology*, Vol. 4., Bajusz, E. and Jasmin, G., Eds., Karger, Basel, 1969, 1.

554. **Trump, B. F., Kim, K. M., Jones, R. T., and Valigorsky, J. M.**, Pathology of organelles in the human hepatic parenchymal cell, in *Progress in Liver Diseases*, Vol. 5, Schaffner, F. and Popper, H., Eds., Grune & Stratton, New York, 1976, 51.

555. **Tulkens, P. and Trouet, A.**, Uptake and intracellular localization of kanamycin and gentamycin in the lysosomes of cultured fibroblasts, *Arch. Int. Physiol. Biochim.*, 82, 1018, 1974.

556. **Tyler, D. D.**, Evidence of a phosphate-transporter system in the inner membrane of isolated mitochondria, *Biochem. J.*, 111, 665, 1969.

557. **Ungewickell, E. and Gratzer, W.**, Self-association of human spectrin. A thermodynamic and kinetic study, *Eur. J. Biochem.*, 88, 379, 1978.

558. **Upton, A. C.**, Radiation carcinogenesis, in *Methods in Cancer Research*, Vol. 4, Busch, H., Ed., Academic Press, New York, 1968.

559. **Upton, A. C.**, Radiation, in *Cancer Medicine*, Holland, J. F. and Frei, E., Eds., Lea & Febiger, Philadelphia, 1973, chap. 1—5.

560. **Van den Besselaar, A. M., de Druijff, B., van den Bosch, H., and van Deenen, L. L.**, Phosphatidylcholine mobility in liver microsomal membranes, *Biochem. Biophys. Acta*, 510, 242, 1978.

561. **Van Hoeven, R. P., Emmelot, P., Krol, J. H., and Oomen-Meulemans, E. P. M.**, Studies on plasma membranes. XXII. Fatty acid profiles of lipid classes in plasma membranes of rat and mouse livers and hepatomas, *Biochim. Biophys. Acta*, 380, 1, 1975.

562. **Vaughan, L. and Penniston, J. T.**, Cation control of erythrocyte membrane shape: Ca^{++} reversal of discocyte to echinocyte transition caused by Mg^{++} and other cations, *Biochem. Biophys. Res. Commun.*, 73, 200, 1976.

563. **Verity, M. A. and Brown, W. J.**, Hg^{2+}-induced kidney necrosis. Subcellular localization and structure-linked lysosomal enzyme changes, *Am. J. Pathol.*, 61, 57, 1970.

564. **Vesell, E. S. and Passananti, G. T.**, Inhibition of drug metabolism in man, *Drug Metab. Dispos.*, 1, 402, 1973.

565. **Vessey, D. A. and Zakim, D.**, Regulation of microsomal enzymes by phospholipids. II. Activation of hepatic uridine diphosphate-glucuronyl transferase, *J. Biol. Chem.*, 246, 4649, 1971.

566. **Vickers, J. D., Rathbone, M. P., and Roses, A. D.**, Alterations of erythrocyte ghost protein phosphorylation in the Duchenne and myotonic muscular dystrophies, *Biochem. Med.*, 20, 434, 1978.

567. **Walborg, E. F.**, Ed., *Glycoproteins and Glycolipids in Disease Processes* (ACS Symp. Ser. 80), American Chemical Society, Washington, D.C., 1978.

568. **Wallach, D. F. H.**, Membrane receptors and diseases involving membrane receptors, in *Plasma Membranes and Disease*, Wallach, D. F. H., Ed., Academic Press, London, 1979, 151.

569. **Walter, G. F., Tassin, S., and Brucher, J. M.**, Familial mitochondrial myopathies, *Acta Neuropathol. Suppl.*, 7, 283, 1981.

570. **Watkins, D. K.**, High oxygen effect for the release of enzymes from isolated mammalian lysosomes after treatment with ionizing radiation, *Adv. Biol. Med. Phys.*, 13, 289, 1970.

571. **Watson, M. L.**, The nuclear envelope: its structure and relation to cytoplasmic membranes, *J. Biophys. Biochem. Cytol.*, 1, 257, 1955.

572. **Wattenberg, L. W.**, Dietary modification of intestinal and pulmonary aryl hydrocarbon hydroxylase activity, *Toxicol. Appl. Pharmacol.*, 23, 741, 1972.

573. **Waxman, S. G.,** Conduction in myelinated, unmyelinated, and demyelinated fibers, *Arch. Neurol. (Chicago),* 34, 585, 1977.

574. **Ways, P., Reed, C. F., and Hanahan, D. J.,** Red-cell and plasma lipids in acanthocytosis, *J. Clin. Invest.,* 42, 1248, 1963.

575. **Weed, R. I.,** Disorders of red cell membrane: history and perspectives, *Semin. Hematol.,* 7, 249, 1970.

576. **Weed, R. I., LaCelle, P. L., and Merrill, E. W.,** Metabolic dependence of red cell deformability, *J. Clin. Invest.,* 48, 795, 1969.

577. **Weibel, E. R. and Paumgartner, D.,** Integrated stereological and biochemical studies on hepatocytic membranes. II. Correction of section thickness effect on volume and surface density estimates, *J. Cell Biol.,* 77, 584, 1978.

578. **Weiner, H. L.,** Multiple sclerosis, in *Current Neurology,* Tyler, H. R. and Dawson, D. M., Eds., Houghton Mifflin, Boston, 1978, 53.

579. **Weissmann, G.,** Studies of lysosomes. VI. The effect of neutral steroids and bile acids on lysosomes in vitro, *Biochem. Pharmacol.,* 14, 525, 1965.

580. **Weiss, L. and Dingle, J. T.,** Lysosomal activation in relation to connective tissue disease, *Ann. Rheum. Dis.,* 23, 57, 1964.

581. **Weissburger, J. H., Hadidian, Z., Fredrickson, T. N., and Weisburger, E. K.,** Host properties determine target, bladder or liver in chemical carcinogenesis, in *Bladder Cancer: A Symposium,* Lampe, K. F., Ed., Aesculapius, Birmingham, Ala., 1967, 45.

582. **Welch, R. M., Harrison, Y. E., Gommi, B. W., Poppers, P. J., Finster, M., and Conney, A. H.,** Stimulatory effect of cigarette smoking on the hydroxylation of 3,4-benzpyrene and the N-demethylation of 3-methyl-4-monomethylaminoazobenzene by enzymes in human placenta, *Clin. Pharmacol. Ther.,* 10, 100, 1969.

583. **Wells, I. C.,** Hemorrhagic kidney degeneration in choline deficiency, *Fed. Proc.,* 30, 151, 1971.

584. **Westerman, M. P., Diloy-Puray, M., and Streczyn, M.,** Membrane components in the red cells of patients with sickle cell anemia. Relationship to cell aging and to irreversibility of sickling, *Biochim. Biophys. Acta,* 557, 149, 1979.

585. **Weston, P. D., Barrett, A. J., and Dingle, J. T.,** Specific inhibition of cartilage breakdown, *Nature (London),* 222, 285, 1969.

586. **Westphal, U.,** Mechanism of steroid binding to transport proteins, in *Pharmacological Modulation of Steroid Action,* Genazzani, E., Di Carlo, F., and Mainwaring, W. I. P., Eds., Raven Press, New York, 1980, 33.

587. **Whelan, D. T. and Scriver, C. R.,** Hyperdibasicaminoaciduria: an inherited disorder of amino acid transport, *Pediatr. Res.,* 2, 525, 1968.

588. **Whitaker, J. N.,** The distribution of myelin basic protein in central nervous system lesions of multiple sclerosis and acute experimental allergic encephalomyelitis, *Ann. Neurol.,* 3, 291, 1978.

589. **Wibo, M. and Poole, B.,** Protein degradation in cultured cells. II. The uptake of chloroquine by rat fibroblasts and the inhibition of cellular protein degradation and cathepsin B_1, *J. Cell Biol.,* 63, 430, 1974.

590. **Wickner, W.,** The assembly of proteins into biological membranes: the membrane trigger hypothesis, *Annu. Rev. Biochem.,* 48, 23, 1979.

591. **Williams, L. T., Lefkowitz, R. J., Watanabe, A. M., Hathaway, D. R., and Besch, H. R., Jr.,** Thyroid hormone regulation of β-adrenergic receptor number, *J. Biol. Chem.,* 252, 2787, 1977.

592. **Williamson, J. R.,** Mitochondrial function in the heart, *Annu. Rev. Physiol.,* 41, 485, 1979.

593. **Wills, E. D.,** Effects of irradiation on subcellular components. I. Lipid peroxide formation in the endoplasmic reticulum, *Int. J. Radiat. Biol. Relat. Stud. Phys. Chem. Med.,* 17, 217, 1970.

594. **Wills, E. D. and Wilkinson, A. E.,** Effects of irradiation on subcellular components. II. Hydroxylation in the microsomal fraction, *Int. J. Radiat. Biol.,* 17, 229, 1970.

595. **Wills, E. D.,** Effects of irradiation on subcellular components. I. Metal ion transport in mitochondria, *Int. J. Radiat. Biol.,* 11, 517, 1966.

596. **Wilson, J. G.,** Current status of teratology. General principles and mechanisms derived from animal studies, in *Handbook of Teratology,* Vol. 1, Wilson, J. G. and Fraser, F. C., Eds., Plenum Press, New York, 1977, 47.

597. **Wilson, J. G.,** Teratogenic effects of environmental chemicals, *Fed. Proc.,* 36, 1698, 1977.

598. **Wilson, J. W. and Leduc, E. H.,** Mitochondrial changes in the liver of essential fatty acid-deficient mice, *J. Cell Biol.,* 16, 281, 1963.

599. **Winters, R. W., Graham, J. B., Williams, T. F., McFalls, V. W., and Burnett, C. H.,** A genetic study of familial hypophosphatemia and vitamin D resistant rickets with a review of the literature, *Medicine,* 37, 97, 1958.

600. **Wisher, M. H. and Evans, W. H.,** Functional polarity of the rat hepatocyte surface membrane. Isolation and characterization of plasma-membrane subfractions from the blood-sinusoidal, bile-canalicular and contiguous surfaces of the hepatocyte, *Biochem. J.,* 146, 375, 1975.

601. **Woese, C. R.,** *The Genetic Code,* Harper & Row, New York, 1967.

602. **Wollman, S. H., Spicer, S. S., and Burstone, M. S.,** Localization of esterase and acid phosphatase in granules and colloid droplets in rat thyroid epithelium, *J. Cell Biol.*, 21, 191, 1964.

603. **Wong, K. P. and Sourkes, T. L.,** Measurement of endogenous UDPGA in tissues, and some hormonal effects on its concentration in liver, *Fed. Proc.*, 26, Abstr. 3235, 835, 1967.

604. **Woolf, L. I., Goodwin, B. L., and Phelps, C. E.,** Tm-limited renal tubular reabsorption and the genetics of renal glycosuria, *J. Theor. Biol.*, 11, 10, 1966.

605. **Woolam, D. H. M.,** Principles of teratogenesis: mode of action of thalidomide, *Proc. R. Soc. Med.*, 58, 497, 1965.

606. **Yachnin, S., Streuli, R. A., Gordon, L. I., and Hsu, R. C.,** Alteration of peripheral blood cell membrane function and morphology by oxygenated sterols; a membrane insertion hypothesis, *Curr. Top. Hematol.*, 2, 245, 1979.

607. **Yaffe, S. J., Levy, G., Matsuzawa, T., and Baliah, T.,** Enhancement of glucuronide conjugating capacity in hyperbilirubinemic infant due to apparent enzyme induction by phenobarbital, *N. Engl. J. Med.*, 275, 1461, 1966.

608. **Yamada, M. and Miyaji, H.,** Binding of sex hormones by male rat liver microsomes, *J. Steroid Biochem.*, 16, 437, 1982.

609. **You, K.,** Salicylate and mitochondrial injury in Reye's syndrome, *Science*, 221, 163, 1983.

610. **Young, D. L., Powell, G., and McMillan, W. O.,** Phenobarbital-induced alterations in phosphatidylcholine and triglyceride synthesis in hepatic endoplasmic reticulum, *J. Lipid Res.*, 12, 1, 1971.

611. **Zakim, D. and Vessey, D. A.,** Regulation of microsomal enzymes by phospholipids. IX. Production of uniquely modified forms of microsomal UDP-glucuronyltransferase by treatment with phospholipase A and detergents, *Biochim. Biophys. Acta*, 410, 61, 1975.

612. **Zampaglione, N., Jollow, D. J., Mitchell, J. R., Stripp, B., Hamrick, M., and Gillette, J. R.,** Role of detoxifying enzymes in bromobenzene-induced liver necrosis, *J. Pharmacol. Exp. Ther.*, 187, 218, 1973.

613. **Zwaal, R. F. A., Roelofsen, B., and Colley, C. M.,** Localization of red cell membrane constituents, *Biochim. Biophys. Acta*, 300, 159, 1973.

614. **Zwaal, R. F. A., Roelofsen, B., Comfrius, P., and van Deenen, L. L. M.,** Organization of phospholipids in human red cell membranes as detected by the action of various purified phospholipases, *Biochim. Biophys. Acta*, 406, 83, 1975.

615. **Borthwick, N. M. and Smellie, R. M. S.,** The effects of oestradiol-17 beta on the ribonucleic acid polymerases of immature rabbit uterus, *Biochem. J.*, 147, 91, 1975.

616. **Estabrook, R. W., Hildebrandt, A., Remmer, H., Schenkman, J. B., Rosenthal, O., and Cooper, D. Y.,** Role of cytochrome P-450 in microsomal mixed function oxidation reactions, in Biochemie des Sauerstoffs, *19 Colloquium der Gesellschaft für Biologische Chemie,* Hess, B. and Staudinger, H. J., Eds., Springer Verlag, Berlin, 1969, 142.

617. **Feuer, G., Miller, D. R., Cooper, S. D., de la Iglesia, F. A., and Lumb, G.,** The influence of methyl groups on toxicity and drug metabolism, *Int. J. Clin. Pharmacol.*, 7, 13, 1973.

618. **Gelboin, H. V.,** Mechanisms of induction of drug metabolism enzymes, in *Handbook of Experimental Pharmacology,* Vol. 2, Brodie, B. B. and Gillette, J. R., Eds., Springer-Verlag, Berlin, 1971, 431.

619. **Hildebrandt, A., Remmer, H., Estabrook, R. W.,** Cytochrome P-450 of liver microsomes — one pigment or many, *Biochem. Biophys. Res. Commun.*, 30, 607, 1969.

620. **Spelsberg, T. C., Thrall, C., Webster, R., and Pikler, G.,** Isolation and characterization of the nuclear acceptor that binds the progesterone-receptor complex in oviduct, *J. Toxicol. Environ. Health*, 3, 309, 1977.

621. **Tulkens, P. and Trouet, A.,** in *Membranes and Disease,* Bolis, L., Hoffman, J. F., and Lead, A., Eds., Raven Press, New York, 1976, 141.

622. **Tulkens, P., Van Hoof, F., and Trouet, A.,** Lysosome overloading and dysfunction induced by streptomycin in cultured fibroblasts, *Arch. Int. Physiol. Biochem.*, 83, 1004, 1975.

623. **Woollam, D. H. M.,** Principles of teratogenesis: mode of action of thalidomide, *Proc. R. Soc. Med.*, 58, 497, 1965.

FURTHER READINGS

Bittar, E. E., *Cell Biology in Medicine,* John Wiley & Sons, New York, 1973.

Bolis, L., Hoffman, J. F., and Leaf, A., *Membranes and Disease,* Raven Press, New York, 1976.

Dingle, J. T., *Lysosomes in Biology and Pathology,* Elsevier, Amsterdam, 1979.

Estabrook, R. W. and Lindenlaub, E., *The Induction of Drug Metabolism,* Schattauer Verlag, Stuttgart, 1979.

Fleischer, S., Hatefi, Y., MacLennan, D., and Tzagoloff, A., *The Molecular Biology of Membranes,* Plenum Press, New York, 1978.

Jakoby, W. B., Bend, J. R., and Caldwell, J., *Metabolic Basis of Detoxication,* Academic Press, New York, 1982.

Plaa, G. L. and Hewitt, W. R., *Toxicology of the Liver* (Target Organ Toxicology Series), Raven Press, New York, 1982.

Richter, G. W. and Scarpelli, D. G., *Cell Membranes. Biological and Pathological Aspects,* Williams & Wilkins, Baltimore, 1971.

Tao, M., *Membrane Abnormalities,* CRC Press, Boca Raton, Fla., 1982.

Weissmann, G. and Claiborne, R., *Cell Membranes, Biochemistry, Cell Biology and Pathology,* H. P. Publishing, New York, 1975.

Chapter 4

GLOSSARY

The purpose of the glossary is to provide a quick guide to the reader in assessing laboratory data. The increased reliability of new methodologies and the more frequent application of biochemical, hematological, and pathological laboratory procedures in medicine has considerably increased the possibility of achieving a correct definite diagnosis in most cases. Laboratory investigations have also sharpened the accuracy of the physician's assessment by supplying exact data which confirms the signs and symptoms of the clinical impressions. Laboratory investigations can occur at early stages of the disease when corrective therapy is more likely to result in recovery. Thus for the physician the laboratory tests become an indispensable aid to his diagnosis in addition to a good physical examination. It is, however, essential that he know what tests to order and how to interpret the results. The glossary attempts to help him in this responsiblity.

Within the past few years there has been a tremendous growth in the use of the biochemical laboratory which has greatly advanced the basic knowledge of the clinician. The application of highly automated laboratory methods made it possible to obtain multiple analysis quickly and reliably. However, although the immense expansion yields a mass of data, sometimes some of these data are irrelevant to the disease and may even obscure the pertinent investigations.

In this chapter, we try to give guidelines to the student and the practising physician regarding the use of laboratory results in the diagnosis of disease. The various headings are organized in alphabetical order for convenience. These headings may be used to relate the initial history and physical examination to the clinical laboratory investigations and to show how to make a definite diagnosis earlier and more correctly. Human diseases are various, but we discuss only the most common clinical conditions and avoid unnecessary details which can be found in many textbooks of clinical biochemistry and medicine.

Abdominal Tumor — Intraabdominal tumors may produce albuminuria, probably by pressure upon the renal veins, causing renal congestion.

Abetalipoproteinemia — Lack of β-lipoprotein in the blood, which leads to a series of deficiencies in cholesterol transport. Accordingly, cholesterol accumulates in tissues, giving rise to Tangier's disease. Investigation:

- Serum cholesterol — very low
- Lipoprotein electrophoresis shows essentially no β-lipoprotein bond

Achlorhydria — Absence of free hydrochloric acid in gastric juice samples during the digestive period. True achlorhydria most commonly occurs in pernicious anemia and severe anemias; chronic gastritis, gastric neuroses, and gastric carcinoma; oral sepsis; subacute degeneration of the cord; occasionally in adrenal insufficiency, hyperthyroidism, diabetes mellitus, pulmonary tuberculosis, generalized arteriosclerosis. False achlorhydria represents neutralization of free hydrochloric acid by duodenal regurgitation.

Acid-base balance — Servomechanism maintaining the blood pH between physiological limits (pH 7.35 to 7.45). The effectors of this regulation are the lungs and kidneys through adjusting the rate of respiration and the reabsorption of bicarbonate. If these organs cannot cope with the disturbance, acidosis or alkalosis develops.

Acid-base balance disturbances are connected with an impairment of homeostasis, representing acidosis or alkalosis. The causes of acidosis are excess hydrogen ion production in ketosis due to diabetes or tissue damage and starvation; absolute anoxia, or relative anoxia due to muscular exercise (physiological); increased catabolism due to starvation; and excessive intake of hydrogen (ammonium chloride toxication). Renal failure, renal tubular insufficiency and retention of carbon dioxide in pulmonary disease, and increase loss of intestinal bicarbonate also lead to acidosis. Dehydration, if extreme, is commonly associated with acidosis. Later stages of normal pregnancy are connected with compensated acidosis due to slight alkali deficit.

Alkalosis is relatively rare. The causes of this condition are ingestion of large amounts of bicarbonate, and excessive loss of potassium or loss of hydrogen ions from the stomach in pyloric stenosis. Investigations:

- Arterial blood gases (pH, pCO_2, and pO_2)
- Serum electrolytes (sodium, potassium, chloride, and bicarbonate)
- Blood urea nitrogen

Acidemia — Blood pH below 7.35.

α_1-Acid glycoprotein — This plasma fraction is also called orosomucoid or α_1-seromucoid. It is considered as an acute phase reactant, and plasma concentration is increased in inflammatory conditions, rheumatoid arthritis, and malignant neoplasias.

Acidosis — Reduction of the body alkali reserve causing acidemia. Acidosis is due to an excess of acid metabolites which are incompletely oxidized or poorly eliminated. The disturbance of the acid-base balance causes a shift of the blood pH below 7.35. Acute or chronic forms may be uncompensated or compensated. Types are metabolic acidosis, respiratory acidosis, and mixed form. When toxic symptoms develop, the condition is called acid intoxication.

Biochemical characteristics of uncompensated respiratory acidosis in the blood are a disproportionate increase in pCO_2 and CO_2 content and bicarbonate, and a decrease of pH. If fully compensated, CO_2 content is high but the increase of bicarbonate is proportionate and pH remains within normal ranges.

Acid phosphatase — This enzyme is widely distributed in various tissues; it is mainly present in lysosomes and becomes activated when lysosomes are broken up. This is useful in diagnosing carcinoma of the prostate or metastases in this organ, when serum acid phosphatase level is elevated. Acid phosphatase is also increased in hemopoietic disturbances and in reticuloendothelioses such as Gaucher's disease.

Acne — Chronic inflammatory condition of the sebaceous glands commonly involving the face, back, and chest. The lesions affect persons between the age of puberty and 30 years and are often associated with menstrual or gastrointestinal disorders. The primary lesion is comedo or blackhead.

Acrodermatitis enterohepatica — Disease of unknown origin with onset in early infancy. Syndrome is characterized by growth retardation, photophobic intermittent diarrhea, and psychic disturbances, in addition to a symmetrical rash.

Acromegaly — Chronic condition resulting from growth hormone-secreting tumors of the anterior pituitary. Characteristic features are increase in size of the viscera (splanchnomegaly), soft tissue, and bones (acromegaly) without increase in height, with the foot, hand, and face showing most change. Other symptoms include joint pain, swelling, shortness of breath, and muscle weakness. Metabolic disturbances are primarily related to a change in sugar tolerance; secondary changes affect endocrine organs. Increased growth hormone secretion alters carbohydrate metabolism and causes insulin resistance and reduction of glucose phosphorylation. Serum phosphate is increased; PBI and urinary steroid level are usually undisturbed. Investigations:

* Fasting blood sugar — increased
* 2-hr Postprandial blood sugar — increased
* Serum phosphate — increased
* Serum thyroxine and triiodothyronine binding — normal
* Thyrobinding protein — normal
* Urinary sugar — increased
* Urinary 17-hydroxycorticosteroids — normal
* 17-Ketosteroids — normal
* Aldosterone — normal
* Gonadotropins — normal
* Radiological examinations

Acute intermittent porphyria — Serious familial disease involving dermal and mental symptoms due to the accumulation of porphobilinogen and δ-amino-levulinic acid. Investigation:

* Urinary porphyrin screen (Watson-Sibworth reaction)

Acute renal failure — Excretion of water and electrolytes is blocked and subsequent body accumulation occurs. The concentration of potassium and chloride increases, and sodium usually decreases, due to the fact that the retention of water is proportionately greater than the retention of sodium. Many products of protein catabolism such as inorganic acids, phosphates, and sulfates are also apparent in greater amounts. These are reduced by plasma bicarbonate, leading to losses of these ions. In acute renal failure the damaged tubular cells cannot replace bicarbonate, and are unable to excrete hydrogen. Metabolic acidosis subsequently occurs. Acute renal failure may develop following transfusions and hemolytic reactions, shock, dehydration, glomerulonephritis, side effect of drugs, and heavy metals. Investigations:

* Serum sodium and bicarbonate — decreased
* Potassium, chloride — increased
* Blood volume — increased
* Urinary sodium, potassium — decreased
* Urine volume — decreased
* Urine pH — normal or alkaline

Addison's disease — Hypofunction of insufficiency of the adrenal cortex mostly due to autoimmune destruction of the cortical tissue. This may be secondary to hypofunction of the anterior pituitary, in which case lack of adrenocorticotropic stimulation causes atrophy of the adrenal cortex with depression of function. Tuberculosis or overwhelming infections also result in atrophy of the adrenal cortex and diminished secretion of glucocorticoids, mineralocorticoids, and androgens with a consequent decrease of plasma levels of these hormones. Lack of corticosteroids causes reduced intestinal glucose absorption and decreased synthesis. Investigations:

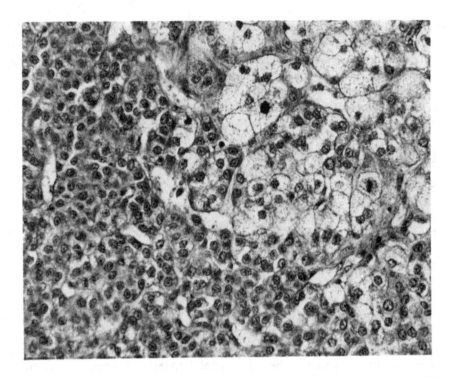

PLATE 1.   Adrenal cortical adenoma.

- Serum sodium and chloride — decreased
- Potassium — increased
- Blood urea nitrogen — may be increased (prerenal azotemia)
- Fasting blood sugar, 2-hr postprandial blood sugar — decreased
- Serum thyroxine and triiodothyronine — normal
- Thyrobinding protein — normal
- Urinary 17-hydroxycorticosteroids, 17-ketosteroids, aldosterone — decreased
- Urinary gonadotropins — normal

Adrenal cortex — Disorders of this endocrine gland are diagnosed by direct and indirect tests. These include the serial determinations of 24-hr urinary 17-ketosteroids and 17-hydroxycorticosteroids. In Addison's disease (hyposecretion) determinations are carried out before and after ACTH administration (ACTH stimulation test). In Cushing's syndrome (hypersecretion) the ACTH test and the cortisone suppression test are useful in differentiating hyperplasia or adenocortical adenoma from carcinoma. Hypo- or hypersecretion of aldosterone can be assessed by measuring the 24-hr urinary level. Aldosterone secretion can be measured by an indirect test using measurements of serum electrolytes and renin level and the waterload test. Light microscopy of adenocortical adenoma is shown (Plate 1).

Adrenal cortical insufficiency — Aldosterone and possible hydrocortisone secretion are insufficient, stimulating the conservation of salts by renal tubules. The urinary excretion of sodium, chloride, and sometimes bicarbonate is increased and the blood level of these electrolytes is reduced. With salt changes a substantial amount of water is also lost so the circulatory volume is decreased. Investigations:

- Serum sodium and chloride — decreased
- Potassium — increased

- Bicarbonate — normal or decreased
- Blood volume — decreased
- Urinary sodium — increased
- Urinary potassium - normal or decreased
- Urine volume — normal or decreased
- Urine pH — normal or decreased
- Urine pH — normal or alkaline

Agammaglobulinemia — Primary disease is due to a congenital deficiency of γ-globulin, resulting in very low immunoglobulin levels; characterized by the inability to respond to artificial immunization by circulating antibody production. It is more common in children; occurrence in adults suggests that the condition may be acquired. Bacterial infections are recurrent; prolonged antibiotic therapy fails to produce protection; isohemaglutinins are absent.

Albumin — This is responsible for osmoregulation and transport of ions, pigments, drugs, and metabolites in the blood. Abnormalities are mainly connected with reduced synthesis, excessive catabolism, losses, or nonspecific acute conditions. Reduced synthesis occurs in chronic liver disease, severe anemias, cachectic states, malabsorption, and malnutrition. Losses occur in protein-losing states such as in nephrotic syndrome, nephrosis enteropathy, in exudation such as extensive burns, or gastrointestinal losses. Excessive breakdown occurs in uncontrolled diabetes mellitus and severe thyrotoxicosis. Nonspecific causes include chronic infections, colds, boils, and some other minor illnesses. During pregnancy there is a steady decrease. Subnormal values are connected with toxemia of pregnancy, particularly eclampsia. There are two rare congenital abnormalities (1) bisalbuminemia with two albumin types with no apparent clinical consequences and (2) analbuminemia with deficient synthesis.

Alcoholism — Chronic alcoholism presents many symptoms such as hepatomegaly, splenomegaly, bilateral parotid gland enlargement, heart-rhythm disturbances, hypertension, anxiety, depression, gastritis, hepatitis, cirrhosis, anorrhexia, peripheral neuropathy, general malnutrition, and withdrawal symptoms upon abstinence. Light microscopy of the liver cell in chronic alcoholism shows the accumulation of neutral fat (Plate 2). Investigations:

- Hyperuricemia
- Hypophosphatemia
- Hypomagnesemia
- Hypocalcemia
- Serum aspartate transaminase—elevated

Alkalemia — Blood pH above 7.45.

Alkaline phosphatase — This enzyme is widely distributed in various tissues. Intestinal mucosa has the greatest activity followed by kidney, bone, thyroid gland, and liver. They are required for the hydrolysis of organic phosphate esters, particularly in mucosal absorptions and digestion. Regenerating and proliferating hepatic tissue contains high levels of this enzyme. It also has an essential role in osteoblastic tissue. Increased alkaline phosphatase activity is associated with varying degrees of metabolic activity of the osteoblasts.

In the normal state alkaline phosphatase is present in the blood. Determination of alkaline phosphatase levels is important in differentiating hepatocellular and obstructive jaundice. It is always elevated in obstructive jaundice, but infrequently raised in hepatocellular jaundice. It is markedly elevated in intrahepatic cholestasis, large duct biliary obstruction, and infil-

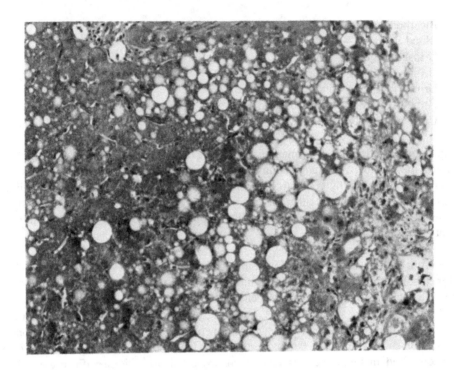

PLATE 2.   Alcoholic liver disease. This is a microscopic image of the liver in chronic alcoholism. The clear vacuoles represent the accumulation of neutral fat. There is also an increased deposition of fibrous tissue which eventually leads to cirrhosis.

trative lesions of the liver. It is raised in increased osteoblastic activity of the bones. Alkaline phosphatase is also useful in distinguishing various bone diseases and hyperparathyroidism in combination with serum calcium and phosphorus determinations and X-ray examinations.

Slightly elevated values have reported in osteomalacia, metastatic carcinoma involving bone, healing fractures, and Gaucher's disease with bone resorption. Alkaline phosphatase is normal in various forms of osteoporosis, osteomyelitis, tumors not involving bone, cretinism, and most cases of chronic arthritis. Physiological increase occurs in late pregnancy, childhood, and particularly in early puberty.

Alkalosis — In this condition bicarbonate content of the blood is relatively high and there is a tendency toward alkalemia. Causes include the ingestion of large amounts of sodium bicarbonate, persistent vomiting with loss of hydrochloric acid, or forced breathing with the reduction of carbon dioxide from the blood. Disturbance of the acid-balance shifts the blood pH above 7.45, acutely or chronically. It may be uncompensated or compensated. Types are metabolic alkalosis, respiratory alkalosis, and mixed forms. Investigation:

* Depending on the cause, blood gases and electrolytes

Alkaptonuria — Inborn error of metabolism resulting in urinary excretion of homogentisic acid due to lack of an enzyme essential for oxidation. This leads to a darkening of urine on standing.

Allergy — Altered responsiveness due to sensitivity to environmental factors. This may be manifested in upper respiratory diseases, asthma, dermatitis, and purpura.

Alopecia — Lack of hair. Most common example is the frontal baldness in adult males.

Amino acid — Blood amino acid content normally rises following the ingestion of proteins and decreases after the administration of insulin, growth hormone, or testosterone, probably due to a stimulation of protein synthesis. Some conditions cause abnormal increase due to impairment of metabolism or elimination. Hepatic insufficiency, effect of hepatotoxic compounds, and eclampsia inhibit urea formation; advanced nephritis and urinary obstruction or suppression reduce elimination. Abnormal amounts of amino acids are excreted in the urine in aminoaciduria. This may occur in the absence of any apparent disturbance of metabolism.

Amino-acidopathies — Inherited disorders, inborn errors of metabolism, related to abnormal amino acid accumulation in blood or tissues and abnormal urinary excretion. Amino-acidopathies are hereditary but some disorders are acquired. Some disorders affect catabolism, others membrane transport of amino acid. The most important amino-acidopathies together with enzyme affected and disease condition are given in the following list:

1. Classical phenylketonuria — phenylalanine 4-monooxygenase
2. Phenylketonuria — dihydropteridine reductase
3. Tyrosinemia — 4-hydroxyphenyl-pyruvate dioxygenase
4. Classical histidinemia — L-histidine ammonia-lyase
5. Branched-chain hyperaminoacidemia (classical maple syrup disease) — branched-chain α-ketoacid oxidase
6. Cystathionuria — cystathionine-γ/lyase
7. Homocysteinuria — cystathionine-β-synthase
8. Hartnup disease — neutral amino acids
9. Alkaptonuria — homogentisic acid metabolism
10. Wilson's disease — unknown, secondary effects on cytochrome oxidase

Some other conditions may affect plasma amino acid levels such as:

1. Prolonged fasting (glycine, threonine — increased; alanine — decreased)
2. Protein-calorie malnutrition (glycine, proline, tyrosine — increased; valine, leucine, isoleucine, typotophan — decreased), with the severity being related to the degree of malnutrition
3. Obesity (valine, leucine, isoleucine, phenylalanine, tyrosine — increased; glycine — decreased); this reflects insulin insensitivity
4. Hepatitis (methionine, tyrosine — increased); this reflects the severity of liver disease

Aminoaciduria — Excretion of amino acids in urine. In neonates this may be physiological, disappearing within 4 to 6 weeks. At later ages, the presence of amino acids in the urine is pathological. Abnormal amounts are excreted in acute hepatic necrosis, in Wilson's disease, eclampsia, poisoning with carbon tetrachloride or chloroform, wasting diseases, protracted fever (tuberculosis, typhoid), or malignancy due to extensive tissue autolysis. Massive tyrosyluria occurs in acute yellow atrophy of the liver, and in degenerating lung tumors. Increasing tyrosyluria indicates a rapidly progressive degenerative process. Aminoaciduria may be restricted to one or several amino acids. In the latter case sometimes glucosuria, ketonuria, and phosphaturia are accompanying symptoms (Fanconi syndrome). Great excess of a single amino acid reflects specific aminoacidopathy. Investigation:

• Urinary amino acid chromatography

Aminotransferases — The activity of these serum enzymes may be connected with many disorders. Increased values of aspartate aminotransferase are usually connected with recent or chronic damage to liver cells, myocardium, skeletal muscle, or hemolysis. Very high values may indicate acute viral or drug-induced hepatitis; moderately increased values may represent cholestasis and hepatitis in most instances. Increase of alanine aminotransferase is more liver specific than aspartate aminotransferase. Usually alanine aminotransferase is greater than aspartate aminotransferase in acute hepatitis; the changes are opposite in alcoholic liver damage.

Amylase — Serum amylase originates from the pancreas, liver, salivary gland, fallopian tubes, and probably from striated muscles. Disorders with abnormal blood levels usually diagnose pancreatic diseases including acute and chronic pancreatitis, pancreatic carcinoma, choledocholithiasis, intestinal obstruction, and duodenal ulcer. Other causes include renal failure, mumps, and infective parotitis related to other enzyme sources. Acute pancreatitis is connected with the inflammation of the pancreas and rupture of cell membranes resulting in increased serum amylase. In chronic pancreatitis serum amylase is only increased in relapse. In choledocholithiasis, gallstone obstruction of the pancreatic duct causes a backup of amylase into the blood. Acute inflammation of the gland may increase further serum and urinary amylase levels. In pancreatic carcinoma, the pancreatic duct is obstructed causing a backup of amylase with moderate elevation. Intestinal obstruction may extend to the duodenum and secondarily cause pancreatic duct obstruction, backing amylase into the blood. Inflamed and necrotic duodenum may also be directly responsible for the increased blood level. Duodenal ulcer may perforate the pancreas resulting in secondary pancreatitis with similar laboratory findings as in primary pancreatitis. In mumps and parotitis infection, salivary amylase leaks into the blood. Renal failure is connected with an inadequate urinary elimination of amylase.

Amyloidosis — Rare disorder characterized by the intracellular accumulation of amyloid in virtually all tissues. Amyloid is a fibrillar protein, which gives the Congo red reaction. Amyloid disease enhances blood urea nitrogen by interference with urinary excretion.

Anaphylaxis — Acute hypersensitivity reaction triggered by drugs, animal serum, insect bites and allergenic agents resulting in bronchospasm, massive edema, collapse, hypotension, wheezing, and cyanosis.

Anastomosis — Linkage of blood vessels through holes and ducts. This intercommunication provides an increased growth of blood supply in the case of an arteriovenous anastomosis when a modified vessel connects an artery with a vein without the intervention of capillaries.

Anemia — Decrease of the hemoglobin level below 12 g/dℓ in females and 14 g/dℓ in males. Types include normocytic, microcytic, and macrocytic anemia. Many enzyme defects in erythrocytes are responsible for hereditary, congenital, and hemolytic anemia such as hexokinase, glucosephosphate isomerase, phosphofructokinase, diphosphofructoaldolase, glyceraldehyde phosphate dehydrogenase, pyruvate kinase, glyceroaldehyde phosphate dehydrogenase, pyruvate kinase, glyceroaldehyde phosphate dehydrogenase, triosephosphate isomerase, glucose 6-phosphate dehydrogenase, and 6-phosphogluconate dehydrogenase.

Anion gap — The quantity of unmeasured anions, mostly proteins. Normal range is often quoted as 4 to 12 meq/ℓ but this is likely to be wider. Increased in certain types of metabolic acidosis. May be low in multiple myeloma.

Anorexia — Loss of appetite. Anorexia nervosa represents a hysterical aversion to food or a psychological effect causing food restriction which may lead to serious malnutrition.

Anterior pituitary insufficiency — Secondary adrenal insufficiency, hypothyroidism, malnutrition, or reduced growth hormone cause this disease associated with low blood sugar. Lack of glucocorticoids causes deficient gluconeogenesis and increased insulin effect. Investigations:

• Fasting blood sugar, 2-hr postprandial blood sugar — decreased
• Urinary 17-ketosteroids and 17-hydroxysteroids before and after ACTH administration

Antibodies — Specialized proteins secreted by immunologically competent cells in response to external immunological stimuli or, rarely, internal ones.

Antigen — Foreign proteins entering the circulation usually evoke antibody formation. Antigen could also be produced by the body in various autoimmune diseases when it is termed autoantigen.

α_1-**Antichymotrypsin** — An acute phase reactant chymotrypsin inhibitor. It is increased in inflammation, rheumatoid arthritis, and malignant neoplasias.

α_1-**Antitrypsin** — This plasma fraction exerts a proteinase action and inhibits trypsin activity. It is an acute phase reactant, increased in inflammation, rheumatoid arthritis, malignant neoplasias and significantly reduced in neonatal cholestasis, cirrhosis, liver cell carcinoma, and emphysema connected with α_1-antitrypsin deficiency.

Anuria — Failure of the kidneys to excrete urine as in acute renal failure. May be due to prerenal, renal, or postrenal (lower urinary tract) causes.

Arrhythmia — Disturbance of the regular cardiac rhythm.

Asphyxia — Lack of oxygen that may lead to cellular or even organic death.

Asthma — Acute or chronic respiratory disease involving attacks of bronchospasm and bronchial edema with increased mucus. It is often triggered by allergic factors. Investigation:

• Arterial blood pO_2 — decreased

Atrophy — Acquired reduction in size of an organ or cell which had reached mature size. Hypoplastic organs may exhibit atrophy. It may be physiologic or pathologic. In the latter case, wasting of an organ may be due to inactivity, loss of innervation, or degenerative disease.

Azotemia — Increase of urea in the circulation due to deficient renal function. Investigations:

• Blood urea nitrogen — increased
• Plasma creatinine — increased
• Creatinine clearance — decreased

Bacterial infections — In both acute and chronic forms the hypertrophied lymph cells produce increased amounts of α- and γ-globulins, but albumin synthesis is decreased.

Occasionally β-globulin is also increased. Parasitic, protozoal, and viral diseases may be associated with similar changes. The increase of globulin levels may be more intense. Investigations:

- Albumin — decreased or normal
- α-Globulin — increased
- γ-Globulin — increased

Bence-Jones protein — This occurs in the urine of some patients with macroglobulinemia and myeloma. The proteins precipitate at pH 7.9 and redissolve on further heating. Actually, the free light chains of immunoglobulins precipitate at 50° and redissolved at 60 to 65°. Investigations:

- Electrophoresis, where Bence-Jones protein appears as a sharp band positioned in the β or γ-globulin zone
- Immunoelectrophoresis, where an immunoglobulin plus a light chain precipitin area appears distorted

Beriberi — Condition due to thiamine deficiency. It is endemic in those living mainly on a polished rice diet; sporadic when the diet is limited in thiamine. Manifestations depend upon the severity and duration of vitamin lack. These are multiple neuritis, general weakness, paralysis, progressive edema, mental deterioration, and finally heart failure.

Biliary obstruction — This disorder represents blocking of the common bile duct and the bile canaliculi. Investigations:

- Increased unconjugated bilirubin and initially elevated alkaline phosphatase; later serum alanine and aspartate aminotransferase are enhanced
- Increased γ-glutamyl transpeptidase and 5′- nucleotidase suggest obstructive jaundice unless tumor is present

Bile acids — Increased bile acid content of serum indicates hepatobiliary dysfunction. Serum trihydroxy to dihydroxy acid ratio determined in jaundiced serum reflects various liver diseases; if greater than 1, it represents hepatocellular disease, if less than 1, cholestasis. Usually pathological bile acid values are found in all forms of hepatitis, cirrhosis, fatty changes of the liver, acute and chronic cholestasis, acute and chronic obstruction of the bile ducts, in idiopathic hyperlipemia and hypercholesterolemia.

Bilirubin — This compound is a by-product of hemoglobin metabolism, mainly derived from the red blood cells (approximately 70-75%) and from myoglobin, cytochromes, and catalases (25-30%). Daily physiological erythrocyte destruction amounts to 7 to 8 g hemoglobin. The cells of the reticuloendothelial system such as liver, spleen, and bone marrow represent the sites of hemoglobin metabolism. In the degradation process, first the globin part is cleared, then the protein-deficient heme is converted to protoporphyrin by the loss of iron. In a further process the ring structure is opened to form bilirubin. This is released from the endothelial cells into the blood stream. Accumulation of bilirubin in abnormal quantities indicates jaundice. Bilirubin is highly insoluble in aqueous media; its solubility in the plasma and transport depend on protein binding to extracellular carrier, chiefly albumin, or protein carrier within hepatocytes, such as ligandin. Plasma bilirubin turnover normally ranges within a narrow limit. Abnormalities are connected with an increase in both unconjugated and conjugated bilirubins. Unconjugated hyperbilirubinemia due to increased production occurs in hemolysis, in increased ineffective erythropoiesis associated with vitamin B_{12} deficiency, and in sideroblastic anemia, erythropoietic porphyria, thalassemia major, or lead poisoning. Unconjugated hyperbilirubinemia due to reduced hepatic clearance occurs in Gilbert's syndrome, in congenital nonhemolytic jaundice type I, and hemolytic jaundice type I (Crigler-Najjar syndrome) and type II (Arias syndrome). Conjugated hyperbilirubinemia is due to abnormal excretory transport connected with Dubin-Johnson syndrome and Rotor syndrome. This condition also occurs in cholestasis and in biliary tract obstruction. Urine is normally negative. Bilirubinuria indicates conjugated bilirubin in serum. It is positive early in viral hepatitis.

Biologically active amines — Include catecholamines (dopamine, epinephrine, and norepinephrine), serotonin and histamine, and important metabolites (homovanillic acid and vanilmandelic acid). Catecholamines and metabolites are increased in hypertension, pheochromocytoma, and neuroblastoma and reduced in collagen diseases. Variations in blood levels reflect catechol-*O*-methyltransferase and monoamine oxidase (isoenzymes) activities, which may be involved in the manifestation of hypertension and related to mental disease. Serotonin level in the blood increases moderately in some diseases of the digestive system (chronic pancreatitis, celiac disease, Crohn's disease), nervous system, and in chronic arteritis. It is highly increased in carcinoid tumors and reduced in rheumatoid arthritis and phenylketonuria. Histamine increases in mastocytoma, urticaria pigmentosa, and other acute urticarias, and in dominantly eosinophil inflammatory pulmonary infiltrates. It is highly raised in pregnancy in preeclampsia.

Blood — Presence of blood in various body fluids usually represents bleeding. The causes of pathological bleeding are coagulation disorder, trauma, inflammation, vascular diseases, and neoplasm. Blood in the urine of females is often due to menstruation. Blood in the urine indicates kidney diseases; infections occurring in glomerulonephritis, pyelonephritis and renal tuberculosis; vascular disorders such as renal vein thrombosis or embolic glomerulitis; kidney stones; neoplasms; or cystitis. Blood in the gastic juice is found in peptic ulcer, gastritis, and esophageal varicies. It is less frequent in stomach carcinoma. Blood in the sputum occurs in carcinoma, tuberculosis, bronchiectasis, pulmonary infarction, and congestive heart failure. Viral influenza and *Klebsiella pneumoniae* also causes bloody sputum.

Blood is found in the feces from bleeding of the lower gastrointestinal tract, due most commonly to neoplasms. Dysentery and ulcerative colitis also cause bloody stools associated with inflammatory conditions. Rectal bleeding is connected with hemorrhoids, anal fissures, and carcinomas. In children, intestinal obstruction causes blood in the stool. Blood in the spinal fluid occurs most commonly in vascular disorders such as ruptured aneurisms, cerebral arteriosclerosis, and arteriovenous anomalies. Trauma of the brain also results in blood in cerebrospinal fluid.

Peritoneal, pleural, and synovial fluids contain blood indicating previous trauma or neoplasm. In women, blood may be present in the peritoneal fluid due to ruptured uterus or ruptured ectopic pregnancy.

Blood gases — In respiratory disturbances the development of acidosis is connected with blood gas levels. The partial pressure of O_2 (pO_2) and CO_2 (pCO_2) are particularly essential. In pulmonary disease due to mechanical or neurologic defects in respiratory movement, or obstruction of large or small airways, blood supply in the alveoli may be normal, but abnormalities in pO_2 and pCO_2 are apparent. Arterial pO_2 is low; pCO_2 is low or normal in pulmonary edema, fibrosis or infiltration, pneumonia, and collapse of the lung. Low arterial pO_2 with high pCO_2 may occur in the same conditions and in severe asthma, chronic bronchitis, emphysema, poliomyelitis, and lesions of the central nervous system due to neurological impairment of respiratory drive, as well as in ankylosing spondylitis, due to impairment of movement of the respiratory cage.

Blood glucose — The end products of dietary carbohydrates are glucose, fructose, and galactose. Fructose and galactose are converted to glucose in the liver. Glucose enters cells and is converted to glucose 6-phosphate. This step is controlled by insulin. It is important to maintain an adequately high blood glucose concentration for cerebral function; several hormones can raise it. Hyperglycemia is not essential, since insulin reduces blood glucose; thus, the control in a simplified manner is connected with several hormones. Blood glucose levels fall by the action of insulin. The secretion of insulin is regulated by the blood glucose level. An increased blood glucose level is brought about by growth hormone, glucocorticoids, epinephrine, and glucagon.

Some diseases are connected with failure in the reduction of blood glucose level caused by insulin deficiency as in diabetes mellitus, excess growth hormone in acromegaly, hormones antagonizing insulin as corticosteroids in stress or Cushing's syndrome, or epinephrine in stress and pheochromocytoma. The presence of glucose in the urine indicates hyperglycemia.

Branched chain aminoaciduria — Also called maple syrup urine disease. The underlying biochemical lesion is a defect in oxidative decarboxylation of keto derivatives of branched-chain amino acids. Accordingly, branched-chain keto amino acids accumulate and are excreted in the urine, having the distinctive odor of maple syrup. Investigation:

- Thin-layer or gas chromatography demonstrates the presence of branched chain keto acids

Cachexia — Weakness and emaciation associated with extreme weight loss caused by some debilitating disease such as tuberculosis, syphilis, pituitary destruction, or chronic mercury poisoning.

Calcium metabolism — Serum calcium phosphate and alkaline phosphatase levels are affected in a parallel fashion. Many factors regulate the concentration of serum calcium and phosphorus, for example, parathyroid hormone, calcitonin, and vitamin D. Parathyroid hormone is the most important regulator. A decrease in serum calcium stimulates parathormone secretion and as a consequence calcium is mobilized from the bones by resorptions. At the same time the absorption of calcium from the intestines and renal tubules is enhanced, the secretion of phosphate by the tubules is increased and its reabsorption is inhibited. An increase in serum calcium stimulates calcitonin secretion from the parathyroid or thyroid gland and serum calcium returns to normal. Vitamin D influences serum calcium by enhancing intestinal absorption. In the absence of this vitamin, serum calcium may decrease, and parathyroid activity is consequently stimulated. The total concentration of calcium in the serum is directly associated with the concentration of serum protein, particularly with that of albumin. The ionizable calcium varies with pH. Alkalosis causes a decrease, acidosis an increase.

A disorder of calcium metabolism may be connected with bone disease, renal calculi, diabetes, convulsions, or tetany. Polyuria, polydipsia, fatigue, or malaise may also suggest calcium anomaly. Increased calcium excretion may be a sign of renal tubular acidosis or of hyperparathyroidism. Decreased excretion may suggest malabsorption, hypoparathyroidism, or hypovitaminosis D.

Carbon dioxide tension (pCO_2) — Partial pressure of physically dissolved CO_2 in blood measured by a special CO_2 electrode.

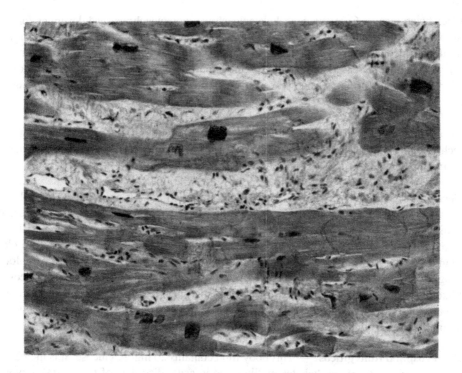

PLATE 3.   Image of a myocardium in the clinical condition recognized as congestive heart failure. There is a dissociation of fibers and increased intercellular space.

Carcinomatosis — Generalized dissemination of cancer tissue in widespread locations of the body.

Cardiac failure — Connected with a marked increase of serum aminotransferases, due to congestion, lesion, or necrosis of the heart muscle. Increase of lactic dehydrogenase plays less significance. In congestive heart failure, muscle fibers are dissociated and the intercellular space is increased (Plate 3). Investigation:

- Aminotransferases — increased
- Lactic dehydrogenase — normal or increased

Cardiomyopathy — Degenerative disease of the heart muscle.

Casts — Found in the urine or sputum. Bronchial casts are present in bronchitis, resolving pneumonia, and cardiovascular diseases. The appearance of bronchial cast is mucous, hemorrhagic, or fibrinous. In bronchial asthma, mucus and fibrils are apparent. Urinary casts originating from renal tubules are important in diagnosis, since in the normal urine only occasional hyaline casts are present. White cell casts indicate acute or chronic pyelonephritis. Red cell casts occurs in glomerular injury resulting from glomerulonephritis, or embolic glomerulitis due to subacute bacterial endocarditis or from mural thrombosis of the heart or various collagenous diseases. Heme pigment casts may be formed in acute tubular necrosis. Epithelial casts are present following toxic or metabolic acute tubular necrosis. These casts may show a waxy appearance due to degeneration. Some casts seen in the nephrotic syndrome containing lipid droplets.

Catecholamines — Compounds include epinephrine, norepinephrine, and dopamine formed from phenylalanine and tyrosine in the brain, postganglionic neurons, sympathetic ganglia, and adrenal medulla secreted into the circulation, activated by the splanchnic nerves, by stress and stimuli. Increased amounts of catecholamines are secreted in pheochromocytoma, resulting in elevated urinary excretion of catecholamines and their metabolite vanilmandelic acid.

Cerebellar ataxia — Disturbance of motor coordination manifested in abnormal gait and speech due to various degenerative, toxic, nutritional, hemorrhagic, and infectious causes affecting the cerebellum.

Cerebrospinal fluid — Examinations should be carried out for appearance, for volume by measuring pressure, and for clotting. The spinal fluid is normally clear and colorless. It is turbid in infections and contains blood in cerebral hemorrhage. Yellow color may indicate lysis of erythrocytes from previous bleeding, tumor, or jaundice. The formation of a clot on standing in a clear spinal fluid may be due to a tumor of the spinal cord. In tuberculous meningitis, poliomyelitis, and syphilitic meningitis, spinal fluid is turbid and produces a weblike clot. In acute bacterial meningitis, the discharging fluid forms a coarse fibrin clot. In subarachnoid hemorrhage the fluid is colored but usually a clot is not formed due to defibrination in the subarachnoid space.

Cerebrospinal fluid glucose — This body fluid contains 20 mg/dℓ glucose. It is increased in primary and symptomatic diabetes mellitus and in acute disorders of the central nervous system. Decreased levels are significant in bacterial of tuberculous meningitis. In hypoglycemia the level of cerebrospinal fluid sugar is also low; glucose determinations, therefore, should be carried out concomitantly in blood and cerebrospinal fluid.

Cerebrospinal fluid volume — The normal volume of cerebrospinal fluid is 60 to 120 mℓ, measured indirectly by the spinal fluid pressure, which is normally between 75 to 200 mm. The volume (pressure) is increased in infections, trauma, and neoplasms of the brain and meninges. Vascular diseases leading to hemorrhage also increase pressure. Some conditions causing increased venous pressure will elevate the pressure of the spinal fluid, such as congestive heart failure and pulmonary emphysema. Decreased pressure is connected with subarachnoid block above the site of the needle.

Ceruloplasmin — This plasma fraction is a copper-binding protein and exerts oxidase activity. It is an acute phase reactant. The plasma concentration of ceruloplasmin is reduced in Wilson's disease (hepatolenticular degeneration) and in severe hepatocellular disease. It shows a physiological increase during pregnancy, and is increased in the plasma of women taking contraceptive hormones. Biliary duct obstruction and estrogen therapy also cause an increase.

Choledocholithiasis — The entrance of gallstones into the common bile duct causes extrahepatic cholestasis resulting in an increase of direct bilirubin in the blood and urine. Although urinary and fecal urobilinogen are decreased, they are still present, since the obstruction is rarely complete. If there is an associated cholangitis urinary urobilinogen may be elevated. Intermittent inflammation and edema of the bile duct and occasional passing of the stones may cause variations in the obstruction and consequent bilirubin values. If the obstruction is due to neoplasms the increase of bilirubin is progressive and remains on a steady level. Investigations:

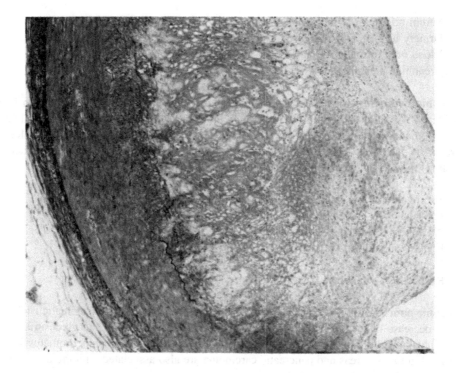

PLATE 4. Atherosclerosis of a coronary artery. The deposition of fatty substances including cholesterol thickens the vessel wall and leads to impairment of the blood flow.

- Serum bilirubin (indirect) — normal
- Serum bilirubin (direct) — intermittently increased
- Urinary bilirubin — increased
- Urinary urobilinogen — decreased
- Fecal urobilinogen — decreased

Cholestasis — Results from a blockage of the bile canaliculi or ducts due to stone, external pressure, allergic reaction to drugs (e.g., phenothiazine), or inflammation. It could be intra- or extrahepatic.

Cholesterol — This is increased in the plasma in intrahepatic cholestasis and large duct biliary obstruction. It is also raised in association with diabetes mellitus, hypothyroidism, nephrotic syndrome, and certain hyperlipidemias. In atherosclerosis fatty substances including cholesterol accumulate in the coronary artery (Plate 4). Cholesterol is reduced in severe hepatocellular disease. Cholesterol ester represents about two thirds to three quarters of total cholesterol. Ester: free cholesterol ratio is reduced in cholestasis due to an increase of free cholesterol, and in hepatocellular disease due to decrease of the ester.

Cholinesterase — This is synthesized in the liver and present in high concentration in the plasma. It is reduced in hepatitis, cirrhosis, and liver metastasis. Low levels are also found in malnutrition. Cholinesterase is elevated in nephrosis and obesity.

Chronic glomerulonephritis — Tubular secretion and renal clearance of uric acid are diminished; consequently, the serum uric acid level is enhanced. Toxic nephritis, diabetic glomerulosclerosis, and collagen disease produce similar effects. Some diuretics may cause uric acid retention, probably by inhibiting tubular secretion. Investigations:

- Serum uric acid — increased
- Urinary uric acid — decreased
- Blood urea nitrogen — increased
- Creatinine clearance

Chronic nephritis — The renal clearance of phosphate is decreased resulting in high blood levels. Most of this phosphate is excreted in the stool. Intestinal absorption of calcium is impaired and more dietary calcium goes into the stool. The reduction of blood calcium activates the parathyroid gland and induces bone resorption. Secondary osteoblastic reaction may develop with increased alkaline phosphatase. Investigations:

- Serum calcium — decreased
- Phosphates — increased
- Alkaline phosphatase — increased
- Urinary calcium — decreased
- Renal function tests
- Response to vitamin D

Chronic pancreatitis — Chronic inflammation and destruction of the pancreatic islet cell cause a decrease of insulin production and subsequent hyperglycemia and glycosuria. The hepatic uptake and peripheral utilization of glucose are decreased. In hemochromatosis iron deposition in the pancreas and pancreatic carcinoma are also associated with the destruction of the islet cells and hyperglycemia.

Chronic renal failure — The regulatory capability of the kidney to maintain electrolyte levels and acid-base equilibrium is impaired. There is often destruction of a large part of glomeruli. Tubular damage impairs water and selective electrolyte reabsorption. Lack of response of the tubules to aldosterone causes failure of salt retention and that of antidiuretic hormone causes failure of water retention. Consequently water, sodium, and potassium are not conserved and their plasma level decreases. Excretion of hydrogen and replacement of bicarbonate in the plasma are inhibited resulting in metabolic acidosis. Sulfate and phosphate excretion is also blocked, contributing to acidosis. Pyelonephritis, glomerulonephritis, nephrosclerosis, and obstructive uropathy cause this condition. Investigations:

- Serum sodium — decreased
- Potassium — normal or decreased
- Chloride — normal or decreased
- Bicarbonate — decreased
- Blood urea nitrogen — increased
- Serum creatinine — increased
- Blood volume — variable
- Urinary sodium, potassium — increased
- Urine volume — variable (usually increased)
- Urine pH — increased

Cirrhosis — In this condition the functional hepatic cells are replaced by fibrotic tissue which destroys the normal tubular architecture of the liver. The microscopic image of postnecrotic cirrhosis is seen on Plate 5. Investigations:

- Increased liver enzymes including serum alanine and aspartate aminotransferase, alkaline phosphatase, and γ-glutamyl transpeptidase

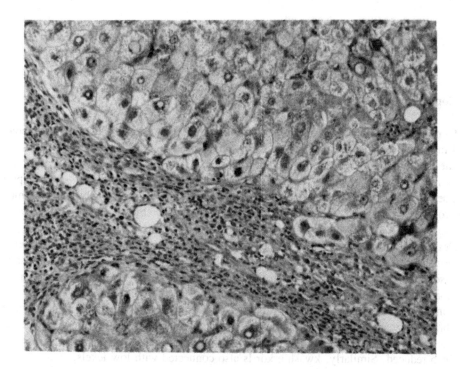

PLATE 5.   Aspect of the liver in postnecrotic cirrhosis. Although the pathogenesis of this condition is not firmly known, it seems that viral hepatitis or hepatotoxins play a very strong role.

- Bilirubin — very highly elevated
- Albumin — decreased
- γ-Globulin — reduced

Coma — This condition indicates an unconscious state of varying depth. Manifestations vary with the cause; for example, serum glucose is increased in diabetic coma, decreased in hypoglycemic coma; ketone bodies are increased in diabetes; blood pH and gases are abnormal in poisonings; osmolarity is increased in methyl alcohol intoxication; serum bilirubin and serum enzymes are enhanced in hepatic coma.

Congestive heart failure — The decreased cardiac output is connected with a significant decrease of the renal blood flow. The decreased glomerular filtration rate results in enhanced blood urea and creatinine. Thrombosis of the bilateral renal artery or aneurysm of the aorta produce similar changes. Circulatory insufficiency causes renal retention of water and sodium and, subsequently, increased total body water and sodium. The retention of water may be secondary to sodium and associated with increased secretion of antidiuretic hormone. This may be responsible for the fact that in late stages more water is retained and hyponatremia may develop. Investigations:

- Blood urea nitrogen — increased
- Creatinine — increased
- Serum sodium — normal or decreased
- Potassium — normal
- Chloride — decreased
- Bicarbonate — normal
- Blood volume — increased

- Urinary sodium — decreased
- Urinary potassium — normal
- Urine volume — decreased
- Urine pH — normal

Conn's disease — see Hyperaldosteronism.

Copper metabolism — Copper is an essential trace element. It is important in hemopoiesis and in the production and function of some transport proteins and enzymes. It is absorbed in the upper part of the jejunum and transported in the plasma as ceruloplasmin (an α_2 globulin). It is also bound to albumin and present in the erythrocytes as erythrocuprein. Copper is stored in the liver, bones, and muscles. Regulation is probably under hormonal control. Estrogens and androgens increase serum copper by increasing ceruloplasmin concentration.

Deviations from normal serum copper level (70 to 130 μg %) indicate various diseases. An increase occurs in cases of acute and chronic infections, myocardial infarction, rheumatic diseases, malignant tumors, and in pregnancy; a decrease occurs in cases of nephrosis and Wilson's disease (hepatolenticular degeneration).

Abnormal amounts of copper accumulate in Wilson's disease in the liver, kidney, cornea, cerebral cortex, and basal ganglia. Plasma-bound copper and ceruloplasmin are reduced and free copper is excreted in the urine. In nephrotic syndrome, plasma copper and ceruloplasmin are also reduced. Similarly, kwashiorkor is also connected with low levels.

Creatinine — Formed from creatine phosphate originating from muscular tissue. It is exclusively excreted by renal glomeruli and not reabsorbed by tubules. Creatinine determination in plasma and urine to calculate clearance constitutes the most reliable method for establishing renal function.

Crohn's disease — Chronic inflammation of the terminal ileum affecting the absorption of many nutrients. This causes malabsorption. A case of Crohn's disease is presented in Plate 6. Investigations:

- Serum albumin, cholesterol, triglycerides — decreased
- Fecal fat content — increased
- Serum carotene, vitamin B_{12} — decreased

These findings are not specific but provide support to a histological diagnosis.

Cryoglobulinemia — Occurs in several diseases. Cryoglobulins are plasma proteins which precipitate below body temperature in vitro or in superficial capillaries on being exposed to cold. In the latter case ulceration of the skin or purpura occurs. Cryoglobulins are present in lymphomas, myelomatosis, collagen diseases, and in any disease connected with elevated γ-globulin, such as chronic infections and cirrhosis.

Crystals — Crystals or calculi may be present in urine, duodenal and synovial fluids, and sputum, and can be useful in diagnosis. In the urine, calcium phosphate, calcium oxalate, cystine or uric acid crystals indicate various types of kidney stones. In the duodenal fluid, calcium salt of bilirubin and cholesterol crystals suggest abnormalities in biliary canaliculi. In the synovial fluid, uric acid crystals can be used for the diagnosis of gout. Calcium pyrophosphate crystals indicate acute arthritis.

PLATE 6.   Crohn's disease. This condition represents a regional granulomatous enteritis of the colon.

Cushing's disease — Adrenal cortical hyperfunction due to a basophil adenoma of the interior pituitary. Differentiation from Cushing's syndrome is made through the dexamethasone suppression test. However, basophil tumors may also be suppressible. Investigations:

- Serum cortisol — increased
- Urinary free cortisol — increased
- Absence of the circadian rhythm in serum cortisol
- Mild hyperglycemia

Cushing's syndrome — Hyperplasia of the adrenal cortex, carcinoma, adenoma, or basophilic adenoma of the pituitary are connected with this syndrome. It is associated with increased gluconeogenesis and reduced peripheral glucose utilization. Glucosuria is present. The production of corticosteroids (cortisol) is excessive, which impairs the formation of new bone structures. Consequently, less calcium is deposited, resulting in osteoporosis. There is usually no change in blood calcium level, but consequently more calcium is excreted in the feces and urine. Cortisol also antagonizes the action of vitamin D on the intestinal mucosa and reduces calcium absorption. Investigations:

- Serum calcium, phosphates — normal
- Serum alkaline phosphatase — normal
- Serum thyroxine, triiodothyronine — normal
- Thyrobinding protein — normal
- Urinary calcium — normal
- Fasting blood sugar, 2-hr postprandial blood sugar — increased
- Urinary sugar — increased
- Urinary 17-hydroxysteroids, 17-ketosteroids — increased
- Urinary aldosterone, gonadotropin — normal

Cystathionuria — Excretion of cystathionine in urine. The metabolic defect involves a defective cystathionase apoenzyme.

Cystic fibrosis — Blocking of the ducts or exocrine glands through increased viscosity of the secreted mucus. It leads most commonly to respiratory and pancreatic insufficiency. Abnormalities in these target organs are presented, an electron microscope image showing airway obstruction (Plate 7), and a micrograph showing progressive loss of acinar tissue in the pancreatic cells (Plate 8). Investigations:

- Sweat chloride test; its chloride content is greater than 60 meq/ℓ
- Pancreatic exocrine enzymes decreased
- Most commonly trypsin is increased

Dehydration — Due to continuing loss of water through kidney, lung, and skin, the reduced water intake causes a depletion of intracellular and extracellular stores of body water. Blood electrolyte levels are markedly increased, particularly sodium and chloride, resulting in hemoconcentration. In addition, renal tubular reabsorption of sodium and chloride is enhanced by the stimulus. Investigations:

- Serum sodium — increased
- Potassium — normal
- Chloride — increased
- Bicarbonate — normal or decreased
- Blood volume — decreased
- Urinary sodium, potassium — increased
- Urine volume — decreased
- Urine pH — decreased

Delirium tremens — Syndrome in chronic alcoholics is presenting with confusion, delusions, hallucinations, restlessness, and marked tremor. Delirium tremens is associated with hypomagnesemia and electrolyte disturbances.

Demyelination — Loss of the outer (myelin) sheath of nerves.

Diabetes insipidus — Water excretion is excessive, connected with decreased production of antidiuretic hormone by the posterior pituitary and the disability of the kidney to reabsorb water. In some cases as much as 10 liters of water is lost daily. Oral intake of water is enhanced, but compensation is not complete. Hemoconcentration reflects an increased serum sodium and chloride. Causes of this condition include pituitary tumor, encephalitis, and syphilis.

Nephrogenic diabetes insipidus, associated with a failure of renal tubules to respond to an adequate secretion of antidiuretic hormone, also modifies blood electrolytes in a similiar fashion. Investigations:

- Serum sodium — normal or increased
- Potassium — normal
- Chloride — increased
- Bicarbonate — normal
- Blood volume — decreased
- Urinary sodium, potassium — normal
- Urine volume — increased
- Urine pH — normal

PLATE 7. Scanning electron micrograph of the airway from a child with cystic fibrosis. Thick tenacious secretions that obstruct the airways reflect an abnormality in the secretion or transport of macromolecules in exocrine glands. (Courtesy of Dr. J. M. Sturgess, Warner-Lambert/Parke-Davis Research Institute, Sheridan Park, Ontario, Canada.)

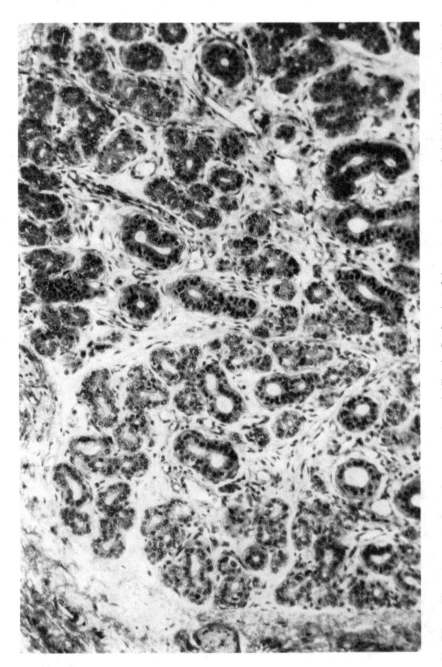

PLATE 8. Photomicrograph of exocrine pancreas in cystic fibrosis. Dysfunction of the pancreas occurs at or before birth. The cystic fibrosis pancreas shows a progressive loss of acinar tissue at 38 weeks gestation; these changes occur in fetal life and progress through early childhood. The ducts are dilated and often obstructed by proteinacious materials. (Courtesy of Dr. J. M. Sturgess, Warner-Lambert/Parke-Davis Research Institute, Sheridan Park, Ontario, Canada.)

Diabetes mellitus — Relative or absolute insulin deficiency causes a decreased hepatic and extrahepatic glucose uptake. Usually the number of islet cells is reduced and often replaced by fibrosis or hyaline substance. In this disease lipid utilization is increased and mobilized from the body stores; subsequently blood levels are increased. Liver converts lipids to ketones which are further metabolized through the Krebs cycle. In diabetic acidosis, ketones accumulate in the blood due to excess production and inadequate elimination. Investigations:

- Fasting blood sugar — normal or increased
- 2-hr postprandial blood sugar — increased
- Urinary sugar — increased
- Serum cholesterol, triglycerides, phospholipids — increased
- Plasma ketones — markedly increased sometimes
- Hemoglobin A_{1c} content — also gives a reliable measure of diabetes

Diabetic acidosis — Concentration of most serum electrolytes is reduced and body water is depleted. This is due to a sequence of deviation from normal metabolism. Insufficient insulin supply causes an inability of the cells to utilize glucose and, subsequently, blood glucose level rises. When this reaches the renal threshold level, glucose is excreted, taking large amounts of water and electrolytes into the urine. Replacing glucose, large amounts of fats are metabolized to acetoacetic acid and β-hydroxybutyric acid which are also excreted in the urine. When renal excretion of these keto acids reaches a maximum, they appear in the blood. To compensate for the loss of water and to maintain a normal circulatory volume, intracellular water moves extracellularly; pH is reduced, stimulating respiration, but more carbon dioxide is exhaled to reduce acidosis. Investigations:

- Serum sodium — decreased
- Potassium — normal or increased
- Chloride, bicarbonate — decreased
- Blood volume — decreased
- Urinary sodium, potassium — increased
- Urine volume — increased
- Specific gravity — increased
- Urine pH — decreased
- Ketones in blood and urine

Diarrhea — Water and electrolytes are excreted by an abnormal route. The bowel has increased peristalsis, and adequate replacement by reabsorption from bile, intestinal, and pancreatic secretion is insufficient. Chronic diarrhea may lead to daily loss of 5 ℓ of water and electrolytes. The levels of sodium, potassium, chloride, and bicarbonate are decreased. Bicarbonate decreases to a greater extent since the intestinal juice contains more of these ions, resulting in subsequent metabolic acidosis. Additionally, lack of carbohydrate intake and accompanying enhanced fat metabolism may produce ketonemia, which aggravates metabolic acidosis. Investigations:

- Serum sodium, potassium, chloride, bicarbonate — decreased
- Blood volume — decreased
- Urinary sodium — decreased
- Potassium — normal or decreased
- Urine volume — decreased
- Urine pH — decreased

Disaccharidase disorders — Sucrase, maltase, and lactase are confined to the brush border surface membrane of the small intestines. A variety of diseases are connected with enzyme deficiencies, mainly with the reduction of lactase, such as gastroenteritis, acute infant diarrhea, pellagra, kwashiorkor, cholera, celiac sprue, and tropical sprue.

Disorders of blood glucose — Normal blood sugar level is fairly constant, regulated by many homeostatic factors. Among these the most important is the liver, which is the primary source of sugar in fasting. When blood sugar level decreases, the liver output of glucose increases. The increase stimulates the pancreatic islet cells, and release of insulin inhibits further hepatic output of glucose. The hepatic homeostatic mechanism is controlled by insulin, adrenocortical hormone, thyroid, epinephrine, glucagon, growth hormone, and possibly glycogen.

Diuresis — Normally represents urination. In practice, it is often used to denote an increased daily urine volume (>2000 mℓ).

Down's syndrome — Complex abnormalities including mental deficiency and a characteristic facial form. Down's syndrome is congenital, most frequently due to the trisomy of chromosome 21. It is associated with mental retardation and occurs more frequently when maternal age is over 35 years. It is also called mongolism or mongolian idiocy.

Drug-induced hepatitis — Various drugs such as chlorpromazine may produce a hypersensitivity reaction and subsequent cholestasis. Side effects of anabolic steroids or sulfonylureas may elicit similar changes. In addition, some other drugs may cause hepatocellular necrosis in scattered areas. Alkaline phosphatase is increased. The diagnosis can be supplemented by case history, therapeutic cortisone trial, and liver biopsy. Eosinophilia and positive skin tests for the particular drug can confirm the diagnosis. Investigations:

- Serum acid phosphatase — normal
- Alkaline phosphatase — increased
- Aminotransferase — may be increased
- Serum bilirubin (indirect) — normal or increased
- Serum bilirubin (direct) — highly increased
- Urinary bilirubin — increased
- Urobilinogen — decreased
- Fecal urobilinogen — decreased

Duodenal fluid pH — This is usually alkaline. Its measurement may be significant in determining total bicarbonate content after intravenous secretin administration. In a normally functioning pancreas this is 90 meq/ℓ in a 20-min specimen. Chronic pancreatitis causes a reduction.

Duodenal fluid volume — This is decreased significantly in carcinomas of the ampulla and following intravenous secretion in chronic pancreatitis.

Dystrophy — Due to defective nutrition or defective or abnormal development or degeneration. Muscular dystrophy represents a progressive familial hereditary disorder, marked by atrophy and stiffness of the muscles. Retarded development of gonads or occasionally diabetes insipidus resulting from impaired function of the pituitary and hypothalamus cause adiposogenital dystrophy, also called Fröhlich's disease.

Eczema — This is an inflammatory skin reaction due to delayed hypersensitivity caused by many endogenous and exogenous factors.

Edema — Conditions leading to abnormal water and sodium retention cause high total extracellular fluid volume and edema formation in the interstitial space. Such conditions are connected with the nephrotic syndrome, liver diseases, protein malnutrition, cardiac failure, and essential, malignant, or renal hypertension.

Emphysema — Increase in the air space occupied by the alveoli due to the destruction of alveolar septa. This will lead to depressed arterial pO_2 and pCO_2.

Endocytosis — Ingestion of endogenous cellular components or foreign substances (bacterial, drug-derived) by primary or secondary lysosomes.

Erythrocyte — Red blood cells are formed normally in the bone marrow in the adult. During various disease conditions they may be formed in the spleen and other organs of the reticuloendothelial system. The average life span of the red blood cells is 120 days. They are stored in the spleen and released under stress conditions. It has not been established whether they are broken down in the blood stream or in the reticuloendothelial system. Disorders of erythrocytes include anemia (connected with reduction in number), or polycythemia (associated with an increase). Anemias are due to decreased synthesis, increased destruction, or loss of blood. Decreased production of red cells is associated with idopathic steatorrhea, pernicious anemia, aplastic anemia, and hypochromic anemia in infancy. Abnormal increase occurs in polycythemia vera. Some disorders of red blood cells are related to clinical disorders. Acetylcholinesterase is decreased in paroxysmal nocturnal hemoglobinuria; adenosine deaminase in immune deficiency disease; catalase in oral ulcerations; galactokinase and galactose 1-phosphate uridylyl transferase in galactosemia, cataracts, neurological disorders, and liver disease; glucose 6-phosphate dehydrogenase in hereditary and drug-induced hemolytic anemia; γ-glutamyl-cysteine and glutathione synthetases in hereditary hemolytic anemia and neurological disease; hexokinase, 6-phosphofructokinase, pyruvate kinase, triosephosphate isomerase in hereditary hemolytic anemia with or without other effects; methemoglobin reductase in hereditary methemoglobinemia; and uroporphyrinogen I synthetase in acute intermittent porphyria. Investigations:

* Determination of hematocrit, blood smear, hemoglobin, serum iron, and bilirubin contents

Erythroblastosis fetalis — Disease involving destruction of red-blood-forming cells in the newborn, producing hemolytic anemia characterized by jaundice and increased numbers of nucleated red blood cells (erythroblasts). It occurs when a mother is Rh negative and develops antibodies against a fetus which is Rh positive. Hemolytic jaundice is due to excessive production of bilirubin in unconjugated form (negative direct van der Bergh reaction).

Extrahepatic biliary obstruction — A blockage of the common biliary duct caused by inflammation, lithiasis, carcinoma of the pancreas, or ampulla of Vater. It is associated with an increase of serum aminotransferases and bilirubin. The levels are lower than produced by hepatitis and remain on a steady level. Acute pancreatitis shows similar changes. Investigations:

* Serum aminotransferases — increased

- Lactic dehydrogenase — normal
- Bilirubin — increased

Familial hypercholesterolemia — Metabolic derangement causes enhanced serum cholesterol. Progress of the disease is connected with xanthomata. Complications include atherosclerosis. Investigations:

- Serum cholesterol — increased
- Triglycerides, phospholipid — normal
- Lipoprotein phenotyping

Fanconi syndrome — Hereditary constitutional infantile anemia, resembling pernicious anemia. Impaired tubular reabsorption of uric acid causes low serum level and high urinary excretion. This syndrome is associated with increased output of phosphates, glucose, and amino acids. Investigations:

- Serum uric acid — decreased
- Urinary uric acid — increased

Fat abnormalities — Lipid droplets in the urine suggest the nephrotic syndrome. Increased fat content of the feces (steatorrhea) indicates malabsorption and may be due to chronic pancreatitis, celiac disease, or biliary obstruction. Lipid in the peritoneal or pleural fluid may represent obstruction and rupture of the lymphatics and exit of chyle into these cavities.

Fatty liver — Accumulation of fat in the cytoplasm of the hepatic parenchyma cells due to various causes. It could be demonstrated with liver biopsy. The changes are reversible at early stages.

Feces — Normal color is brown. It is green in severe diarrhea due to biliverdin excretion; red color represents bleeding in the lower gastrointestinal tract; black color indicates upper gastrointestinal bleeding; clay color is found in obstructive jaundice due to lack of bile pigments.

Feces volume — Normal daily production is 100 to 200 g. The volume is increased in diarrhea due to enteritis, ulcerative colitis, or toxic actions. In malabsorption and fibrocystic disease of the pancreas the volume of feces is elevated.

α-Fetoprotein — This plasma protein shows a physiological increase during the first year of life and pregnancy. It is greatly increased in liver cell carcinoma and gonadal tumors and moderately increased in acute hepatitis and chronic active hepatitis.

Fibers — Elastic or muscle fibers may be present in feces and sputum. In the feces, the presence of fibers indicates that the defective digestion may be due to chronic pancreatitis or fibrocystic disease of the pancreas. The presence of fibers in the sputum may be associated with destruction of alveoli, bronchi, or blood vessels due to chronic infection, abscess, or pulmonary tuberculosis.

Fibrosis — This disorder represents an accumulation of fibrous connective tissue. This occurs in cirrhosis; epithelial tissue is replaced by connective tissue fibers.

Fistula — Narrow tube or canal formed by incomplete closure of an abscess, wound, disease process, or congenital conditions. It transmits some fluid, secretions, or contents of some organ into body cavity. Fistulas opening from a viscus to the exterior of the body are named according to the viscus involved, e.g., biliary, gastric, vesical, or cecal.

Fructosuria — Benign congenital disease involving the urinary excretion of fructose. Investigations:

- Specimen positive on reducing tests
- Specimen negative on glucose oxidase tests
- Fructose identification by paper or thin layer chromatography

Galactosemia — Congenital disease characterized by the accumulation of galactose in the blood of young infants. Failure to eliminate galactose from the blood before 6 weeks of age leads to hepatomegaly, jaundice, and mental retardation. In congenital galactosemia uridyldiphosphogalactose transferase is absent in the liver and the metabolism of galactose to glucose is deficient. Investigations:

- Fasting blood sugar, 2-hr postprandial blood sugar — normal
- Blood — galactose present
- Urine — galactose present

Gastric juice — Examinations should be carried out for appearance, volume, and pH. Gastric juice is normally clear. It is red after fresh bleeding and dark brown or black in earlier bleeding associated with heme formations. The production of gastric juice is normally 60 to 80 mℓ/hr. It is increased in duodenal ulcer and in noninsulin secreting islet cell adenoma (Zollinger-Ellison syndrome) and reduced in pernicious anemia, atrophy, and carcinoma. In gastric ulcers the volume is similar to or lower than that of a normal person, but the acid concentration is usually lower. In atrophy, acid secretion is very low. The output of acid is the smallest in carcinoma and the subnormal secretory activity may precede the development of tumor by many years. In pernicious anemia acid is virtually absent. In chronic peptic ulcer acid it is in great excess.

Gastric juice pH — Normally below 3.5. Abnormalities occur in peptic ulcer and in Zollinger-Ellison syndrome when hydrochloride concentration and secreted volume are high causing an acid shift. In contrast, in gastric atrophy, pernicious anemia, and gastric carcinoma, usually hypochlorhydria and frequently achlorhydria are associated with an alkaline shift.

Gastric secretion — Important components of gastric secretion are hydrochloric acid, pepsin, mucin, and intrinsic factor, stimulated mainly by the vagus nerve, gastrin (a hormone produced by the pyloric glands), and secretagogues (substances present in food). Abnormalities are connected with hyper- or hyposecretion. Hypersecretion of gastric juice may be associated with duodenal ulceration or with the pancreatic tumor of Zollinger-Ellison syndrome, which produces large amounts of gastrin. Hyposecretion occurs in pernicious anemia, chronic gastritis, and carcinoma of the stomach.

Gastric ulcer — In bleeding gastric ulcer, blood urea nitrogen may be increased associated with increased breakdown of blood. A micrograph of benign peptic ulcer is presented (Plate 9). Investigations:

- Blood urea nitrogen — increased
- Urinary urea nitrogen — increased

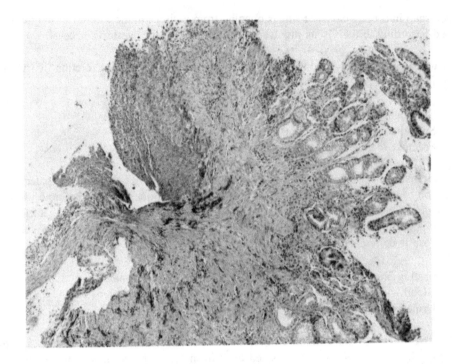

PLATE 9. Microscopic aspects of a common benign peptic ulcer.

Gastrinoma — Tumor in the upper gastrointestinal tract, mostly in the antrum, secreting gastrin. The excessive secretion of this hormone gives rise to the Zollinger-Ellison syndrome. Investigations:

* Excessive gastric acid content
* Very high, usually nonsuppressible immunoreactive gastric levels
* Malabsorption

Gaucher disease — Lipid storage decrease due to the accumulation of glucocerebroside in liver, spleen, and bones. The metabolic defect involves a deficiency of glucocerebrosidase.

α-Globulins — Abnormalities are connected with increases occurring in inflammations, trauma, collagen diseases, neoplasm, and nephrotic syndrome. Low levels are connected with hepatic parenchymal disease, malabsorption syndrome, and malabsorption due to reduced synthesis.

β-Globulins — Major components of this fraction are β-lipoprotein and transferrin. Increased β-globulin primarily represents an increase in β-lipoprotein. This may occur in obstructive jaundice and nephrotic syndrome. Changes in transferrin level are usually undetected by electrophoresis.

γ-Globulins — This fraction contains circulating antibodies (immunoglobulins) and increased levels are connected with chronic inflammatory processes. The increase can be diffuse when all antibodies are elevated, or may be a localized band of a paraprotein. Diffuse elevation occurs in chronic infection, rheumatoid arthritis, cirrhosis, sarcoidosis, and systemic lupus erythematosus. A discrete band occurs in the γ-globulin region in macroglobulinemia, myelomatosis, and essential nonmalignant and transient paraproteinemia. γ-Globulin is decreased in malabsorption and malnutrition, nephrotic syndrome, and hypogammaglobulinemia.

Glomerular dysfunction — Connected with reduced glomerular filtration rate. Causes are reduction of differential hydrostatic pressure in the glomerulus due to low systemic blood pressure (dehydration, hemorrhage, or shock), congestive heart failure, stenosis of renal artery, acute tubular necrosis, obstruction of the ureters or urethra, and to diseases of the glomerulus such as acute or chronic glomerulonephritis.

Increased glomerular permeability occurs in nephrotic syndrome and in autoimmune disease. Consequences of glomerular dysfuction are uremia, acidosis with low plasma bicarbonate, and elevated uric acid, phosphate, potassium, and calcium levels. Oliguria increases urea and low sodium excretion in the urine.

Glomerulonephritis — In chronic form in this kidney disease urea and creatinine excretion is blocked and their level rises in the blood. Other renal parenchymal diseases such as nephrosclerosis, diabetic nephrosis, gout, chronic pyelonephritis, and polycystic kidney disease may show similar changes. Investigations:

* Blood urea nitrogen, creatinine — increased
* Urinary urea nitrogen, creatinine — decreased

Glucose — Abnormal glucose metabolism causes a change in the glucose concentration of blood, urine, and spinal fluid which provides an important diagnostic tool in several disorders such as primary or symptomatic diabetes mellitus and acute central nervous system disorders.

Glucose tolerance test — Administration of 50- to 75-g glucose orally to fasting people kept at least 3 days previously on a high carbohydrate diet. The test often does not give reproducible results but is still the best in diagnosing diabetes mellitus and hypoglycemia. Before the oral glucose load, a blood sample is taken for baseline. Next, blood and urine specimens are taken at 1/2, 1, 1 1/2, 2, and 3 hr and analyzed for glucose. If serum glucose exceeds 180 mg/dℓ at 1 hr or fails to return to the baseline by 2 hr, glucose tolerance is impaired.

γ-Glutamyltranspeptidase — This enzyme derives essentially from the liver — partly from the cytosol and partly from the plasma membrane. Increased serum levels may represent diseases which are accompanied by changes in γ-glutamyltranspeptidase levels. Conditions such as the action of drugs (representing an increase) or impairment brought about by toxic compounds affect the release of γ-glutamyltranspeptidase from other tissues reflects cardiac, neurological, and renal disorders, and the presence of some kidney tumor. Best indicator in metastasis of the liver, biliary tract and pancreatic diseases, or liver condition of chronic alcoholism.

Glycogen-storage disease — Due to some enzyme deficiency involved in glycogenolysis, glycogen and glucose are stored in the liver (lysosomes) and the release of the latter into blood is impaired. Hyperglycemic response to glucagon or epinephrine is poor. Glycogen-storage diseases are grouped into various types. Some of these types are limited to a few tissues, and some are systemic. Type I, glucose 6-phosphatase deficiency affects the liver, kidney, and intestine. Signs include convulsions, hepatomegaly, growth retardation, severe hypoglycemia, hyperlacticacidemia, and hyperlipemia. Signs of Type II, α-glucosidase deficiency, include muscular weakness, sometimes hepatitis and cardiomegaly, and neurological symptoms. Type III ab, debranching enzyme (amylo-1,6-glucosidase) deficiency, is indicated by hepatomegaly, growth retardation, and hypoglycemia. Type IV, branching enzyme (amylo-1,4 → 1,6 transglucosidase) deficiency, is indicated by liver cirrhosis. Indications of Type V, muscle phosphorylase and phosphofructokinase deficiency, are muscular weakness and the fact that exercise does not produce lactic acid. Investigation:

- Fasting blood sugar, 2-hr postprandial blood sugar — decreased

Gout — Hereditary condition of uric acid metabolism often associated with fever and leukocytosis. In primary gout uric acid production is enhanced. Due to the increased pool, uric acid is deposited in the joints, but there is a rise in the serum and urine. Investigations:

- Serum uric acid — increased
- Urinary uric acid — increased
- Radiological investigations

Haptoglobin — This plasma protein fraction binds hemoglobin and exerts peroxidase activity. Its concentration in the blood is reduced in hemolytic anemias and in liver diseases.

Hemodialysis — Process of linking the circulation with an apparatus called an artificial kidney, where urea and other constituents are dialyzed out of the blood stream. Used in acute and chronic renal failure.

Hemochromatosis — Hereditary disease of unknown origin characterized by deposition of iron in the form of complexes comprising ferritin into the liver, pancreas, reticuloendothelial system, and skin. Investigations:

- Serum iron — increased
- Total iron-binding capacity — decreased
- Radioimmunoassay of ferritin (most specific)

Hemoglobinopathies — Diseases involving mutation in the primary structure of hemoglobin. The occurrence of some of the abnormal hemoglobins are common in sickle cell anemia (hemoglobin S and C), atypical hemolytic anemia (hemoglobin C, D, and E), thalassemia (hemoglobin F), and methemoglobinemia (hemoglobin M). Abnormal hemoglobins are reported in nutritional anemias, infantile anemia, leukemia, and spherocytic jaundice. Investigation:

- Hemoglobin electrophoresis

Hemolytic disorders — May be associated with the production of bilirubin stones which can cause obstructive jaundice. Constitutional hepatic dysfunctions such as Gilbert's syndrome may also be connected with similar changes. Investigations:

- Serum bilirubin (indirect) — increased
- Serum bilirubin (direct) — normal
- Serum enzymes (alanine or aspartate aminotransferase) — normal

Hemophilia — Sex-linked hereditary disease occurring only in males but transmitted by females characterized by prolonged coagulation time and abnormal bleeding due to lack of coagulation factors (factors VIII and IX).

Hepatic diseases — If associated with necrosis such as cirrhosis, viral hepatitis, hepatomas, or metastatic malignancy, the necrotic cells release intracellular aminotransferases and γ-glutamyltranspeptidase into the blood. Investigation:

- Aminotransferases — increased
- γ-Glutamyltranspeptidase — increased

Hepatitis — Inflammation of the liver parenchyma due to bacterial or viral infections, toxins, or sometimes drugs and steroids. Investigations:

- Bilirubin — increased, mostly direct
- Albumin — decreased
- Alanine and aspartate aminotransferase — increased
- Alkaline phosphatase — normal

Hereditary fructose intolerance — Congenital disease occurring first in infancy. It manifests with vomiting, anorexia, hypoglycemia, and hepatomegaly, and is associated with renal dysfunction similar to Fanconi syndrome. The biochemical deficiency involves fructose-1-phosphate aldolase and fructose-1,6-diphosphate aldolase, the latter defect probably being secondary. Investigation:

- On fructose load, prolonged elevation of serum fructose level, hypoglycemia, hypophosphatemia, and increased serum lactate

Hereditary spherocytosis — Rapid hemolysis of intrinsically abnormal red cells leading to an enhanced production of bilirubin. Hepatic compensation occurs to some degree, resulting in the excretion of increased amounts of urobilinogen in the urine and stool. Severe hemolysis exceeds the hepatic capacity of bilirubin metabolism and is associated with enhanced serum indirect bilirubin. Investigations:

- Serum bilirubin (indirect) — increased
- Serum bilirubin (direct) — normal
- Urinary bilirubin — normal
- Urobilinogen — increased
- Fecal urobilinogen — increased
- Serum enzymes
- Alanine or aspartate aminotransferase — normal

High-density lipoprotein (HDL) — This plasma fraction is also called α-lipoprotein. It is important in lipid transport and transport of lipid-soluble vitamins and hormones. HDLs are reduced in liver diseases and chronic kidney disease and are important in the assessment of atherosclerosis. Genetic defect occurs in Tangier disease.

Hirsutism — Inappropriately increased body hair in females caused by hereditary factors, endocrine disorders, drugs, or masculinizing tumors. Investigations:

- 17-Ketosteroids — increased
- Cortisol — increased or normal
- Dihydroandrosterone — increased

Histidinemia — A genetic disease characterized by increased histidine levels in blood. The metabolic defect involves the absence of histidase activity converting histidine to urocanic acid. Investigation:

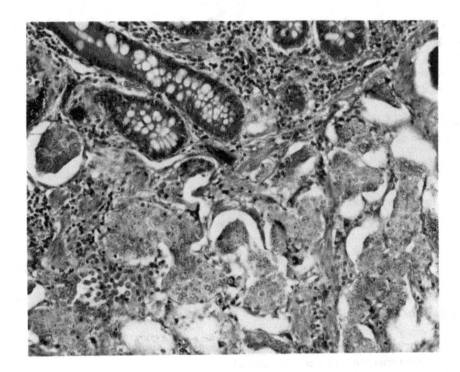

PLATE 10.   Typical aspect of an endocrine cell tumor localized in the small intestine diagnosed as carcinoid. The distinguishing feature of this neoplasm is that it can secrete histamine, 5-hydroxytryptophan, serotonin, and derivatives, as well as prostaglandins and some peptides.

- Assay of urocanic acid in sweat — missing

Holocrine activity — Secretory activity of a gland denoted by excretion of disintegrated cellular materials.

Homeostasis — Servomechanism using feedback control which regulates the level of the given constituent within physiological limits.

Homocystinuria — Excretion of homocystine in the urine, associated with β-cystathionase deficiency.

Hormone production — The endocrine system consists of seven essential glands: pituitary, thyroids, parathyroids, adrenals, gonads (testes in males, ovaries in females), pineal and islet cells of the pancreas. These glands secrete many hormones, and disturbances of secretions are connected with various diseases. Accordingly, measurements of these hormones provide essential data in diagnosing endocrine diseases.

5-Hydroxyindolacetic acid — A metabolite of serotonin formed in the liver, lungs, and in minor amounts in the brain. It is secreted into the blood and functions as a vasopressor substance. Increased serotonin breakdown and enhanced secretion of 5-hydroxyindolacetic acid occur in malignant carcinoid tumors arising in the intestinal tract (Plate 10), in metastasis of the liver, and in bronchial adenomas. Urinary 5-hydroxyindolacetic acid is increased in adult celiac disease due to an increased amount of chromaffin cells in the intestines.

Hyperaldosteronism (Conn's disease) — Condition due to increased aldosterone secretion by a functioning cortical adenoma and causing mild hypertension. Investigations:

- Serum sodium, bicarbonate — increased
- Potassium — decreased
- Immunoreactive aldosterone — increased in serum and urine
- Basal aldosterone level in recumbent position is not affected by assuming the erect one

Hyperbilirubinemia — Connected with increased serum bilirubin level. Investigation:

- Total/direct bilirubin — if elevated, measurements of serum alanine and aspartate aminotransferase, γ-glutamyl transpeptidase, and 5′ nucleotidase may provide an indication of the origin of hyperbilirubinemia

Hypercapnia — Represents elevated pCO_2 in the blood.

Hyperglycemia — Increased serum glucose concentration, usually due to diabetes mellitus and only rarely to other causes. Investigations:

- Fasting blood sugar
- 2-hr postprandial glucose
- Oral glucose tolerance test

Hyperoxaluria — Increased urinary excretion of oxalic acid. Primary hyperoxaluria is present in nephrolithiasis, nephrocalcinosis, and early renal failure. In some types of this disorder, enzymatic defects in glyoxalate metabolism are involved. Investigation:

- Patients excrete 150 to 650 mg oxalate per day (normal <50 mg/day)

Hyperparathyroidism — Caused by adenoma (90%), hyperplasia (10%), or carcinoma (rare), and associated with enhanced secretion of parathyroid hormones. This hormone increases serum calcium by enhanced bone resorption and enhanced renal tubular and intestinal absorption of calcium. Due to the accelerated renal tubular secretion, serum phosphates are reduced and their resorption may cause an elevation of alkaline phosphatase and urinary calcium. The latter change may also be connected with increased intestinal absorption. Investigations:

- Serum calcium — increased
- Phosphates — decreased
- Serum alkaline phosphatase — increased
- Serum immunoreactive parathyroid hormone level — increased
- Urinary calcium — increased
- Turnover rate with ^{45}Ca isotope or cortisone suppression test in doubtful cases

Hyperplasia — Increase in the number of cells in a given organ. It may be physiological, compensatory, or pathological in origin.

Hyper-, hypokalemia — Connected with increased or decreased serum potassium levels, respectively. Investigations:

- Serum potassium, urinary potassium, blood gases (further investigation depends on the underlying cause)

Hypertension — Represents an increased blood pressure. Investigations:

- Serum sodium increased, potassium decreased in Conn's disease
- Catecholamine excretion — increased in pheochromocytoma
- Cortisol — increased in Cushing's syndrome
- Renin level elevated, creatinine clearance reduced in kidney diseases

Hyper-, hypotonicity — Represents increased or decreased osmolality of body fluids.

Hypertrophy — Increase in the size of cells in an organ without an increase in their number, as in the muscles of athletes or in heart muscles subjected to increased load (e.g., hypertension).

Hyperthyroidism — In this disease, total body muscle mass production is diminished and less creatine is converted to creatinine. The excess thyroid hormone also reduces the formation of creatine to creatinine in the remaining muscle. Increased thyroid hormone secretion causes enhanced intestinal glucose absorption, peripheral glucose utilization, and glycogenolysis indirectly. Investigations:

- Urinary creatine increased
- Creatinine — decreased
- Fasting blood sugar — normal or increased
- 2-hr postprandial blood sugar — increased
- Urinary sugar — increased

Hypogammaglobulinemia — Characterized by a primary immunological deficiency. Usually occurs in infancy with recurrent infections. The condition is associated with decreased levels of circulating immunoglobulins or impaired cellular immunity. Hypogammaglobulinemia can be transient, congenital, or acquired. The synthesis of γ-globulins is decreased irrespective of whether it is primary (congenital, idiopathic, or physiological), or secondary due to leukemia, Hodgkins disease, or multiple myeloma. In people with O blood type, anti-A and anti-B isoagglutinins are absent; their antibody titer to various vaccines and their response to bacterial infections are fairly poor. Investigations:

- Albumin — normal
- γ-Globulins — decreased

Hypoglycemia — Represents a decreased serum glucose concentration usually due to a functioning islet cell tumor. An example of insulinoma is illustrated in Plate 11. Investigations:

- Serum glucose — decreased
- Immunoreactive serum insulin — increased

Hyper-, hypoventilation — Represents an increased or decreased rate of breathing.

Hypogonadism — Insufficient development of the gonads resulting in deficiency of gonadal function and of secondary sex characteristics. It may be due to hypothalamic, pituitary, and gonadal (testis, ovary) causes. Investigations:

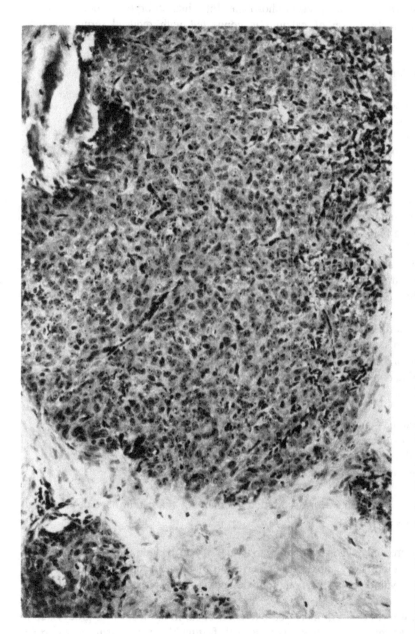

PLATE 11. Uncontrolled proliferation of β-cells of the endocrine pancreas leads to the formation of insulinomas. Patients with this tumor show hypoglycemic attacks which quickly recede following the administration of glucose.

- Level of peripheral sex hormones (testosterone and estradiol)
- Level of pituitary gonadotropic hormones (luteinizing, follicle-stimulating hormones)
- Stimulatory tests after the administration of follicle-stimulating or luteinizing hormone releasing factor

Hypoparathyroidism — Lack of parathormone causes a decrease in renal tubular secretion of phosphate resulting in elevated serum phosphate. Intestinal absorption, bone resorption, and renal tubular reabsorption of calcium is connected with reduced serum calcium. Investigations:

- Serum calcium — decreased
- Phosphates — increased
- Serum alkaline phosphatase — normal
- Urinary calcium — decreased

Hypoplasia — Decrease in the number of cells of an organ. It might be due to developmental causes or, more often, to inactivity.

Hypothyroidism — Deficiency in thyroid function. Hepatic cholesterol synthesis is reduced. Tissue utilization, storage, and excretion are also reduced; serum cholesterol and cholesterol ester are high. Investigations:

- Serum cholesterol, triglyceride, phospholipids — increased
- Thyroxine and triiodothyronine — reduced
- Thyrotrophic hormone — increased

Hypovitaminosis D — The intake of calcium may be normal, but the lack of vitamin D in sufficient amounts prevents the absorption of calcium in adequate amounts and it is excreted in the stool. Serum calcium is low, causing secondary hyperparathyroidism. X-ray examination of long bones and a good response to vitamin D can supplement the biochemical tests. Investigations:

- Serum calcium — normal or decreased
- Phosphates — decreased
- Serum alkaline phosphatase — increased
- Urinary calcium — decreased

Hypovolemia — Represents decreased blood volume; e.g., in hemorrhage or after treatment with diuretics.

Hypoxia — Decreased oxygen availability to tissues. Types include anemic anoxia (hemoglobin decreased); circulatory anoxia (poor perfusion of tissues); histotoxic anoxia (toxic compounds, e.g., cyanide poisoning); and anoxic anoxia (atmospheric oxygen reduced).

Idiopathic steatorrhea — An increased excretion of fat in the feces (>7 g/day) due to unknown causes. Absorption of glucose is abnormal, fasting blood sugar is usually normal, but blood sugar fails to rise after a carbohydrate meal due to impaired absorption of glucose from the intestines. Submucosal malabsorption manifests in intestinal lipodystrophy (Whipple's disease; see Plate 12). Hepatic lipoprotein synthesis is also diminished due to the impairment of amino acid absorption through the small intestines. Consequently, serum lipid concentration is reduced. Generally the impaired absorption of amino acids results in a

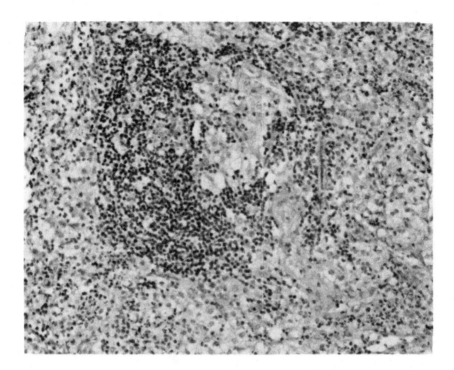

PLATE 12. Manifestation of Whipple's disease in the lymph node. Clear cells in the picture correspond to macrophages characteristic of this disease. Similar phagocytic macrophages can be seen in other organs including the central nervous and digestive systems.

decrease of plasma protein synthesis. Chronic pancreatitis and other disorders associated with malabsorption show similar changes. Investigations:

- Fasting blood sugar — normal
- 2-hr postprandial blood sugar — decreased
- D-Xylose absorption test
- Serum carotene — decreased
- Stool fat — increased
- Serum cholesterol, cholesterol esters, triglycerides, phospholipids, lipoproteins — decreased
- Plasma ketones — normal
- Albumin, α-, β-, γ-globulin — decreased

Immunoglobulins — There are three major classes (IgG, IgM, and IgA) and two minor classes (IgD and IgE). Synthesis of these immunoglobulins is increased in myelomatosis, macroglobulinemia, cryoglobulinemia, heavy chain disease, and essential paraproteinemia. Synthesis is increased due to congenital disorders.

IgG forms about 75% of immunoglobulins and contains most normal plasma antibodies. Deficiency is connected with pyogenic infections, inherited hypogamma globulinemia, IgA myeloma, Waldenström's macroglobulinemia, malabsorption syndrome, and also occurs during extensive protein loss. Excess occurs in myelomatosis. IgA forms about 20% of immunoglobins. Deficiency is associated with intestinal disease, protein loss, certain leukemias, and respiratory tract infection. Excess occurs in myelomatosis. IgM is about 7%. Deficiency is seen in septicemia and in some IgG and IgA myelomas.

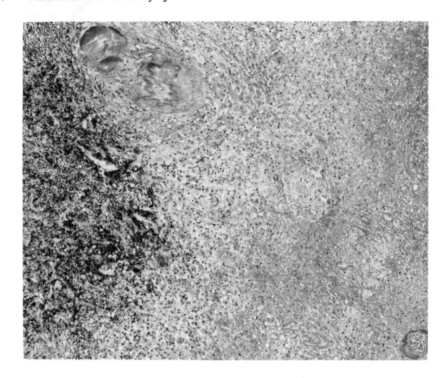

PLATE 13.   Infarction of the spleen as shown in this illustration can represent a complication of hemodynamic disorders.

Interstitial cells — Cells making up the interstitium, i.e., connective tissues. Cell types vary from organ to organ.

Ischemia — Represents a deficiency of blood supply to cells or organ. Complications lead to hemodynamic disorders such as infarction of the spleen (Plate 13), heart, and other organs.

Juvenile diabetes mellitus — Diabetes mellitus occurring in young people, probably due to autoimmune reaction against the pancreatic β-cells. In contrast to maturity-onset diabetes, the syndrome is characterized by an absolute lack of insulin.

Juxtaglomerular apparatus — Epithelial cells situated at the vascular pole of Bowman's capsule. These cells contain many granules and secrete renin.

Keto acids — In the liver the production of keto acids depends on the concentrations of free fatty acids, insulin, and glucagon. Major representatives are acetoacetate and β-hydroxybutyrate. A combination of overproduction and underutilization leads to pathogenic concentrations in diabetic ketoacidosis. In prolonged fasting keto acids are the primary protein sparing compounds. Prolonged starvation in obese people reduces keto acids.

Ketone bodies — These are β-hydroxybutyric acid, acetoacetic acid, and acetone. Nitroprusside reagent detects these compounds, by reacting preferentially with acetoacetic acid.

Ketonemia/ketosis — Represents an accumulation of ketone bodies in the circulation.

Ketonuria — Represents an excretion of ketone bodies in the urine.

Kidney diagnostic tests — These include measurements of glomerular filtration rate, renal plasma flow, and tubular function. Glomerular filtration rate can be assessed by creatinine or urea clearance. Renal plasma flow can be determined by standard clearance methods or by the phenolsulfophthalein test. Tubular function tests include exclusion of renal aminoaciduria or glycosuria, and measurements of phosphate reabsorption, renal acid, and water excretion.

Klinefelter's syndrome — Characterized by eunochoidism, azospermia, gynecomastia, and increased urinary gonadotropins. It is due to the presence of two or more X chromosomes in the male karyotypes. Investigations:

* Follicle stimulating and luteotrophic hormone — increased
* Testosterone — decreased

Kwashiorkor — Disease rampant in underdeveloped countries due to lack of adequate intake of essential amino acids causing a protein-calorie malnutrition. This condition leads to widespread atrophy of the absorptive area and marked decrease in intraluminal digestion caused by deficiency of the peptide hydrolases.

Lactate dehydrogenase — Lactate dehydrogenase is a mixture of five tetrameric isoenzymes derived from two subunits termed H (heart) and M (muscle). These isoenzymes occur in cardiac and skeletal muscle, erythrocytes, liver, testes, and seminal fluid. Serum lactate dehydrogenase isoenzyme assays have clinical value in myocardial infarction and megaloblastic anemia. These also increased in the serum in hepatocellular disease, muscular dystrophy, and hemolysis and highly elevated in hepatic neoplasia. Frequency and extent are correlated with dissemination.

Laennec's cirrhosis — This is connected with chronic alcoholism, and the degeneration, necrosis, and fibrosis of the liver tissue leading to an inhibition of bilirubin conjugation results in an increased indirect bilirubin. Fibrosis also causes intrahepatic biliary obstruction and increased direct bilirubin level. With the advancement of the disorder, many further functions of the liver deteriorate, particularly protein and cholesterol synthesis. Finally, storage capability is destroyed and intermittent hypoglycemia is apparent. Liver biopsy can confirm the diagnosis. Investigations:

* Serum bilirubin (indirect) — increased
* Serum bilirubin (direct) — increased
* Urinary bilirubin — increased
* Urobilinogen — normal or increased
* Fecal urobilinogen — normal
* Serum alkaline phosphatase — increased
* Serum cholesterol — increased
* In the advanced state, albumin synthesis and cholesterol esterification impaired

Leukemia — Malignant disease of the white blood cells characterized by permanent increase in their numbers. May be acute or chronic, leukocytic, lymphocytic, or monocytic. Increased production and breakdown of white cells causes an increased nucleic acid turnover and elevated uric acid synthesis. Multiple myeloma, polycythemia vera, lymphoblastomas, and other malignancies may be associated with similar changes. Investigations:

* Complete blood count

- Serum uric acid — increased
- Urinary uric acid — increased

Leukocyte — Normally monocytes and granulocytes are synthetized in the bone marrow, lymphocytes in the lymph nodes, and plasma cells in the lymphocytes or primitive connective tissue cells. They are stored in the bone marrow, spleen, and lymph nodes. Neutrophils are found occasionally in relatively large amounts in the lungs. The average life span of leukocytes is 9 to 13 days. Lymphocytes have a life span of 100 to 200 days. They are broken down in the reticuloendothelial system. In adults the white cell count is around 8,000 to 10,000 cells per cubic millimeter of blood. During the first day of life they may be elevated to 20,000 to 30,000 cells per cubic millimeter and remain elevated throughout childhood. The production of leukocytes is increased abnormally in acute leukemia and chronic myelogenous leukemia. Variable changes characterize multiple myeloma, Hodgkin's disease, Addison's disease, Cushing's syndrome, and Gaucher's disease. Acute bacterial infections and infectious mononucleosis cause an enhanced total white cell count; in the chronic stage it may be normal. Acute viral infection is associated with a reduction.

Leukodystrophy — Degeneration of the white matter of the brain involving demyelination and glial reaction. This entity appears to be related to disorders of lipid metabolism.

Lipase — Responsible for the lipolytic activity of the serum. Pancreatic acinar cells are the major source of this enzyme. These cells secrete lipase into the duodenum via the pancreatic duct and common bile duct. Serum levels of lipase are increased in acute pancreatitis, pancreatic carcinoma, intestinal obstruction, and other diseases causing secondary pancreatitis. Elevation of lipase is more specific to pancreatic disease than amylase.

Lipemia — Represents an increased concentration of cholesterol and primarily triglycerides, giving rise to a turbid, occasionally milky serum.

Lipidosis — Connected with enzyme defects causing sphingolipid storage. Heterozygous carriers can be detected by monitoring pregnancies where the fetus is at risk. From the amniotic fluid, cells can be cultured and enzyme activity measured. Gaucher disease is connected with glucocerebroside accumulation due to β-glucosidase deficiency; Niemann-Pick disease, with sphingomyelin accumulation — sphingomyelinase deficiency; Krabbe disease, with galactocerebroside accumulation — β-galactosidase deficiency; ceramide lactoside lipidosis, with ceramide lactoside accumulation — β-galactosidase deficiency; metachromatic leukodystrophy, with ceramide galactose 3-sulfate accumulation — sulfatidase deficiency; Tay-Sachs disease, with ganglioside GM_2 accumulation — hexosaminidase A deficiency; Fabry's disease, with ceramide trihexoside accumulation — α-galactosidase deficiency; generalized gangliosidosis, with ganglioside GM_4 accumulation — β-galactosidase deficiency.

Lipoprotein — Carrier proteins bind chylomicrons and various lipids in a complex form and carry them in the circulation. Lipoprotein electrophoresis allows the visualization of several groups of lipoproteins, such as chylomicrons, β- , preβ- , and α-lipoproteins. Each fraction contains variable amounts of lipids attached to specific apoproteins.

Lipoproteinemias — Lipoproteins are separated by flotation by analytical ultracentrifuge or by electrophoresis. Classes include chylomicrons with exogenous triglycerides as major lipid components; very-low-density lipoproteins (VLDL) with endogenous triglycerides (pre β-mobility); low-density lipoproteins (LDL) with cholesterol, cholesterol esters, and phospholipids (β-mobility); high-density lipoproteins (HDL) with phospholipids and cholesterol esters (α-mobility).

Hyperlipoproteinemias are groups as follows by phenotyping. *Type I*: Hyperchylomicronemia with absolute deficiency of post-heparin lipase and diacylglycerol lipase. Secondary associated diseases are diabetic acidosis, hypothyroidism, and dysglobulinemia. These may cause the lipoprotein defect.

Type IIa: Increased LDL; Type IIb: Increased LDL and VLDL. Secondary associated diseases include hypothyroidism, nephrosis, obstructive liver disease, Cushing's syndrome, acute intermittent porphyria, and dietary excess lipid intake.

Type III: Characterized by floating lipoproteins. Secondary causes are rare, and include hypothyroidism, renal insufficiency, and diabetes mellitus.

Type IV: Increased VLDL. VLDL is also increased in some secondary diseases including diabetes mellitus, obstructive liver disease, nephrotic syndrome, uremia, glycogen storage disease, alcoholism, pancreatitis, and dysglobulinemia. It is raised due to side effects of various drugs and steroid hormones.

Type V: Increased VLDL with hyperchylomicronemia, reduced post-heparin lipase, and diacylglycerol lipase. Secondary associated diseases are similar to Type IV.

Hypolipoproteinemia occurs in Tangier disease with low plasma cholesterol, phospholipid, and HDLs. In contrast, cholesterol deposits occur in other tissues (reticuloendethelial system). Hypobetalipoproteinemia is connected with reduced cholesterol and LDLs. Abetalipoproteinemia is connected with low plasma level of all lipids and lipoprotein fractions, with the exception of HLDLs, which are present. Clinical symptoms in the latter disease include malabsorption of fat, atoxic neuropathy, retinitis pigmentosa, and acanthocytosis.

Liver diagnostic tests — These include the determination of bile pigments (bilirubin and urobilinogen), bile acids, serum enzymes (alanine and aspartate aminotransferase, alkaline phosphatase, γ-glutamyl transferase, lactate dehydrogenase, and 5′-nucleotidase), plasma proteins (albumin, immunoglobulins, prothrombine-time, α-antitrypsin, α-fetoprotein, and ceruloplasmin), other components (glucose, ammonia, cholesterol, and vitamin B_{12}), and sulfobromophthalein retention time.

Liver cirrhosis — In the diseased liver glycogen stores are depleted and gluconeogenesis is inadequate. With severity, hyperglycemia cannot be produced in response to epinephrine. Severe hepatitis, obstructive jaundice, and other hepatic disorders as well as carcinomatosis may be connected with similar changes. In severe cases conversion of ammonia to urea is inhibited. Consequently blood ammonia is raised and urea is decreased. Anemia may occur in cirrhosis due to impairment of iron, folic acid, and vitamin B_{12}. The intake of these factors may be reduced and the reduced production of erythrocytes is related to their lack. Investigations:

- Fasting blood sugar, 2-hr postprandial blood sugar — increased
- Liver function tests, liver biopsy
- Blood urea nitrogen — decreased
- Ammonia — increased

Liver profile — A series of laboratory tests which provide information on the functional status of the liver and the biliary drainage system. These tests are not specific for any one disease; nevertheless, they are very useful in making the diagnosis in pathological conditions involving the liver the biliary tract.

Low-density lipoprotein (LDL) — This plasma fraction is also called β-lipoprotein and is important in lipid transport, mainly cholesterol. There are age-dependent varieties in plasma concentration of this fraction. It is elevated in nephrosis and in Type-II hyperlipidemias. Genetic defects in the production of LDLs include decrease such as abetalipoproteinemia or increase such as familial hypercholesterolemia.

Lymphoma — Malignant solid tumor of lymphoid cells.

Lymphosarcoma — Sarcoma originating from lymphotic tissue.

Lysozyme — Plasma fraction responsible for the lysis of bacteria. It is strongly increased in monocytic leukemias.

Macroglobulinemia — In this disease the proliferating cells resemble lymphocytes which produce large amounts of IgM. Vascular changes and anemia dominate the symptoms; hemorrhagic manifestations in the retina produce visual disturbances.

Malabsorption — Strictly defined, it denotes insufficient absorption of foodstuffs from the intestinal lumen due to lesions affecting the absorptive surface. In common usage, malabsorption comprises maldigestion due to deficient enzymatic activity as well. Biochemical tests cannot identify the cause of malabsorption.

Malabsorption syndrome — Abnormal amounts of fat retained in the alimentary canal, as a consequence of reduced absorption of broken fat or inadequate breakdown connected with pancreatic disease. They form an insoluble, nonabsorbable soap with calcium. Consequently serum calcium is decreased and secondary hyperparathyrodism is produced. Vitamin D absorption is also decreased limiting further calcium absorption. Poor water and electrolyte absorption leads to hyponatremia and hypokalemia. Investigations:

- Serum alkaline phosphatase — increased
- Blood volume — decreased
- Serum sodium, potassium chloride — normal
- Bicarbonate — normal or decreased
- Fecal fat excretion — increased
- Xylose absorption — decreased
- Vitamin B_{12}, folate, calcium, carotene — decreased
- Urinary sodium, potassium, calcium — decreased
- Urine volume — normal
- Urine pH — normal or decreased

Maldigestion — Form of malabsorption due to the deficiency of pancreatic enzymes. Accordingly, digestion of food in the duodenum is inadequate, resulting in malabsorption of carbohydrates, proteins, and fats alike. Investigations:

- Fecal fat — increased
- Xylose tolerance test — normal
- Pancreatic exocrine secretion — decreased

Malnutrition — Represents an inadequate intake of essential nutrients resulting in syndromes due to various deficiencies such as beriberi (vitamin B_1 deficiency), pellagra (vitamin B_6 deficiency), or protein-calorie malnutrition. Although glucose intake is decreased, peripheral utilization still continues. Hepatic stores are depleted and blood sugar level falls. Following a carbohydrate meal, blood sugar rises but remains at abnormal levels due to the long impairment of the glycolysis machinery. The cause of some symptoms is connected with a lack of shortage of dietary calcium ingestion. It is associated with decreased serum calcium which increases parathyroid activity and, in turn, mobilization of calcium from bone and renal tubular secretion of phosphate.

In malnutrition, caloric intake may be sufficient and lipids are not mobilized from stores; but due to insufficient dietary protein, lipoprotein synthesis in the liver is reduced. Investigations:

- Fasting blood sugar — decreased
- 2-hr postprandial blood sugar — increased
- Serum calcium — normal or decreased
- Serum phosphates — decreased
- Alkaline phosphatase — increased
- Urinary calcium — decreased
- Serum cholesterol, cholesterol esters, phospholipids, lipoproteins — decreased
- Triglycerides — normal
- Plasma ketones — normal or increased

Maturity-onset diabetes — This syndrome occurs only in adults. It is due to the relative deficiency of insulin. The efficiency of the hormone is, however, impaired due to tissue resistance. Whereas the disease proper is not life-threatening, its complications involve diabetic neuropathy, nephrosclerosis, and atherosclerosis.

Melanine metabolism — Pigment formed from tyrosine by oxidation in skin melanocytes. Tyrosine comes directly from the diet or is derived from phenylalanine catalysed by *p*-phenylalanine hydroxylase. Abnormalities occur in phenylketonuria, connected with the absence of phenylalanine hydroxylase. Phenylalanine accumulates and large amounts are excreted in the urine and as other derivatives by *o*-hydroxylation. In alkaptonuria homogentisic acid oxidase is absent and homogentisic acid is excreted in the urine. In albinism there is a congenital tyrosinase deficiency resulting in impaired melanin formation. In melanotic sarcoma melanocytes produce excess melanine, and large amounts of precursors (mainly conjugated 5,6-dihydroxyindole) are excreted in the urine.

Menkes' syndrome — Also known as "kinky hair syndrome". It is a congenital disease associated with defective copper metabolism, manifesting in the neonatal period with hypothermia, feeding difficulties, and occasional jaundice. Seizures begin at 2 to 3 months. The hair is short, stubby, and feels like steel wool.

Metabolic acidosis — This disorder is not associated with an increased anion gap but there is an elevated serum chloride. Hyperchloremic metabolic acidosis occurs following ingestion of organic acids, bicarbonate loss (diarrhea, gastrointestinal fistula, or renal tubular acidosis), and in the presence of abnormal proteins which are cationic at pH 7.4.

Metastasis — Seeding of tissues with tumor cells originating in the primary tumor.

Metastatic bone tumors — Metastatic neoplasms of bone are connected with enhanced osteoblastic activity and increased alkaline phosphatase. This is very useful together with X-ray examinations and bone biopsy. Metastatic carcinoma of the bone is connected with the invasion and breakdown of bone (Plate 14), and with an associated release of calcium into the blood. Calcium is excreted and therefore serum calcium usually remains normal. Occasionally the invasion is so massive that the serum calcium is enhanced and can be controlled by steroid therapy.

PLATE 14.   Representation of metastatic invasion of carcinoma into the rib (dark area).

Carcinoma of the breast, thyroid, and the kidney is associated with osteoblastic metastases and consequent alkaline phosphatase changes. Metastatic carcinomas originate from lung, kidney, and rectum may produce osteolytic lesions without any change of alkaline phosphatase activity. Investigations:

- Serum acid phosphatase — normal or increased occasionally
- Serum alkaline phosphatase — increased
- Serum calcium — increased or normal
- Serum phosphate — normal
- Urinary calcium — increased

Metastatic hepatic carcinoma — In this disorder if a primary tumor has been diagnosed elsewhere in the body, a marked increase occurs in serum alkaline phosphatase. In contrast, primary hepatomas are not connected with any remarkable change of this enzyme. Liver biopsy scan or exploratory laparatomy can confirm the diagnosis. Investigations:

- Serum acid phosphatase — normal
- Alkaline phosphatase — increased
- Serum bilirubin — normal

Metastatic prostatic carcinoma — Often connected with high levels of serum acid phosphatase. Alkaline phosphatase is also elevated. Acid phosphatase may remain normal in the highly anaplastic type of metastatic prostatic carcinoma. The diagnosis is confirmed by X-ray examinations of the bone and biopsy of the prostate. Investigations:

- Serum acid phosphatase — increased
- Alkaline phosphatase — increased

Methemoglobinemia — Presence of methemoglobin in the blood in excess. Chemical detection of methemoglobinemia requires spectrophotometric analysis.

Methylmalonic aciduria — This syndrome involves an increased excretion of methylmalonic acid in the urine (>5 mg/day up to 5 g/day) as a consequence of methylmalonic acidemia. Several enzyme defects might be responsible for the disorder.

Mosaicism — An anomaly of chromosomal patterns in the same organism when two or more types of chromosomal numbers are present in different cells. This is exemplified in some patients with Klinefelter's syndrome who may have a 46, XY or 47, XXY mosaicism.

Multiple myeloma — Excessive proliferation of plasma cells characterized by the secretion of a specific immunoglobulin and, above all, grossly increased production of light chain. These are then excreted as free κ or λ chains. Protein electrophoresis shows M peak. Immunoelectrophoresis shows distorted immunoglobulin and κ or λ-chain bands. Thus multiple myeloma is associated with the presence of abnormal proteins in the serum which migrate with γ-globulin, and occasionally with β- or α_2-globulin. Plasma protein level is usually increased with the exception of albumin which is decreased. Other abnormal proteins are present less frequently. Bence-Jones protein appears in the urine. Investigations:

- Albumin — decreased
- γ-Globulin — increased

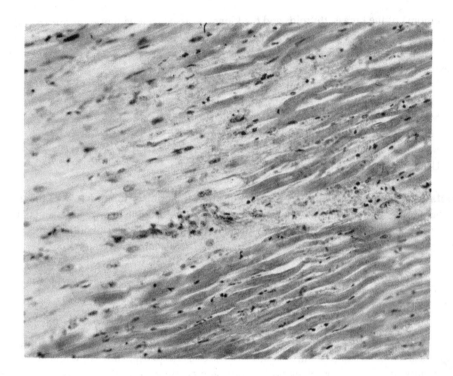

PLATE 15. Heart infarct, representing ischemic damage to the myocardium (lower part of the figure). It occurs as the result of coronary inclusion.

Muscular dystrophy — Degenerative disease, probably of autoimmune origin, resulting in the loss of the functional muscle fibers. Subsequently the total body muscle mass is reduced, and creatinine production from creatinine is decreased. Blood creatine and urinary excretion is increased, whereas creatinine is diminished in both pools. Investigations:

• Blood creatine — increased
• Blood creatinine — decreased
• Urinary creatine — increased
• Urinary creatinine — decreased
• Lactate dehydrogenase increased (particularly LDH isoenzyme)
• Creatine phosphokinase and aldolase increased

Myelomatosis (multiple myeloma) — This disease represents a generalized malignant proliferation of plasma cells (myeloma tumor) throughout the bone marrow and occasionally in the soft tissues in multiple locations. There is a gradation of cell types from normal plasma cells to multinucleate and abnormal forms with large nucleoli representing myeloma cells. These cells produce globulins in most cases.

Myocardial infarction — Ischemic damage to the myocardium represents heart infarct (Plate 15). It is associated with cellular injury and necrosis leading to release of intracellular aminotransferases and dehydrogenases into the blood. There is a significant correlation between the extent of infarct and the degree of increased enzyme level. Aspartate aminotransferase is elevated 6 to 12 hr after the infarction, and returns to normal within 4 to 7 days. Lactic dehydrogenase is elevated 12 to 24 hr after the infarction and returns to normal within 1 to 2 weeks. One lactic dehydrogenase isoenzyme (LDH_5) is increased to the highest level. Investigations:

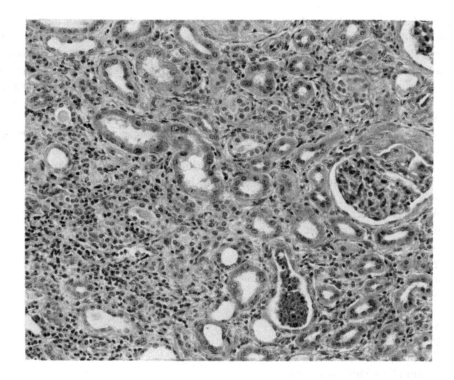

PLATE 16.   Chronic interstitial nephritis. This tubulointerstitial damage is frequently produced by drugs and toxins.

- Amino transferases — increased
- Lactic dehydrogenases — increased
- Isoenzyme separation

Necrosis — Death of cells, tissues, or even organ without involving death of the whole organism.

Neoplasia — New growth of cells meaning tumor production that might be either benign or malignant.

Nephritis — Inflammation of the kidney parenchyma that might involve the glomeruli (glomerulonephritis) or the calices (pyelonephritis). Plate 16 illustrates the morphological appearance of chronic interstitial nephritis. Investigations:

- Glomerulonephritis — presence of erythrocytes in the urine, red cell counts, protein in the urinary sediment
- Pyelonephritis — white cell counts
- Increased protein in the urine
- Mainly leukocytes in the urinary sediment

Nephrocalcinosis — Presence of calcium deposits in the renal parenchyma, but with the collection system excepted. It may be due to any cause underlying hypercalcemia, renal tubular acidosis, renal parenchymal disease, and urolithiasis.

Nephrotic syndrome — Syndrome presenting with gross edema, massive proteinuria, hypoalbuminemia, and hyperlipidemia. This is caused by various conditions such as glomerulonephritis, amyloidosis, glomerulosclerosis, or renal vein thrombosis of lupus erythematosus, and is connected with a decreased serum level of albumin and γ-globulin. The glomeruli show increased permeability and proteins are lost in the urine; α_2- and β-globulins are normally increased, and the latter are associated with hyperlipemia. Diseased glomeruli lead to protein loss and therefore plasma proteins are reduced. Through compensation, hepatic synthesis is enhanced, including the synthesis of lipoproteins. Serum calcium is reduced due to reduced serum protein lost together with bound calcium in the urine. Ionizable calcium concentration remains normal and the syndrome is not accompanied by secondary hyperparathyroidism. In cases of uremia, ionizable calcium is also lost and hyperparathyroidism ensues. Hypoproteinemia brought about by malnutrition, malabsorption syndrome, or cirrhosis of the liver may also produce hypocalcemia. Investigations:

- Albumin — decreased
- α_2-Globulins — increased
- β-Globulins — increased
- γ-Globulins — decreased
- Serum cholesterol and cholesterol esters — increased
- Serum triglycerides — increased
- Serum phospholipids — increased
- Serum lipoproteins — increased
- Plasma ketones — normal
- Serum calcium — decreased
- Serum phosphates — normal
- Serum alkaline phosphatase — normal
- Serum proteins — decreased
- Urinary calcium — normal
- Serum protein electrophoresis
- Renal function tests
- Excessive proteinuria and lipuria also present

Neurotransmitter — Low-molecular-weight compounds mediating the transmission of impulse in the nerve synopses. These transmitters are as follows: adrenergic (norepinephrine and dopamine), cholinergic (acetylcholine), serotoninergic (serotonin).

Niemann-Pick disease — Congenital disease due to the accumulation of sphingomyelin. Patients have hepatomegaly, splenomegaly, and extensive central nervous system damage, and frequently a cherry-red spot in the macular region. Antenatal detection determines sphingomyelinase activity in cultured amniotic cells from mothers assumed to be carriers.

Nitrogen balance — The relationship between ingested and excreted protein nitrogen.

Nonprotein nitrogen compounds — Small-molecular-weight nitrogen-containing compounds in the blood including ammonia, creatine, creatinine, urea, uric acid, and various amino acids. Most of these substances are products of protein metabolism. Creatine is formed in muscle, and creatinine in the kidney and liver. Blood and urinary urea level is dependent on dietary intake and metabolism of proteins. Creatinine concentration is, however, independent from any endogenous source of nitrogen and proportional to the total muscle mass.

Obstructive jaundice — Obstruction of bile ducts impairs the excretion of cholesterol. In primary biliary cirrhosis abnormal lipoproteins are produced and serum triglycerides, phospholipids, and lipoproteins are also increased. Investigations:

- Serum cholesterol — increased
- Serum triglycerides — increased
- Serum phospholipids — increased
- Serum lipoproteins — increased
- Plasma ketones — normal

Oliguria — Represents a decrease of the daily urine output below 500 mℓ.

Oncotic pressure — Contribution of serum proteins to the total intravascular osmotic pressure.

Ornithine-carbamyl transferase — Important enzyme of the urea cycle. Normal serum level is very low; an increase, therefore, represents specific mitochondrial or cytolytic damage due to degeneration, infection, or toxic effects. Determination of the enzyme is very useful, particularly in chronic liver diseases when serum aminotransferases do not show the functional changes.

Osmotic pressure — Pressure exerted by the solutes present in body fluids.

Osteoblastoma — Tumor of bone cells.

Osteogenic sarcoma — An example of this disorder is seen on Plate 17. It occurs in two forms, an osteoblastic and an osteolytic form. The osteoblastic type is associated with a highly increased serum alkaline phosphatase. Enhanced alkaline phosphatase is also present in the serum in some other types of bone tumors, such as malignant giant-cell tumors and chondrosarcomas. Investigations:

- Acid phosphatase — normal
- Alkaline phosphatase — highly increased
- Serum calcium — normal
- Serum phosphorous — normal

Osteomalacia — Literally, means softening of the bone. Actually, osteomalacia involves the loss of calcium and phosphate from the bone trabeculae due to increased osteoblastic activity. Characteristic low blood calcium levels occur, which stimulate the activity of parathyroid glands. This in turn induces bone reabsorption which compensates for the low blood calcium. Increased osteoblastic activity is associated with enhanced serum alkaline phosphatase levels. Primary hyperparathyroidism produces similar abnormalities. Investigations:

- Serum acid phosphatase — normal
- Serum alkaline phosphatase — increased

Osteoporosis — Enlargement of the spaces of bone, producing a porous appearance. The loss of bony substance results in brittleness or softness of the bones.

Ovaries — Urinary gonadotropin or estrogen and progesterone measurements in the urine and serum can be used for diagnosis of disorders. Vaginal smears provide an indirect test for ovarian function.

Paget's disease — Paget's disease or osteitis deformans is characterized by idiopathic localized bone destruction and reabsorption, followed by compensatory new but abnormal

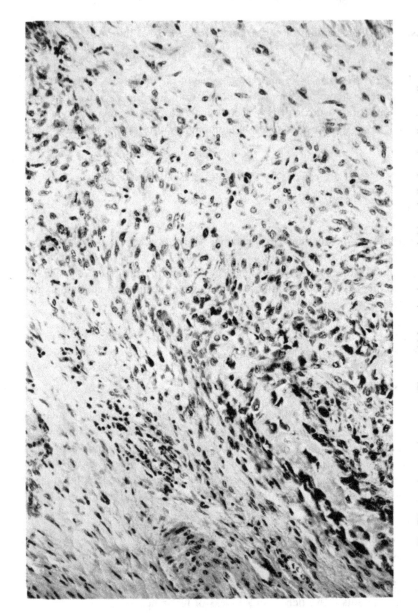

PLATE 17. Osteogenic sarcoma. The osteoid formation is surrounded by spindle-shaped malignant cells.

bone formation. The osteoblastic activity of certain areas of the skeleton is extremely high, e.g., long bones and skull, is extremely high and it is associated with highly increased serum alkaline phosphatase levels. This may be a reflection of osteogenic sarcoma formation, a frequent complication of the disorders. Investigations:

- Serum calcium and phosphate — usually normal
- Serum acid phosphatase — normal or increased
- Serum alkaline phosphatase — highly increased
- Urinary hydroxyproline — increased
- Hypercalcinosis

Pancreas carcinoma — Carcinoma of the head usually compresses and obstructs the common bile duct resulting in an increased level of serum alkaline phosphatase. This rise is probably connected with the backflow of hepatobiliary phosphatase, although regurgitation of alkaline phosphatase of bone origin has also been considered. Other disorders causing obstruction of the common duct, such as stones or carcinoma of the ampulla of Vater, produce similar effects on biochemical parameters. Extrahepatic biliary obstructions cause an increase of serum bilirubin and its appearance in the urine. Complete obstruction blocks the bilirubin excretion into the duodenum resulting in the absence of urobilinogen in the stool and urine. Excretion of enzymes and cholesterol is inhibited, but some parameters may remain normal unless there is an accompanying liver disease. Lack of bile in the intestines reduces the absorption of fat-soluble vitamins, including vitamin K which may lead to an abnormal synthesis or prothrombin. Investigations:

- Serum acid phosphatase — normal
- Serum alkaline phosphatase — increased
- Serum cholesterol and phospholipids — increased
- Plasma prothrombin — decreased
- Serum bilirubin, indirect — normal
- Serum bilirubin, direct — increased
- Urinary bilirubin — increased
- Urinary urobilinogen — decreased
- Fecal urobilinogen — decreased

Pancreatitis — Acute pancreatitis is connected with the obstruction of the pancreatic duct or with regurgitation of bile along this duct. It also occurs as a complication of extremely high serum triglyceride level (Plate 18). Acute or chronic inflammation of the pancreatic tissue due to necrosis of the cells causes a release of enzymes into the peritoneal cavity and blood stream. Predisposing factors are alcoholism, trauma, and biliary tract disease. Investigations:

- In the acute form: serum amylase — increased; serum is lipemic; calcium and potassium — decreased
- In the chronic form: enzymes — reduced; bicarbonate — decreased

Parakeratosis — Disorders affecting the stratum corneum of the epidermis.

Paraproteinemia — Appearance of pathological proteins, most often monoclonal ones, in the circulation. Investigation:

- Serum electrophoresis when an M-peak may occur; in case of an M-peak, immunoelectrophoresis may establish the type of the paraprotein present

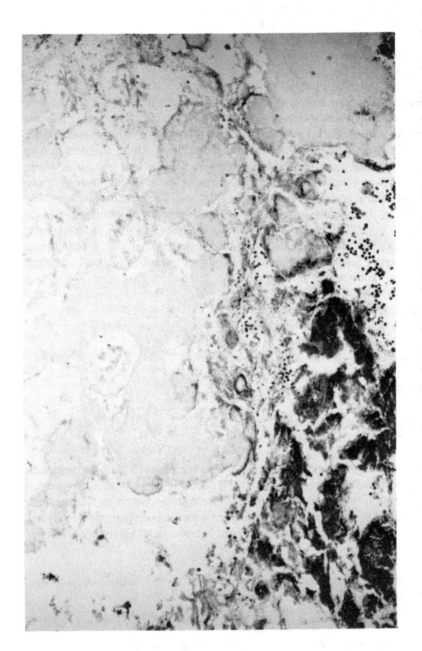

PLATE 18.   Acute pancreatic necrosis occurring as a complication of extremely high hypertriglyceridemia as can be found in patients with Type I or Type V hyperlipoproteinemia.

Parkinson's disease — Syndrome of progressive rhythmic tremor, mask-like facial expressions, and slowing of movements with increasing rigidity. Excessive creatinine is present in the urine.

Pellagra — Deficiency disease occurring endemically in some countries due to deficient diet or lack of assimilation of nicotinic acid. In the early stages it is characterized by debility, spinal pains, and digestive disturbances. In later stages there is erythema and drying and exfoliation of the skin. In severe cases nervous manifestations and mental disturbances result. Nicotinic acid deficiency is associated with inadequate protein intake. Hypoproteinemia and the development of edema are associated with diminution of plasma albumin concentration.

Pentosuria — Benign disease characterized by the appearance of pentosis in the urine.

Peritoneal fluid — Normally negligible, but increased in trauma, infections, and neoplasms. Portal cirrhosis, ruptured ectopic pregnancy and certain ovarian tumors, or obstruction of the hepatic or portal vein may occasionally cause an increase.

Peritoneal and pleural protein — In neoplasms and tuberculosis the protein content of these fluids is significantly increased. Pulmonary infarcts, lupus erythematosus, empyema, and congestive heart failure are connected with smaller elevations in the pleural fluid. In portal cirrhosis, the protein content of the peritoneal fluid is raised.

pH — Measurements of the pH provide an important diagnostic aid. Specimens in which pH is measured include urine, gastric and duodenal fluid, vaginal discharge, and sputum.

Phagocytosis — Ingestion process of foreign particulate or other matter principally by bacteria and by certain cells such as phagocytes and monocytes.

Phenylketonuria — Genetic disease due to the deficiency of phenylalanine *p*-hydroxylase. Accordingly, phenylalanine accumulates and its metabolites are excreted in increased amounts.

Pheromones — Ectohormones secreted to the surface of animals which change the sexual or social behavior of a second animal of the same species towards another.

Phosphatases — Many phosphatases occur in the human body, and are categorized according to the pH at which they exert maximum activity. Alkaline and acid phosphatases have clinical significance. The optimum activity of alkaline phosphatase is at pH 9 to 10, and acid phosphatase at pH 4 to 5.

Pigments — In several disease conditions pigment metabolism is disturbed. Increased pigmentation occurs in the skin in acromegaly, Fanconi's syndrome, hyperthyroidism, and metabolic disturbances of liver function (diffuse melanosis). In Addison's disease and Cushing's syndrome the increased melanin production is due to enhanced melanocyte stimulating hormone and adrenocorticotropic hormone production from functioning pituitary adenoma arising after bilateral adrenalectomy; in ectopic ACTH syndrome malignant tumors of nonendocrine origin produce these hormones. Many metabolic disturbances of the liver, such as hematochromatosis, porphyria cutanea tarda, progressive hepatolenticular degeneration (Wilson's disease), and biliary cirrhosis occasionally cause diffuse melanosis. Melanism is present at birth. The intensity of pigment sedimentation is enhanced during early childhood. During pregnancy darkening of the nipple and genitalia and melanoma occur due to increased estrogen and progesterone production, respectively. Decreased pigmentation develops in several amino acid disorders (due to defective or absent enzymes) such as albinisms (monophenol monooxygenase), histidinemia (histidine ammonia lyase), homocystinuria (cystathionine β-synthase), and phenylketonuria (phenylalanine 4-monooxygenase). Kwashiorkor and other chronic protein deficiency states, vitamin B_{12} deficiency, and Fanconi's syndrome cause decreased pigmentation in hair. Piebaldism is congenital; vitiligo shows some familial tendency. In both conditions discrete areas of absent pigmentation are characteristic. Several drugs can cause abnormal pigmentation, as can some anticonvulsants (diphenylhydantoin or mephenytoin), antimalarial drugs (chloroquin, amodiaquin, or atabrin), hormones (estrogen or progesterones), antineoplastic agents (bleomycin, myleran, cyclophosphamide, or adriamycin), or heavy-metal-containing agents (gold preparations, silver preparations, bismuth, or potassium arsenite). Psoralens cause an increased general pigmentation. On the other hand, local depigmentation occurs at the site of application of some hydroquinones, antimalarial drugs (chloroquine, hydroxychloroquine), fluorinated steroids, or triparanol. Depigmentation may also appear at sites independent from the application.

Pinocytosis — Ingestion process of surrounding liquid droplets by cells or minute vacuoles.

Pituitary — Most functions can be determined indirectly by measuring peripheral hormones. TSH secretion can be assessed by the determination of PBI, ACTH secretion by urinary 17-hydroxycorticosteroids and 17-ketosteroids, and FSH and LH secretion by urinary gonadotropins. When TSH is decreased, PBI is also decreased. When ACTH is increased, urinary steroids are increased. Conversely, when ACTH is diminished, urinary steroids are reduced. Decreased TSH secretion can be confirmed by the TSH stimulation test, diminished ACTH secretion by the ACTH stimulation test, and decreased TSH production using PBI content after TSH administration.

Plasma ammonia — This is increased in severe acute and chronic hepatocellular disease, large portal-systemic shunts, encelopathy, and congenital hyperammonemia syndromes.

Plasmacytoma — Tumor originating from plasma cells.

Plasma proteins — Liver is the major site of the synthesis of most proteins circulating in the blood such as albumin, α- and β-globulins, fibrinogen, and prothrombin. γ-Globulins are also produced in all reticuloendothelial tissues. The main functions of these proteins are the maintenance of plasma osmolarity. They play important roles as antibodies, and carriers of many hormones, vitamins, lipids, trace metals, and drugs; constituents include blood clotting factors and enzymes. Plasma proteins are essential in the plasma buffering system and as sources of nutrient proteins for the tissues. The regulation of plasma protein synthesis depends on various factors such as the quality and quantity of dietary amino acids and metabolic rate, as well as on certain hormones such as growth hormone, insulin, and androgens which enhance protein synthesis, and corticosteroids and thyroid hormone which increase protein catabolism. Causes of increased protein concentration are dehydration and stasis during venopuncture. Certain chronic diseases raise γ-globulin, such as some cases of cirrhosis, some collagen diseases such as systemic lupus erythematosis, certain chronic inflammatory conditions, or sarcoidosis. Causes of reduction are nephrotic syndrome, dietary protein deficiency such as in Kwashiorkor, liver disease (mainly albumin), or severe malabsorption.

Pleural fluid — This is normally negligible, but increased in infections, neoplasms, and trauma, and in diseases associated with hypoalbuminemia and congestive heart failure. Pulmonary infarct and other vascular disorders infrequently cause enhanced volume.

Pneumonitis — Inflammation of the pulmonary interstitium, mostly of viral origin.

Pneumothorax — Separation of the inner and outer pleural membrane by air.

Polycythemia (vera) — Increase in the number of circulating erythrocytes over the normal range.

Poliomyelitis — Viral disease attacking the anterior horn of the spinal cord resulting in the paralysis of motor neurons.

Porphyrin metabolism — Porphyrins are precursors of hemoproteins. Disorders include hepatic diseases such as acute intermittent porphyria hepatica and cutaneous type porphyria hepatica; bone marrow disorders, such as porphyria erythropoietica; and the side effects of many drugs and toxic compounds such as griseofulvin and lead poisoning.

Porphyrinuria — Excessive amounts of prophyrins are excreted in the urine. They are grouped into congenital and acquired categories. In the hereditary type of porphyrinuria, major urinary porphyrins are as follows: in erythropoietic prophyria uroprophyrin I and coproprophyrin I are increased due to bone marrow defect. In hepatic prophyria mainly coproprophyrin III, uroprophyrin I, uroprophyrins III, and porphobilinogen in small amounts are present; porphyrins are also increased in the liver. In acquired type porphyrinuria a mixture of prophyrins are present in the urine. In acute alcoholism, alcoholic cirrhosis, and following the action of toxic chemicals, coproporphyrin III is in excess. In infectious hepatitis and nonalcoholic cirrhosis: coproprophyrin I; in obstructive jaundice: coproprophyrin I; in aplastic anemias, poliomyelitis, and Hodgkin's disease: coproporphyrin III, are the major porphyrins.

Portal cirrhosis — The number of functioning hepatocytes is reduced, and the reduction is associated with reduced albumin production. The reticuloendothelial system is, however, overactive. Hepatic lymphoid and plasma cells show metaplasia and hyperplasia and therefore γ-globulin synthesis is increased. In postnecrotic cirrhosis, plasma protein changes are similar. In biliary cirrhosis, however, albumin is normal, particularly in the early phases, although β- and γ-globulins are increased in the blood. Conversion of proteins and carbohydrates to fat is inhibited due to damaged hepatic cells. Consequently, serum lipids are reduced. Investigations:

- Albumin — decreased
- γ-Globulins — increased
- Serum cholesterol
- Cholesterol esters — decreased
- Serum triglycerides — decreased
- Serum phospholipids — decreased
- Plasma ketones — normal

Potassium disturbances — Disturbances of potassium metabolism are connected with changes in extracellular concentration. Hypokalemia causes muscular weakness and hypotonia by interfering with the neuromuscular transmission. Intracellular potassium losses lead to extracellular alkalosis. Chronic depletion causes lesions in renal tubular cells and may lead to muscle cramps and tetany. Hyperkalemia occurs in anoxia and in acidosis of any kind, particularly when glomerular filtration rate is low. It can be associated with aldosterone deficiency as in Addison's disease. Any abnormalities of potassium metabolism cause characteristic changes in the electrocardiogram. Severe hyperkalemia may lead to cardiac arrest.

Prealbumin — This fraction contains the tryptophan-rich prealbumin and throxine-binding prealbumin. Biological functions are related to binding of vitamin A and thyroxine. Plasma level is reduced in severe hepatic disease and inflammations and increased during the administration of adrenal steroids.

Pregnancy toxicemia — Renal clearance of uric acid is impaired and blood uric acid increases. Urea clearance is usually unaltered. This differentiates pregnancy toxicemia from other inflammatory conditions affecting the kidney. Investigations:

- Serum uric acid — increased
- Urinary uric acid — decreased

Prenatal diseases — These are connected with inborn errors of metabolism. In some cases prenatal diagnosis is reported, expressed in fibroblasts, amniotic fluid or cells, or fetal erythrocytes. Central nervous system damage is the most common consequence of prenatal diseases. Some examples of this class of disorder are chromosome disorders, hemoglobinopathies (β-thalassemia and sickle cell anemia), myotonic dystrophy, collagen synthesis, gangliosidosis, glycogen storage disease, aminoacidopathies, and porphyrias.

Primary aldosteronism — The cause is normally a benign adenoma, hyperplasia, or carcinoma of the glomerulosa layer of the adrenal cortex. Excessive secretion of aldosterone activates the distal tubule of nephrons to reabsorb large amounts of sodium in exchange for potassium and hydrogen. Tubular reabsorption of water is, however, inhibited. Consequently serum potassium decreases and bicarbonate increases. Metabolic alkalosis further enhances the loss of potassium. Investigations:

- Serum sodium, bicarbonate — increased
- Serum potassium, chloride — decreased
- Blood volume — normal
- Urinary sodium — decreased
- Urinary potassium — increased
- Urine volume — increased
- Urine pH — normal or decreased
- Urinary aldosterone — increased
- Plasma renin

Proliferation — Growth with reproduction of a line of cells.

Propionic aciduria — Congenital disease characterized by the excretion of propionic acid in the urine.

Protein abnormalities — Increased protein content of various body fluids indicates some abnormalities with the exception of gastric and duodenal juices, vaginal discharge, sputum, and feces, where the protein content shows normal variations. Daily urinary protein excretion is less than 100 mg; spinal fluid contains 15 to 45 mg/dℓ; peritoneal, pleural, and synovial fluids, 200 to 400 mg/dℓ.

Protein-losing enteropathy — This syndrome is due to many diseases involving disorders of the enteric mucosa with or without ulceration and to abnormalities in the lymphatics draining the intestinal tract. It is due to enteric loss of protein as manifested in hypoproteinemia.

Proteinuria — Excretion of more than 150 mg/day protein in the urine.

Proteoglycan — The name proteoglycan represents native glycosaminoglycans occurring in connective tissue; these include hyaluronic acid, chondroitin, chondroitin sulfate, keratan sulfate, dermatan sulfate, heparin, and heparin sulfate. Secretion of proteoglycans into the urine depends on the biosynthesis and degradation of the total connective tissue. Pathological changes representing mucopolysaccharidosis are connected with increased excretion of glycosaminoglycans, as in rheumatoid arthritis, lupus erythematosus, psoriasis, leukemia, chronic hepatitis, and florid cirrhosis. Chondroitin sulfate is raised in progressive sclerosis, diffuse sclerodermia polyserositis, and dermatomyositis. Decreased excretion is found in primary hepatoma. Mental retardation is found in most mucopolysaccharidoses.

Primary glycosaminoglycanurias are related to enzyme defects resulting in increased excretion. In Hurler's syndrome (mucopolysaccharidosis I), dermatan sulfate, heparin sulfate, and chondroitin sulfate are excreted due to an α-L-iduronidase defect. In Hunter's syndrome (mucopolysaccharidosis II) the same proteoglycans are excreted, due to sulfoiduronate sulfohydrolase deficiency. Sanfilippo A and Sanfilippo B syndrome, also called mucopolysaccharidosis III or polydystrophic oligophrenia, are connected with an increase excretion of heparan sulfate and dermatan sulfate due to heparin sulfate sulfohydrolase or N-acetyl-α-D-glucosamidase deficiency, respectively.

Pulmonary emphysema — Carbon dioxide excretion is impaired and blood concentration is increased which lowers pH. Kidney compensates for the respiratory acidosis by retaining and producing more bicarbonate. Chloride is excreted to compensate for the increased bicarbonate. Respiratory acidosis occurs in other lung diseases such as pneumonia, pulmonary edema, and pneumothorax, and in conditions of hypoventilation connected with anesthesia or poliomyelitis. Investigations:

- Serum sodium, potassium — normal
- Serum chloride — decreased
- Serum bicarbonate — increased
- Blood volume — normal or increased
- Urinary sodium — decreased
- Urinary potassium — normal
- Urine volume — normal
- Urine pH — decreased

Purpura — Hemorrhages into skin and mucous membranes due to increased capillary permeability.

Pyelonephritis — Inflammation of the renal papillae resulting in marked proteinuria.

Pyloric obstruction — Associated with losses of fluid and major electrolytes with the exception of bicarbonate. The constricted pylorus blocks the passage of gastric juice into the intestines and it is vomited. Consequently, large amounts of water, potassium, and hydrogen and chloride ions, as well as smaller amounts of sodium are lost. Loss of hydrogen ions causes metabolic alkalosis with depressed respiratory rate. Carbon dioxide is retained as carbonic acid which partly compensates for the alkalosis. Alkalosis inhibits carbohydrate metabolism and lipid breakdown, and results in excess ketone formation and a shift of pH to acid ranges. Similar changes may manifest in prolonged gastric losses, administration of alkaline salts such as antacids, and may also be due to potassium deficits. Investigations:

- Serum sodium, potassium — decreased
- Serum chloride — decreased

- Serum bicarbonate — increased
- Blood volume — decreased
- Urinary sodium — decreased
- Urinary potassium — normal
- Urine volume — increased
- Urine pH — decreased

Reducing substances in urine — Urine contains several reducing substances such as glucose, fructose, galactose, lactose, pentoses, glucuronates, and homogentisic acid. Glucose is present in diabetes mellitus and renal glucosuria; fructose, in essential fructosuria and hereditary fructose intolerance; galactose, in galactosemia; pentoses, in essential or alimentary pentosuria (both diseases are rare); lactose, in congenital lactose deficiency, in late pregnancy, and during lactation; glucuronate is present in combination with drugs and their metabolites; and homogentisic acid is present in alkaptonuria.

Renal calculi — Minor changes in urinary composition may cause precipitation of normal products of metabolism as crystals or calculi. Conditions favoring stone formation are high urinary concentration of one or more constituents of the glomerular filtrate, low urinary volume or abnormally high rate of excretion, changes of urinary pH, or stagnation of urine due to obstruction. Calculi are mainly composed of calcium oxalate or phosphate, uric acid, xanthine, and rarely cystine. These stones precipitate on a nucleus (nidus) in the kidney calices and wedged in the ureters.

Renal failure — It occurs in acute and chronic form; the former may be reversible. Loss of kidney function is apparent in the acute form. It is accompanied by nitrogen retention (increased blood urea nitrogen), and creatinine retention, acidosis, fluid retention, and hyperuricemia. In the chronic form, further signs of the condition are hyperphosphatemia and hypocalcemia.

Renal function tests — These can be used to detect any renal damage; the cause of any kidney disease cannot be established. The tests are based on the glomerular filtration and tubular reabsorption. Determination of the glomerular filtration is based on the concentration of a substance in the blood and in the urine, and on urinary volume. The glomerular filtration rate or renal clearance of a substance can be calculated as follows:

$$G.F.R. = \frac{U.V}{B}$$

where B = concentration in blood, U = concentration in urine, and V = excreted urine volume in milliliters at a given period. These are substances, such as inulin, which are excreted only by glomerular filtration. In practice, however, urea and creatinine clearance are used. The rate of endogenous creatinine excretion is very close to inulin, but in urea clearance there are some difficulties. Not all urea filtered by the glomeruli is excreted in the urine; about one third is reabsorbed passively and the amount excreted is dependent on the urinary volume.

Abnormal clearance values indicate defects of glomerular filtration. Defective tubular reabsorption can be detected by the presence of amino acid or glucose in the urine, alkaline pH, and changes in specific gravity. In chronic nephritis and invariably in nephrogenic diabetes insipidus the specific gravity is low. Defects in tubular secretion indicate renal tubular disease. It may occur in heart failure and shock.

Renal glycosuria — Tubular reabsorptive capacity of the kidney is impaired causing glycosuria with normal blood sugar levels. Nephritis and the de Toni-Fanconi syndrome cause similar changes. Investigations:

* Fasting blood sugar, and 2-hr postprandial blood sugar — normal
* Urine sugar — increased

Renal tubular acidosis — A defect occurs in the renal tubular reabsorption of calcium and phosphates and consequent urinary loss of these constituents. Serum calcium and phosphate may decrease and secondary parathyroidism ensues associated with an enhanced serum alkaline phosphatase level, which in turn readjusts serum calcium level to normal. If dietary calcium intake is limited, osteomalacia or rickets develop. Fanconi syndrome may be associated with similar changes. The excretion of hydrogen and reabsorption of bicarbonate from the glomerular filtrate is defective in this disorder. Consequently serum bicarbonate is reduced and chloride is increased. In exchange for sodium the tubules secrete potassium; thus serum potassium is decreased. Hydrogen ions are retained, blood pH is reduced, and metabolic acidosis ensues. Investigations:

* Serum sodium, serum potassium — decreased
* Serum chloride — increased
* Serum bicarbonate — decreased
* Serum calcium — normal or decreased
* Serum phosphates — decreased
* Serum alkaline phosphatase — increased
* Blood volume — decreased
* Urinary sodium, potassium, and calcium — increased
* Urine volume — increased
* Urine pH — increased

Rheumatic fever — Febrile disease characterized by painful migratory arthritis and predilection to heart damage leading to chronic valvular disease. Development is influenced by hereditary susceptibility and is related probably to streptococcal infection of the throat and climatic conditions.

Investigations reveal an increase in α- and γ-globulins. The rise of γ-globulin is parallel with increasing titers of streptococcal agglutinating and antistreptolysin factor. Hyperglobulinemia and albuminuria are associated with the febrile state. Plasma glycoprotein is increased, accompanied by tissue reduction. Marked hypochloremia may occur.

Rheumatoid arthritis — Systemic disease of unknown origin involving a chronic proliferative inflammation of the synovial membrane. This in turn destroys the cartilage of the affected joint. Another feature is the rheumatoid nodule in various locations. Investigations:

* Decreased albumin and increased immunoglobulin levels

Rickets — The laboratory symptoms are similar to those of osteomalacia. In this condition X-ray investigations and the response to vitamin D therapy help to confirm the diagnosis. A similar laboratory finding is obtained in other disorders associated with low blood calcium resulting from inadequate intake or absorption or enhanced urinary elimination. Investigations:

* Serum acid phosphatase — normal
* Serum alkaline phosphatase — increased
* Serum calcium — low

Sarcoidosis — This represents a chronic granulomatous inflammatory disease of unknown origin, possibly due to an immunologic imbalance with wide-ranging signs and symptoms. Investigations:

- Hypercalcemia
- Increased alkaline phosphatase with minimal or no hyperbilirubinemia
- Occasional hyperuricemia
- Hyperuricuria in 25% of patients

Sebaceous gland — Exocrine gland excreting sebum onto the skin surface.

Sebotrophic hormones — Hormones facilitating sebum secretion.

Sebum — Oily secretion of certain subcutaneous (sebaceous) glands.

Serum electrolytes — Electrolytes present in serum, such as sodium, potassium, chloride, and bicarbonate, frequently show reciprocal changes in disease conditions. Some changes are connected with water changes. Electrolyte disorder has to be correlated with a clinical diagnosis of the state of hydration, intake and output, and acid-base equilibrium. Vomiting, diarrhea, hyperventilation, polyuria, or oliguria may help in establishing fluid and electrolyte gains or losses.

Serum enzymes — Serum enzymes originate from various tissues and their level reflects the turnover and physiological cellular catabolism. Three groups can be distinguished: enzymes required for normal function in the blood (coagulation and pressure regulation); enzymes secreted from exocrine glands (pancreas, gastric mucosa, and prostate); enzymes with site of action in various tissues and occurring in plasma as waste products. For clinical diagnosis, serum amino transferases, γ-glutamyl transpeptidase, lactate dehydrogenase, alkaline phosphatase, and creatine phosphokinase are most important. Some enzymes are widely distributed in various tissues; some such as creatine phosphokinase are only found in high levels in specific organs. The presence of these enzymes indicates the disease of the original tissue. Measurement of these enzymes is clinically important in the diagnosis of several diseases. Some are present in blood cells, some in plasma or serum. Investigations:

- Acetylcholinesterase — low in paroxysmal nocturnal hemoglobinuria
- Acid phosphatase — raised in prostatic carcinoma, hemolysis, and thromboembolic disease
- Adenosine deaminase — low or absent in immune deficiency disease
- Adenylate kinase — absent in congenital hemolytic anemia
- Alanine aminotransferase — increased in many diseases
- Alkaline phosphatase — raised in hepatic and bone diseases
- Amine oxidase — increased during pregnancy
- α-Amylase — raised in acute pancreatitis and in mumps due to salivary gland inflammation
- Arginase — raised in pernicious anemia and thalassemia major
- Aspartate aminotransferase — increased in many diseases
- Catalase — absent in congenital acatalasia
- Creatine kinase — increased immediately after cardiac infarct
- Cystyl-aminopeptidase — increased during pregnancy
- Dopamine-β-monooxygenase — reflect the activity of the peripheral nervous system
- Fructose-diphosphate aldolase — raised in hemolytic disease

- Glucose 6-phosphate dehydrogenase — reduced in congenital and drug-induced hemolytic anemia
- Glucose 6-phosphate isomerase — absent in nonspherocytic congenital hemolytic anemia
- β-Glucuronidase — defective in some primary glycosaminoglycanurias (mucopolysaccharidosis)
- γ-Glutamyltransferase — increased in many diseases; inductive action of drugs associated with an increase, elevated in experimental tumor models
- Hexokinase — activity decreases during cell aging
- Hypoxanthine phosphoribosyltransferase — absent in Lesch-Nyhan syndrome
- Lactate dehydrogenase — increased in many diseases
- Pyruvate kinase — absent in nonspherocytic congenital hemolytic anemia
- Triacylglycerol lipase — increased in pancreatic diseases and may be in severe renal disease (uremia) due to diminished renal excretion

Serum iron — Iron is essential in hemoglobin, myoglobin, cytochromes, and various enzymes (catalase, peroxidase, and cytochrome oxidase). It is absorbed in the duodenum and upper jejunum, transported probably as an iron-amino acid complex bound to β-globulin called transferrin. It is stored in the liver, spleen, and bone marrow as ferritin and hemosiderin. Iron utilization and hemoglobin synthesis are under hormonal control; erythropoietin and variations in oxygen supply exert part of the control. Disorders are related to increased absorption or depleted stores. In hemochromatosis plasma transferrin is saturated and plasma iron rises. Excess iron is accumulated in parenchymal tissues such as liver, pancreas, endocrine glands, gonads, and skin. If in the pancreas is involved in the storage, it may lead to diabetes. In hemosiderosis excess iron is stored in the reticuloendothelial tissue. Iron deficiency is associated with depleted stores due to relative or absolute deficiency of iron intake and results in anemia. This condition may develop in growing children, pregnant women, or following gastrointestinal bleeding. Measurement of serum iron content is becoming increasingly important in biochemical diagnosis. Deviations from normal level (man, 90 to 140 μg%; woman, 80 to 120 μg%) characterize various diseases. Increases occur in cases of anemia, pernicious anemia, hemochromatosis, and hepatitis. Decreases are connected with loss of blood, iron-deficiency anemia, acute and chronic infections, and malignant tumors.

Serum lipids — Three classes of lipids are present in serum: triglycerides, phospholipids, and steroids, mainly cholesterol. The regulation of serum levels of various lipids is dependent on an equilibrium between input/intake on one side and production/output on the other, involving storage, utilization, and excretion. Various factors controlling their levels have not been clearly established. The major regulatory organ is the liver. Increased intake usually decreases hepatic synthesis and enhances biliary excretion. Dietary fats only affect serum lipids after a meal. Several hormones influence lipids. Insulin increases synthesis of fatty acids in the liver and adipose tissue; cortisol increases breakdown and mobilization. Estrogen reduces plasma half-life of cholesterol; ovariectomy increases it. Thyroxine also reduces plasma cholesterol and phospholipids. Epinephrine mobilizes fatty acids from adipose tissue.

Serum magnesium — This is essential in activating many enzymes and in bone formation. Magnesium is transported in the blood in protein-bound form. Regulation depends on parathyroid and growth hormones. Growth hormone increases intestinal absorption; parathyroid hormone reduces renal excretion. Serum magnesium level is increased in Addison's disease and in hyperthyroidism, and reduced occasionally in diabetes. In diarrhea or ulcerative colitis serum magnesium is reduced due to decreased intestinal absorption and loss of intestinal secretions. There is a reduction of serum level in hypoparathyroidism due to diminished tubular reabsorption, and in Laennec's cirrhosis due to lack of absorption.

Sickle cell anemia — This is a disorder of hemoglobin synthesis. It is due to the presence of a mutant hemoglobin, hemoglobin S. Erythrocytes contain this abnormal hemoglobin which decreases survival time. Investigation:

- Involves hemoglobin electrophoresis. Demonstration of hemoglobin S with variable amounts of HbF and no HbA

Sodium disturbances — Disturbances of sodium metabolism are due to changes in its extracellular concentration. Hyponatremia causes cellular overhydration. The effect of this condition on cerebral cells is headache, confusion, and fits. Hypernatremia causes cellular dehydration and thirst. The effect on cerebral cells is mental confusion and coma.

In the well-regulated homeostatic mechanism, sodium deficiency is the consequence of vomiting, diarrhea, or excessive sweating. The homeostatic mechanism fails in Addison's disease due to the absence of aldosterone. Excess sodium is predominant in Cushing's syndrome and Conn's syndrome (primary aldosteronism) due to an excess of aldosterone or mineralocorticoids.

Special cerebrospinal fluid components — Bilirubin is increased in jaundice and in cerebrovascular diseases connected with hemorrhages. The presence of amino transferases and lactic dehydrogenases suggests vascular or neoplastic diseases of the brain. Chloride content may be reduced in acute and tuberculous meningitis in association with low serum chloride.

Special fecal components — Trypsin is present in chronic pancreatitis; urobilinogen and stercobilin changes indicate various types of jaundice.

Special gastric juice components — In gastric carcinoma certain mucoproteins are present.

Special peritoneal fluid components — Amylase may be present in large amounts in acute pancreatitis and bilirubin in the presence of a biliary-peritoneal fistula.

Special urinary components — Amylase is increased in various pancreatic disorders, gallstone obstructing pancreatic duct and in high intestinal obstruction. Bilirubin and urobilin are raised in disorders of erythrocytes such as hereditary spherocytosis, constitutional and metabolic liver diseases, and in hepatic cirrhosis. Catecholamines are increased in tumors of the adrenal medulla or in tumors of extraadrenal ganglionic tissue called pheochromocytomas.

Homogentisic acid is increased in alkaptonuria. Melanin (5,6-dihydroxyindole conjugate) is raised in melanotic sarcoma. Phenylpyruvic, phenyllactic, and phenylacetic acids are increased in phenylketonuria. Metabolism of some coumarins also results in the formation of these aromatic acids. Porphyrins are increased in porphyrias of hepatic and erythropoietic origin, as well as in lead poisoning and other toxic changes. Some steroids are elevated, such as 17-ketosteroids and 17-hydroxycorticosteroids in Cushing's syndrome, and 17-ketosteroids in Stein-Leventhal syndrome (polycystic ovaries) and during pregnancy together with gonadotropin.

Specific gravity — Changes in the specific gravity of urine or peritoneal and pleural fluids are important signs of disease conditions. Urine specific gravity is abnormal in kidney diseases, diabetes insipidus, and primary aldosteronism; the specific gravity of peritoneal, pleural, and synovial fluids is abnormal in congestive heart failure, hepatic or renal disorders and infections, or bleeding or in the presence of tumor cells. Low specific gravity of the stool (floating in water) indicates malabsorption.

Spinal fluid protein — Increased protein level is mostly connected with hemorrhage. Bacterial meningitis or neoplasms enhance spinal fluid protein, but there are no cells in the latter condition. High protein content without cells is found in certain metabolic diseases such as hypothyroidism and diabetes. In multiple sclerosis γ-globulin fraction is increased.

Sputum — Examinations should be carried out for appearance, pH, and blood contamination. The color is normally clear, occasionally slightly turbid. If there is bleeding in the respiratory tree, it becomes red. It is brown in pneumococcal pneumonia, and green or brown in lung abscesses. The sputum is green in *Pseudomonas* infections of the respiratory tract.

Sputum volume — Normal production is negligible. It is, however, increased in infections of the respiratory tree, bronchiectasis, lung abscess, and pulmonary edema.

Starvation — The major effect of starvation is the presence of organic acids in blood as in diabetic keto acidosis. Lack of carbohydrate intake causes glycogen depletion and increased metabolism of proteins and lipids which produce ketones (keto acids). With starvation or in diabetic acidosis keto acids are synthesized in excess and plasma level increases. Since they are neutralized by bicarbonate, plasma level of these anions is reduced, but the by-product carbon dioxide is eliminated by the lungs so the metabolic acidosis is compensated. When protein intake is reduced, protein catabolism is increased, producing enhanced blood and urinary urea and creatine. Diabetes mellitus and diseases connected with high fever frequently show similar changes. Protein intake is insufficient and the production of plasma proteins is reduced. Investigations:

- Serum sodium, potassium, chloride — normal
- Serum bicarbonate — decreased
- Blood volume — normal or decreased
- Urinary sodium — normal or decreased
- Urinary potassium — normal or increased
- Urinary ketones; urine volume — increased
- Urine pH — decreased
- Blood urea nitrogen, creatine — increased
- Urinary urea, creatinine — increased
- Albumin — decreased
- α-Globulin, β-globulin, γ-globulin — decreased

Steatorrhea — Represents an increased excretion (>7 g/day) of fat in the feces. Investigation:

- Determination of fecal fat in the timed (72-hr) stool collection

Stein-Leventhal syndrome — It is also called polycystic ovary syndrome, and involves anovulation, amenorrhea, infertility, and hirsutism. Investigations:

- Increased luteotrophic hormone
- Low or low-normal follicle stimulating hormone
- Increased dihydroandrosterone and testosterone

Synovial fluid — This is normally clear and has an amber yellow color; it is turbid in infections and red following trauma. Normal synovial fluid forms no clots; clotting on standing represents inflammation. In the presence of 1% acetic acid hyaluronic acid precipitates and mucin is formed. Normal synovial fluid contains mucin. In rheumatoid arthritis there is no coagulate or the precipitation is transient. In gouty arthritis mucin formation is very little.

Synovial fluid volume — This is normally negligible, but may be increased in inflammatory, traumatic, and neoplastic conditions of the joints such as in rheumatoid arthritis and osteoarthritis.

Systemic lupus erythematosis — Inflammatory disease of probably autoimmune etiology affecting most commonly the skin, mucous membranes, joints, and kidney. It is associated with abnormalities of the immunoglobulins, and often manifested as the presence of antinuclear factor and lupus erythematosis cells.

Tangier disease — Genetic disease due to α-lipoprotein deficiency resulting in very low cholesterol and triglyceride levels. Clinical features are fat malabsorption, fatty liver, erythrocyte abnormality (acanthosis), neurological symptomes, and growth retardation.

Tay-Sachs disease — Congenital disease manifested in infancy due to hexosaminidase deficiency and causing blindness, mental retardation, and eventually death.

Teratologic effects — Effects either by drugs, chemical, or physical factors in the environment causing intrauterine deformities in fetuses.

Testes — Hypo- or hypersecretions of testosterone can be determined indirectly by the 24-hr urinary gonadotropins. Through regulatory feedback mechanism decreased production of testosterone is connected with an enhanced pituitary secretion of the interstitial cell stimulating hormone and increased urinary levels of gonadotropins. Conversely, increased secretion of testoterone is associated with reverse changes. Decreased testoterone production is often paralleled by reduced spermatogenesis.

Thalassemia — Congenital disease occurring mainly in the Mediterranean region due to decreased globin synthesis involving either α- or β-chain deficiency and resulting in hemoglobin F production. It is characterized by splenomegaly, anemia, changes in the bones, and skin pigmentation.

Thyroid gland — Disorders of this endocrine gland are diagnosed by direct tests such as the protein-bound iodine or radioactive iodine uptake. If a patient received radiocontrast material, iodine-containing compounds, or thyroid hormone preparations, the uptake of ^{131}I-thyronine or thyrobinding index can be applied. Hypothyroid patients may be given the thyroid-stimulating hormone test. Hyperthyroid patients may be given the triiodothyronine suppression test. Hashimoto thyroiditis (Plate 19), is independent of iodine supply.

Transferrin — This plasma fraction is also called siderophilin and is responsible for binding and transport of iron. It is reduced in nephrosis and in diseases associated with inflammatory or necrotic diseases. A genetic defect causing atransferrinemia has also been described.

Trauma — This is characterized by a sympathetic central-nervous-system-mediated response, connected with the release of mobilizing mediators and hormones such as epinephrine, cortisol, glucagon, growth hormone, and insulin, a relative inhibitor. The ensuing

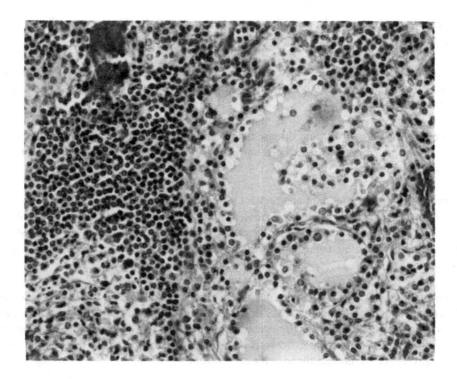

PLATE 19. Hashimoto thyroiditis characterized by heavy lymphocytic infiltration. This condition
is independent from the iodine supply.

status of their interaction is glycogenolysis, gluconeogenesis, ketogenesis, lipolysis, and
muscle proteolysis producing a hypercaloremia. The severity of the trauma can be measured
by the presence of metabolic constituents derived from these processes. There are two phases
of injury, acute and ketoadaptive. The latter can be established by the presence of ketonemia
and ketonuria. Mild trauma follows elective surgery; moderate trauma, major surgery; and
severe trauma occurs following burns to over more than 35% of the body surface.

Triglyceridemia — Due to lipoprotein lipase deficiency and increased lipoprotein rate
fat removal is inhibited. Development of xanthomata on the skin and enlarged spleen are
fairly common. In essential triglyceridemia development of atherosclerosis is less frequent
than in familial hypercholesterolemia. Investigations:

- Serum cholesterol — normal
- Serum triglycerides — greatly increased
- Serum phospholipids — increased
- Serum lipoproteins — increased
- Lipoprotein phenotyping

Tubular dysfunction — Generalized tubular damage is caused by acute tubular necrosis
due to prolonged renal circulatory insufficiency, various poisons, and progressive damage
to the tubules associated with Wilson's disease, galactosemia, hypercalcemia, hypokalemia,
or various heavy metal poisons. Consequences of tubular dysfunction are acidosis, hypokalemia, hypophosphatemia, polyuria, and low urea and high sodium concentration in the
urine.

Turner's syndrome — Due to aneuplody, there are only 45 chromosomes. Some of the sex chromosomes, either X or Y, are missing.

Tyrosinemia — Represents the persistence of tyrosine in blood beyond the neonatal period. Unusual amounts of tyrosine and oxidation products are excreted in the urine, indicating tyrosyluria.

Ulcerative colitis — Chronic disease due to an unknown cause involving inflammation of the colonic mucosa. It is associated with rectal bleeding and diarrhea and is connected with malabsorption.

Urea — The most important end product of protein metabolism. It is formed in the liver and excreted through the kidneys. Increased blood levels represent primary or secondary renal impairments and absolute or relative renal insufficiency. Prerenal, renal, and postrenal abnormalities can be distinguished. Urea clearance is used for the diagnosis of kidney function.

Uric acid — Uric acid is a waste product of nucleic acid (purine) metabolism. It is present in the serum and excreted mainly in the urine. Uric acid is formed from adenine and guanine primarily in organs showing a high metabolic turnover, such as liver and bone marrow, and to a smaller extent in muscle. Exogenous purines are absorbed from the diet, but this does not usually affect the normal blood level of uric acid.

Urine — Examinations include appearance (color and clarity); pH; solid content (specific gravity, osmolality, and refractive index); the presence of protein, glucose or reducing substances, ketone bodies, hemoglobin, bile and urobilinogen; and microscopic examination of formed elements of sedimentation. The normal color of the urine is amber. Abnormal color may occur after the ingestion of various drugs and in some disease conditions. Almost colorless urine indicates a dilution due to large fluid intakes, reduction of perspiration, chronic intestinal nephritis, diabetes insipidus, untreated diabetes mellitus, diuretic therapy, or alcohol ingestion. An orange color indicates concentrated urine due to restricted fluid intake, fever, excess sweating, or small amounts of bile pigments. Red indicates blood, hemoglobin, hematin, or porphyrins. Deep red indicates a mixture of methemoglobin and oxyhemoglobin or porphyrins. Dark brown indicates methemoglobinuria or melanin. Brown-black urine is a sign of excess hemoglobin or homogentisic acid. Smokey urine indicates erythrocytes. Yellow foam bilirubin, and green biliverdin and other bile pigments. In congenital porphyria urine is red; in acute intermittent porphyria hepatica the urine turns brownish red on standing; in alkaptonuria it turns black. In obstructive jaundice the color of urine is brown, in glomerulonephritis smokey brown, and in *Pseudomonas* infection green.

Concentration of solids varies from the solid content of plasma (excluding proteins), but urinary solute concentration can be used as an index for the concentrating ability of the kidney. In pathological conditions comparison of serum and urine osmolality may reflect the nature of disease. Serum osmolality decrease and urine osmolality increase in hepatic cirrhosis and congestive heart failure due to decreased glomerular filtration rate or secondary hyperaldosteronisms, and following surgery due to increased adrenal stimulation or decreased antidiuretic hormone secretion. Serum osmolality increase and urine osmolality decrease in hypokalemia and hypercalcemia due to increased water loss, and in diabetes insipidus and nephrogenic diabetes due to decreased antidiuretic hormone or decreased response. Both types of osmolality are decreased in compulsive water drinking or iatrogenic water loading due to dilution. Both are increased in high-protein diet. Retention of sodium in primary aldosteronism causes a slight increase of serum osmolality and a slight decrease of urine osmolality. In Addison's disease, sodium excretion is increased, serum osmolality decreased, and urine osmolality normal to increased. Other measurements are given under adequate headings.

Urinary amino acid — Elevated amounts of amino acids are present in the urine in diseases with renal tubular defects such as Fanconi syndrome and Wilson's disease, and in a wide variety of inborn errors of metabolism. In phenylketonuria, phenylalanine and phenylpyruvic acid are excreted in excess; in maple sugar disease leucine, isoleucine, and valine; in cystinuria cystine, arginine, ornithine, and lysine; in homocysteinuria, homocysteine; in Hartnup disease, all common amino acids (with the exception of arginine, methionine, proline, and hydroxyproline), as well as indoleacetic acid and indican. β-Aminoisobutyric aciduria and argininosuccinic aciduria are rare conditions associated with the excess elimination of these compounds.

Urinary catecholamines — Catecholamines and phenolic acid metabolites are derived from the adrenal medulla, brain, heart, and catecholamine-secreting tumors. Increased quantities are excreted in the urine in pathological conditions; vanilmandelic acid, homovanillic acid, metanephrine and normetanephrine in pheochromocytoma and ganglioneuroblastoma; norepinephrine, dopamine, vanillic acid, homovanillic acid, and vanilmandelic acid in neuroblastoma; *p*-hydroxyphenylacetic acid in malignant melanoma, Wilm's tumor, ascorbic acid deficiency, and tyrosinosis; and *o*-hydroxyphenylacetic, *o*-hydroxyphenyllactic, and *o*-hydroxyphenylpyruvic acid in phenylketonuria.

Urinary glucose — Normally only a negligible amount of glucose is excreted in the urine. Glucosuria is apparent in primary diabetes mellitus and in symptomatic diseases such as thyrotoxicosis, Cushing's syndrome, acromegaly, and pheochromocytomas. Increased urinary glucose excretion is also found in acute disorders of the central nervous system such as cerebrovascular accidents and meningitis, and in alimentary glucosuria, renal glucosuria, glucose-galactose malabsorption, and galactose-fructose intolerance.

Urinary protein — Normal daily protein elimination is insignificant (i.e., less than 100 mg) due to incomplete tubular reabsorption of the small amount of protein present in the glomerular filtrate. In most cases this protein is albumin, and part of it represents uropepsinogen. Urinary protein excretion is increased in inflammation, vascular, metabolic, and toxic diseases of the kidney, and is associated with hematuria. Kidney infection causes elevated urine protein. It is most significant in glomerulonephritis. Infections of the lower urinary tract, however, usually do not raise urinary protein excretion. Collagenous diseases are also connected with proteinuria, often with casts and increased erythrocyte excretion. Carbon tetrachloride, mercury, and other poisoning cause renal tubular necrosis and proteinuria. Preeclampsia is also associated with increased urinary protein. Diabetes mellitus, gout, and amyloidosis as metabolic conditions frequently show proteinuria. Vascular diseases, namely congestive heart failure and essential hypertension associated with nephrosclerosis also cause increased protein excretion. In multiple myeloma a specific protein called Bence-Jones protein is excreted in large amounts.

Urinary sugar — Some carbohydrate disorders are characterized by excess urinary excretion of sugars which gives positive reducing tests. Glucosuria is given under a separate heading. Other sugar disorders show negative results for glucose oxidase tests the sugar found in various disorders. Fructose is present in essential and hereditary fructosuria; xylulose in primary pentosuria; arabinose in alimentary pentosuria; galactose in galactosemia and galactose-fructose intolerance; lactulose in lactulosemia; lactose and galactose in disaccharide intolerance; and glucose and galactose in monosaccharide malabsorption.

Urine pH — This is normally between 6 and 7.8. Changes reflect blood pH changes such as metabolic alkalosis or acidosis which shift the pH to more alkaline or acid, respectively. If tubular reabsorption is defective, as in chronic glomerulonephritis, renal tubular acidosis, or even in the presence of acidosis, the urine is alkaline due to loss of bicarbonate. Some infecting organisms such as *Proteus* and *Pseudomonas*, which metabolize urea to ammonia also produce alkaline urine. Infections with tubercle bacilli cause acid urine.

Urine specific gravity — The normal specific gravity is 1015 to 1025. This reflects the degree of renal reabsorption of solutes. It is reduced in chronic nephritis due to general loss of reabsorption capability, in pituitary diabetes insipidus due to antidiuretic hormone insufficiency and consequent reduction of distal tubular reabsorption, and in renal diabetes insipidus due to loss of renal tubular responses to antidiuretic hormone. Primary aldosteronism may also reduce specific gravity. It is increased in diabetes mellitus and dehydration due to increasing solutes in the urine.

Urine volume — The daily urine output is dependent on water intake, glomerular filtration rate, and tubular reabsorption. The daily volume is normally between 1 and 2 ℓ daily, representing about 0.5% of the glomerular filtrate. The urine volume is increased by increased fluid intake as in hysterical diabetes insipidus and by increased electrolyte load as in diabetes mellitus. In hyperthyroidism the increased urine volume is due to an increased glomerular filtration rate; in chronic nephritis and diabetes insipidus it is due to decreased tubular absorption. Urine volume is decreased when fluid intake is reduced as in dehydration or due to abnormal loss of body fluids as in diarrhea, vomiting, or hemorrhagic shock; or due to decreased glomerular filtration rate as in congestive heart failure, nephritis or shock, or metabolic diseases or toxic actions on the kidney. Obstructions of the urethra and bladder cause decreased urine volume as in prostatic hypertrophy, urethral strictures, and neurogenic bladder.

Urobilinogen — Derives from bilirubin in the intestines by microbial transformation. Antibiotic therapy causes unreliable measurements. Very low fecal values indicate cholestasis. An increase is connected with increased bilirubin production.

Vaginal bleeding — Vaginal discharge contains blood for 3 to 5 days in every menstruation cycle, which may have 21 to 35 days duration. The average loss is 50 mℓ but may be as much as 150 mℓ. The presence of blood is pathological between periods, or if menstrual bleeding is prolonged or exceeds the maximum volume. Abnormal bleeding is due to hormone disorders or complications of pregnancy, or is associated with neoplasms such as carcinoma of the endometrium or cervix.

Vaginal discharge — This is normally minimal but may be elevated during ovulation and due to bacterial and parasitic infections.

Vaginal discharge pH — The normal pH is 3.5 or 4.5 *Gonococcus* or other infections and atrophic vaginitis shift it to a more alkaline region.

Vernix caseosa — Sebaceous deposit covering the fetus consisting of secretion by fetal skin glands.

Very-low-density lipoprotein (VLDL) — This plasma fraction is also called pre β-lipoprotein. It is responsible for the binding of some lipids, mainly triglycerides. It is elevated in Type IV hyperlipidemias.

Viral hepatitis — Necrotic liver cells release large amounts of amino-transferases. The rise in the blood is more pronounced than in myocardial infarction; however, it is extremely high in carbon tetrachloride poisoning. Aminotransferase changes and a characteristic pattern indicate the acute disease process.

Sudden elevation or a persistent high level suggest relapse or other complications, or postnecrotic cirrhosis. Lactic dehydrogenases are usually not increased in viral hepatitis. Due to necrosis the production of albumin in the liver cells is also impaired. However, γ-globulin synthesis is enhanced associated with the immunological response to the virus and to the necrotic cells. Albumin changes usually manifest in the late preicteric or early icteric stage and return to normal levels in 5 to 6 weeks from the beginning of the disease. If hypoalbuminemia is prolonged, massive necrosis, postnecrotic cirrhosis, or chronic hepatitis is apparent. In these cases, plasma albumin levels indicate the extent of the destructive process. Increase of blood γ-globulin frequently occurs before the onset of jaundice; it often returns to a normal level in 3 to 4 months. γ-Globulin level marks the severity and progression together with serum aminotransferases and have great prognostic significance.

Hepatocellular damage and intrahepatic cholestasis further cause an increase of both indirect and direct bilirubin (the latter being highly raised) and appears in the urine. The intensity of changes progresses for several weeks and then slowly subsides. Diagnosis may be confirmed by liver biopsy. Investigations:

- Serum aminotransferases — increased
- Lactic dehydrogenases — normal
- Albumin — decreased
- γ-Globulin — increased
- Serum bilirubin — increased
- Urinary bilirubin — present
- Serum bilirubin, indirect — increased
- Serum bilirubin, direct — increased
- Urinary bilirubin — increased
- Urinary urobilinogen — normal or increased
- Fecal urobilinogen — normal or decreased
- Serum enzymes — alanine and aspartate aminotransferases are increased unless cholestasis is severe

Waldenstrom's disease — Characterized by uncontrolled production of IgM which in turn results in the "hyperviscosity syndrome" (bleeding, diathesis with internal hemorrhages, and a diffuse brain syndrome leading to coma). Investigations:

- Serum electrophoresis — M-peak
- Immunoelectrophoresis — IgM paraproteins

Water disturbances — These depend on the circulating volume. Water deficiency causes hypovolemia. If cellular hydration is not altered by changes in the effective osmolarity across cell membranes, hypotension and collapse follow, as does death from circulatory insufficiency. Water excess causes hypertension, edema occurs if albumin levels are low, and overloading of the heart may lead to cardiac failure. The homeostatic mechanism regulating water retention fails in diabetes insipidus due to pituitary or hypothalamic damage.

Water intoxication — Syndrome resulting from a water excess more than 7 to 10% which is primarily due to marked hyponatremia. Significant symptoms (headache, drowsiness, weakness, or disorientation) rarely occur above a serum sodium level of 125 mg/ℓ. Investigations:

- Serum sodium (<125 meq/ℓ), serum osmolality (<260 mOsm)
- Hematocrit

Wilson's disease — Also called hepatolenticular degeneration. It is caused by accumulation of copper in liver, central nervous system, cornea, and kidney. Investigations:

- Serum copper — decreased (<70 mg/dℓ)
- Urinary copper — increased (100 mg/ℓ)
- Serum ceruloplasmin — decreased (<20 mg/dℓ)
- Possible increased serum alanine and aspartate aminotransferases

APPENDIX

The International System of Units

The International System of Units (SI Units) uses the mole as a measure of the amount of all substances for which molecular weights are known (rather than the mass units), and the liter as unit of volume. Some conventional units are already based on the mole or equivalents. Many plasma electrolytes are expressed in milliequivalents per liter. Conversion of these values to SI units, millimoles per liter, will alter the numerical value only if the valency of the electrolyte is different from one. In the case of some biological substances, the concentration can be expressed if the molecular mass of one component is known. For example, SI units for triglycerides can be given as moles of glycerol. For substances where the molecular mass cannot be identified, the units will be in grams or milligrams per liter. Exceptions to SI conversion are enzymes, which are given in units per liter and blood gases, expressed in mmHg.

The tables below contain the reference ranges of most frequently applied laboratory procedures in conventional and SI units. These reference ranges should only be considered as guidelines, since they are modified by several factors such as age, sex, genetic inheritance, diurnal variations, diet, socio-economic conditions, and previous diseases.

Table 1
CLINICAL BIOCHEMISTRY — REFERENCE RANGES

Substance	Source	Reference ranges	
		Conventional units	SI units
Albumin	Serum	3.5—5.0 g/dℓ	35—50 g/L
Ammonia	Plasma		
Males		<6 μmol/dℓ	<60 μmol/L
Females		<5 μmol/dℓ	<50 μmol/L
Amylase	Plasma	40—160 Somogyi units (SU)/dℓ	70—300 U/L
	Urine	<7000 SU/day	<1300 U/day
Bicarbonate	Plasma	21—28 mmol/ℓ	21—28 mmol/L
Bilirubin	Serum		
Conjugated		0—0.3 mg/dℓ	0—5 μmol/L
Unconjugated		0.1—1.0 mg/dℓ	1—17 μmol/L
Total		0.1—1.2 mg/dℓ	2—20 μmol/L
Blood, occult	Urine	Negative	
	Feces	Negative	
Calcium			
Protein dependent	Plasma	8.8—10.3 mg/dℓ[a]	2.2—2.6 mmol/L[a]
Ionized	Serum	2.2—2.5 meq/ℓ	1.1—1.2 mmol/L
Calcium	Urine	<300 mg/day	<7.5 mmol/day
Carotenes	Serum	60—300 μg/dℓ	1.1—5.6 μmol/L
Chloride	Serum	96—106 meq/ℓ	96—106 mmol/L
	Sweat	<60 mmol/ℓ	<60 mmol/L
Cholesterol	Serum, plasma	<270 mg/dℓ[b]	<7.0 mmol/L[b]
HDL	Plasma		
Males		30—66 mg/dℓ[b]	0.8—1.7 mmol/L[b]
Females		35—80 mg/dℓ[b]	0.9—2.1 mmol/L[b]
LDL	Plasma	80—180 mg/dℓ	2.1—4.7 mmol/L[b]
CO_2	Blood, arterial	22—30 mmol/ℓ	22—30 mmol/L
Coproporphyins	Urine	Negative, or trace	
Creatine	Urine		
Males		<40 mg/day	<300 μmol/day
Females		<80 mg/day	<600 μmol/day
Creatinine[c]	Plasma		
Males		0.7—1.4 mg/dℓ	60—120 μmol/L
Females		0.6—1.2 mg/dℓ	50—110 μmol/L
Fetal maturity	Amniotic fluid	>1.8 mg/dℓ	>160 μmol/L
	Urine	1.0—1.7 g/day	8.8—15.0 mmol/day
Creatinine clearance	Plasma and urine	95—130 mℓ/min	1.58—2.16 mL/sec
Creatine phosphokinase	Serum, plasma		
Males		<105 U/ℓ	<105 U/L
Females		<45 U/ℓ	<45 U/L
Creatine phosphokinase	Serum, plasma		
MB isoenzyme		Negative	
Cyclic AMP	Urine	3.0—5.0 μmol/g	340—570 nmol/mmo creatinine
Cystine	Urine	Negative	
Fat	Feces		
100 g fat/day		<5.0 g/day	<18 mmol/day
α-Fetoprotein	Serum	<0.5 μg/dℓ	<5 μg/L
Nonpregnant	Amniotic fluid	Depends on gestation time	mg/L
Gastric acid	Gastric fluid	<3 meq/hr	<3 mmol/hr
Stimulation tests			
Basal acid output		<3 meq/hr	<3 mmol/hr
Maximum acid output			
Males		18—30meq/hr	18—30 mmol/hr
Females		15—26 meq/hr	15—26 mmol/hr
Insulin (Hollander)		<2 meq/hr	<2 mmol/hr

Table 1 (continued)
CLINICAL BIOCHEMISTRY — REFERENCE RANGES

Substance	Source	Reference ranges: Conventional units	Reference ranges: SI units
Globulins	Plasma	2.3—3.5 g/dℓ	23—35 g/L
Glucose	Plasma		
Normal fasting		72—108 mg/dℓ	4.0—6.0 mmol/L
2 hr postprandial		<140 mg/dℓ	<8.0 mmol/L
Glucose tolerance test	Plasma		
Fasting		<115.0 mg/dℓ	<6.4 mmol/L
0.5—1 hr		<200.0 mg/dℓ	<11.1 mmol/L
2 hr		<140.0 mg/dℓ	<7.8 mmol/L
Glucose	Spinal fluid	45.0—76.0 mg/dℓ	2.5—4.2 mmol/L
Glucose 6-phosphate dehydrogenase	Erythrocyte	5.0—10.4 U/gHb	5.0—10.4 U/gHb
β-Hydroxybutyrate dehydrogenase	Plasma	<145 U/ℓ	<145 U/L
5-Hydroxyindole acetic acid	Urine	Negative	
Hydroxyproline	Urine	14—45 mg/day	110—340 μmol/day
Iron (total)	Serum		
Males		75—165 μg/dℓ	13—30 μmol/L
Females		65—145 μg/dℓ	12—26 μmol/L
Iron, TIBC	Serum	280—400 μg/dℓ	50—72 μmol/L
Iron, saturation	Serum	20—50%	0.20—0.50
Ketones	Serum	Negative	
	Urine	Negative	
Lactate	Blood	5—20 mg/dℓ	0.5—2.0 mmol/L
Lactate dehydrogenase	Serum	80—129 units at 25°C (lactate → pyruvate)	38—62 U/L at 25° C
		185—640 units at 30°C (pyruvate → lactate)	90—310 U/L at 30° C
Lecithin/sphingomyelin ratio	Amniotic fluid	>3	>3
Magnesium	Plasma	1.7—2.6 mg/dℓ	0.70—1.20 mmol/L
Magnesium	Urine	73—102 mg/day	3.0—4.2 mmol/day
5′-Nucleotidase	Serum	0—14 U/ℓ	0—14 U/L
Osmolality	Serum	280—300 mOsm/kg	280—300 mmol/kg
Osmolality	Urine	300—1090 mOsm/kg	300—1090 mmol/kg
Oxalate	Urine	17.0—43.0 mg/day	190—480 μmol/day
pCO_2	Blood, arterial	35—45 mmHg[b]	35—45 mmHg[b]
pO_2	Blood, arterial	80—100 mmHg[b]	80—100 mmHg[b]
pH	Blood, arterial	7.35—7.45	7.35—7.45
Phosphatase, alkaline	Serum	20—90 IU/ℓ at 30°C (*p*-nitro phenylphosphate)	29—90 U/L at 30°C
Phosphatase, prostatic	Serum	0.3—1.1 U/ℓ	0.3—1.1 U/L
Phosphate, inorganic	Plasma	2.6—4.5 mg/dℓ	0.85—1.45 mmol/L
	Urine	0.9—2.4 g/day[d]	29—77 mmol/day[d]
Potassium	Plasma	3.2—4.7 meq/ℓ	3.2—4.7 mmol/L
Potassium	Urine	26—123 meq/day[d]	26—123 mmol/day[d]
Protein	Plasma		
Total		6.9—8.4 g/dℓ	69—84 g/L
Albumin		3.5—5.0 g/dℓ	35—50 g/L
Globulin		2.4—3.8 g/dℓ	24—38 g/L
Protein	Serum		
Total		6.5—8.0 g/dℓ	65—80 g/L
Protein	Urine	<100 mg/day	<0.10 g/day
Pyruvate	Blood	5—15 μmol/dℓ	50—150 μmol/L

Table 1 (continued)
CLINICAL BIOCHEMISTRY — REFERENCE RANGES

Substance	Source	Reference ranges Conventional units	Reference ranges SI units
Sodium	Plasma	135—147 meq/ℓ	135—147 mmol/L
Sodium	Urine	40—220 meq/day[d]	40—220 mmol/day[d]
Specific gravity	Urine	1.002—1.030	1.002—1.030
Transferases	Serum		
AST (SGOT)		16—60 U/mℓ (Karmen) at 30°C	8—29 U/L at 30°C
ALT (SGPT)		8—50 U/mℓ (Karmen) at 30°C	4—24 U/L at 30°C
Triglyceride	Plasma	<160 mg/dℓ[b]	<1.80 mmol/L[b]
Urea	Plasma	8—20 mg/dℓ[d]	3.0—7.0 mmol/L[d]
Urate	Plasma		
Male		<7.6 mg/dℓ	<450 μmol/L
Female		<6.0 mg/dℓ[c]	<360 μmol/L[c]
Urate	Urine	170—500 mg/day	1—3 mmol/day
Urobilinogen	Urine	0—4 mg/day	0.68 μmol/day
Uroporphyrins	Urine	Negative	
Vanilmandelic acid	Urine	0.8—7.0 mg/day	4—35 μmol/day
Xylose	Blood		
Absorption test			
1 or 2 hr, 25-g dose		33.0—56.0 mg/dℓ	2.2—3.7 mmol/L
Xylose	Urine		
25-g dose		5.0—8.3 g/5 hr	33—55 mmol/5 hr

[a] Protein dependent
[b] Age dependent
[c] Sex dependent
[d] Diet dependent
[e] Pre-menopause

Table 2
LABORATORY DATA OF ENDOCRINE STATUS - REFERENCE RANGES

Substance	Source	Reference ranges Conventional units	Reference ranges SI units
Androstenedione	Serum		
Males, <50 years		0.28—0.70 nmol/dℓ	2.8—7.0 nmol/L
Females, cycling		0.20—0.8 nmol/dℓ	2.0—8.0 nmol/L
Adrenocorticotropin (ACTH)	Plasma	<100 pg/mℓ	<22 pmol/L
Aldosterone	Serum	50—145 pg/mℓ	135—405 pmol/L
Catecholamines, free (as norepinephrine)	Urine	<100 μg/day	<600 nmol/day
Cortisol, 0800 hr	Plasma, fasting	6.2—24 μg/dℓ	170—660 nmol/L
Cortisol	Urine	36—145 μg/day	100—400 nmol/day
Dehydroepiandrosterone sulfate	Serum		
Males and females			
20—30 years		0.26—1.15 μmol/dℓ	2.6—11.5 μmol/L
30—40 years		0.19—0.77 μmol/dℓ	1.9—7.7 μmol/L
40—50 years		0.13—0.51 μmol/dℓ	1.3—5.1 μmol/L
>50 years		0.03—0.13 μmol/dℓ	0.3—1.3 μmol/L
Estriol, pregnancy	Serum	3—11 nmol/dℓ	30—110 nmol/L
Estradiol, prepubertal	Plasma	2.45 ng/dℓ	<90 pmol/L

Table 2 (continued)
LABORATORY DATA OF ENDOCRINE STATUS - REFERENCE RANGES

Substance	Source	Reference ranges	
		Conventional units	SI units
Nonpregnant, adult males		4.90 ng/dℓ	<180 pmol/L
Follicle stimulating hormone (FSH)	Serum		
Males		1—4 mU/mℓ	1—4 IU/L
Females		1—4 mU/mℓ	1—4 IU/L
Postmenopausal		9—47 mU/mℓ	9—47 IU/L
Gastrin	Serum, fasting	<50 pg/mℓ	<50 ng/L
Growth hormone	Serum	2—10 ng/mℓ	2—10 µg/L
17-Hydroxyprogesterone	Serum		
Males		0.1—0.6 mmol/dℓ	1.0—6.0 mmol/L
Females		<0.9 nmol/dℓ	<9.0 nmol/L
Insulin (fasting)	Serum	7—35 µU/mℓ	50—250 pmol/L
Insulin antibodies	Serum	Not detectable	
17-Ketosteroids	Urine		
Males		8.6—22.0 mg/day	30—75 µmol/day
Females		5.8—14.4 mg/day	20—50 µmol/day
Luteinizing hormone	Serum		
Males		1—4 mU/mℓ	1—4 IU/L
Females		1—6 mU/mℓ	1—6 IU/L
Postmenopausal		12—30 mU/mℓ	12—30 IU/L
Parathyroid hormone	Serum	0—9 pmol/dℓ	0—90 pmol/L
Progesterone	Serum		
Males		0.05—0.2 nmol/dℓ	0.5—2 nmol/L
Females			
Follicular		0.05—0.5 nmol/dℓ	0.5—5 nmol/L
Luteal		0.8—9.0 nmol/dℓ	8—90 nmol/L
Pregnancy 3rd trimester		15.0—80.0 nmol/dℓ	150—800 nmol/L
Prolactin	Serum	0—30 ng/mℓ	0—30 µg/L
Renin activity, peripheral	Plasma		
Normal sodium intake (100—180 mmol/day)			
Recumbent (overnight or 6 hr)		0—2.09 ng/mℓ/hr	0—0.58 ng/L/sec
Upright (2 hr)		1.01—4.54 ng/mℓ/hr	0.28—1.26 ng/L/sec
Low sodium intake (10 mmol/day for 4 days)			
Recumbent (overnight or 6 hr)		1.15—5.40 ng/mℓ/hr	0.32—1.50 ng/L/sec
Upright (2 hr)		1.87—10.22 ng/mℓ/hr	0.52—2.84 ng/L/sec
Upright (2 hr) + diuretic		2.59—19.29 ng/mℓ/hr	0.72—5.36 ng/L/sec
Secretin test, volume	Duodenal fluid	≥0.18 mℓ/dkg body wt	≥1.8 mℓ/kg body wt
Bicarbonate		>8.0 mmol/dℓ	>80 mmol/L
Testosterone	Plasma		
Prepubertal		<30 ng/dℓ	<1.0 nmol/L
Adult male (< 50)		400—1000 ng/dℓ	14.0—35.0 nmol/L
Adult female: cycling		30—72 ng/dℓ	1.0—2.5 nmol/L
Preg. 3rd trimester		58—230 ng/dℓ	2.0—7.5 nmol/L
T3 Resin uptake	Serum		
Normal, Toronto range		25—35%	0.25—0.35
Hypothyroid		<25%	<0.25
Hyperthyroid		>35%	>0.35
Triiodothyronine (T3)	Serum	0.78—2.40 ng/mℓ	1.2—3.7 nmol/L
Thyroid stimulating hormone (TSH)	Serum	<10 µU/mℓ	<10 mU/L

Table 2 (continued)
LABORATORY DATA OF ENDOCRINE STATUS - REFERENCE RANGES

Substance	Source	Reference ranges	
		Conventional units	SI units
Thyroxine, free	Serum	0.7—1.8 ng/dℓ	9—23 pmol/L
Thyroxine, total	Serum		
Toronto range		4.7—12.0 μg/dℓ	60—155 nmol/L
Thyroxine binding globulin	Serum	1.2—3.0 mg/dℓ	12—30 mg/L

Table 3
HEMATOLOGY — REFERENCE RANGES

Substance	Source	Reference ranges	
		Conventional units	SI units
Albumin	Spinal fluid	20—40 mg%	200—400 mg/L
Alkaline phosphatase	Leukocytes	On report	
Alpha-1-antitrypsin	Serum	190—350 mg/dℓ	1.9—3.5 g/L
Alpha-2-macroglobulin	Serum	150—350 mg/dℓ	1.5—3.5 g/L
Antinuclear factor	Serum	Negative	
Aspergillus	Serum	Negative	
Carboxyhemoglobin	Blood		
Nonsmokers		<2.0% of Hb	<0.02
Smokers		5.0—9.0% of Hb	0.05—0.09
Complement, C3	Serum	50—120 mg/dℓ	0.5—1.2 g/L
Complement, C4	Serum	20—50 mg/dℓ	0.2—0.5 g/L
Cryoglobulin	Serum	Negative	
Eosinophil count	Blood	40—440/mm³	0.04—0.44 10^9/L
Erythrocyte sedimentation rate	Blood		
Males		0—10 mm/hr	0—10 mm/hr
Female		0—15 mm/hr	0—15 mm/hr
Erythrocytes	Blood		
Male		4.5—6.5 × 10⁶/mm³	4.5—6.5 × 10^{12}/L
Female		3.9—5.6 × 10⁶/mm³	3.9—5.6 × 10^{12}/L
Erythrocytes	Spinal fluid	0 × 10⁵/dℓ	0 × 10^6/L
Erythocyte indices	Blood		
MCV		80—95 μm³	80—95 × 10^{15}/L
MCH		27—36 pg	27—36 pg
MCHC		32—36 g/dℓ	320—360 g/L
Ferritin	Serum		
Male		40—480 ng/mℓ	40—480 μg/L
Female		9—190 ng/mℓ	9—190 μg/L
Fibrinogen	Plasma	150—350 mg/dℓ	1.5—3.5 g/L
Folate	Serum		
Low		<1.4 ng/mℓ	<3.2 nmol/L
Indeterminate		1.4—2.0 ng/mℓ	3.2—4.5 nmol/L
Normal		>2.0 ng/mℓ	>4.5 nmol/L
Folate	Erythrocytes	>120 ng/mℓ	>270 nmol/L
Haptoglobin	Serum	100—300 mg/dℓ	1—3 g/L
Hemoglobin	Blood		
Male		14.0—18.0 g/dℓ	140—180 g/L
Female		12.0—16.0 g/dℓ	120—160 g/L
Hemoglobin A$_{1c}$	Blood	6.0%	<0.06
Hemoglobin A$_2$	Blood	<3.5%	<0.035
Hemoglobin F	Blood	0—3.0%	0—0.03

Table 3(continued)
HEMATOLOGY — REFERENCE RANGES

Substance	Source	Reference ranges	
		Conventional units	SI units
Hemoglobin H	Blood	Negative	
Heinz bodies	Blood	Negative	
Immunoglobulins	Serum		
IgG		800—1800 mg/dℓ	8—18 g/L
IgA		90—450 mg/dℓ	0.9—4.5 g/L
IgM		60—280 mg/dℓ	0.6—2.8 g/L
Immunoglobulins IgG	Spinal fluid	2.0—4.0 mg/dℓ	20—40 mg/L
Iron, Total	Serum	50—180	9—32 μmol/L
Iron, TIBC	Serum	250—420	45—75 μmol/L
Iron Saturation	Serum	25—50%	0.25—0.50
LE cells	Serum	Negative	
Leukocytes	Blood	4.0—11.0 × 10^8/dℓ	4.0—11.0 × 10^9/L
Leukocytes	Spinal fluid	0.3/mm^3	0—3 × 10^6/L
Leukocytes differential	Blood		
Neutrophils		40—75%	0.40—0.75
Lymphocytes		20—45%	0.20—0.45
Monocytes		2—10%	0.02—0.10
Eosinophils		1—6%	0.01—0.06
Basophils		<1%	<0.01
Methemoglobin (proportion of total Hb)	Blood	<2.0%	<0.02
Packed cell volume			
Male		42—52%	0.42—0.52
Female		37—47%	0.37—0.47
Partial thromboplastin time (activated)	Plasma	22—37 sec (activated)	
Platelet count	Blood	150—350 × 10^8/dℓ	150—350 × 10^9/L
Prothrombin time	Plasma	On report	
Reticulocyte count	Blood	50—100 × 10^8/dℓ	50—100 × 10^9/L
Sickle cell	Blood	Negative	
Vitamin B$_{12}$	Serum		
Low		<120 pg/mℓ	<90 pmol/L
Indeterminate		120—190 pg/mℓ	90—140 pmol/L
Normal		>190 pg/mℓ	>140 pmol/L

Table 4
CONCENTRATION RANGES OF FREQUENTLY APPLIED DRUGS

Substance	Source	Reference ranges	
		Conventional units	SI units
Acetaminophen	Serum		
Therapeutic		2—13 mg/ℓ	13—86 μmol/L
Toxic <4 hr		>90 mg/ℓ	>600 μmol/L
Toxic <12 hr		>30 mg/ℓ	>200μmol/L
Carbamazepine	Serum		
Therapeutic		4—12 mg/ℓ	17—50 μmol/L
Digoxin	Serum		
Therapeutic, Toronto range		0.8—2.0 ng/mℓ	1.0—2.6 nmol/L
Disopyramide	Serum		
Therapeutic		2—5 mg/ℓ	6—15 μmol/L
Toxic		>6 mg/ℓ	>18 μmol/L

Table 4 (continued)
CONCENTRATION RANGES OF FREQUENTLY APPLIED DRUGS

Substance	Source	Reference ranges	
		Conventional units	SI units
Ethosuximide Therapeutic	Serum	40—100 mg/ℓ	280—710 µmol/L
Lithium Therapeutic, Toronto range	Serum	0.5—1.5 meq/ℓ	0.50—1.50 mmol/L
Phenobarbital Therapeutic	Serum	20—40 mg/ℓ	85—170 µmol/L
Phenytoin Therapeutic	Serum	10—20 mg/ℓ	40—80 µmol/L
Primidone Therapeutic	Serum	5—12 mg/ℓ	23—55 µmol/L
Procainamide Therapeutic Toxic	Serum	4—8 mg/ℓ >12 mg/ℓ	17—34 µmol/L >50 µmol/L
Procainamide + N-acetylprocainamide Therapeutic	Serum	8—16 mg/ℓ	34—68 µmol/L
Salicylate Therapeutic	Serum	15—30 mg/dℓ	1.10—2.15 mmol/L
Theophyline Therapeutic	Serum	10—20 mg/ℓ	55—110 µmol/L

INDEX

A

Abdominal tumor, 147
Abetalipoproteinemia, 147, 187—188
Acetaminophen, 223
Acetoacetate, 184
Acetone, 184
Acetylcholine, 8, 15, 194
Acetylcholine receptors, 68, 73
Acetylcholinesterase, 171, 206
N-Acetylgalactosamine, structure, 60
N-Acetylneuraminic acid, structure, 60
N-Acetylprocainamide, 224
N-Acetyltransferase, 72
Achlorohydria, 147
Acid-base balance (equilibrium), 70, 147—148, 152, 206
Acid-base metabolism, 8—15
Acidemia, 148
α_1-Acid glycoprotein, 148
Acid hydrolases, 80—82, 85
Acid intoxication, 148
Acid load, excess, deposition in body, 10—11
Acidosis
    characteristics of, 7—8, 10—15, 148, 189
    diabetic, 11, 169, 209
    diseases and disorders associated with, 147—149, 154, 157—158, 169, 175, 189, 201, 203—205, 209, 214
    lactic, 11
    metabolic, see Metabolic acidosis
    normochloremic, 11
    renal tubular, 12, 65—67, 70, 205, 214
    respiratory, see Respiratory acidosis
    water and electrolyte metabolism studies, 7—16
Acid phosphatase, 82—84, 148, 206
    pH, 199
Acid urine, 214
Acinar tissue loss, in cystic fibrosis, 166, 168
Acne, 148
Acquired porphyrinuria, 201
Acrodermatitis enterohepatica, 148
Acromegaly, 32, 148—149, 158
ACTH, see Adrenocorticotropic hormone
Actin, 53—54, 116, 118
Actinomycin D, 102
Active membrane transport, 63—65
Actomyosin, 15, 64
Acute alcoholism, 76
Acute intermittent porphyria, 149, 201
Acute nephritis, 6, 20
Acute pancreatitis, 154, 171, 197—198, 208
Acute phase, trauma, 211
Acute phase reactant proteins, 33
Acute radiation sickness, 119
Acute renal failure, 8, 149, 155, 176, 204
Addison's disease, 5—6, 8, 20, 32, 84, 149—150, 186, 199, 207—208, 213

Adenosine deaminase, 206
Adenosine diphosphate, 71—72
Adenosine triphosphatase, 7, 15, 17, 34, 55, 64, 76, 78, 80, 117—118, 122
    deficiency, erythrocytes, 66
Adenosine triphosphate, 7, 17, 53, 62—63, 65, 71, 75, 77—78, 95, 116—118, 122
Adenylate kinase, 206
Adenylcyclase, 40
Adiposogenital dystrophy, 170
ADP, see Adenosine diphosphate
Adrenal cortex, disorders of, 149—151
Adrenal cortical carcinoma, 32
Adrenal cortical insufficiency, 150—151
Adrenal corticosteroids, 32
Adrenal glucocorticoids, 33—34
Adrenal steroids, 17, 19
β-Adrenergic receptors, 72
Adrenocortical adenoma, 150
Adrenocorticotropic hormone, 13, 30, 199—200
    reference ranges, 220
Adrenocorticotropic hormone stimulation test, 150
Aflatoxin, 115—116
Agammaglobulinemia, 151
Age effects, water metabolism, 1—2, 5
Airway obstruction, in cystic fibrosis, 166—167
Alanine aminotransferase, 104, 154, 206
Albinism, 189
Albocuprein I and II, 36
Albumin, 6, 96—97
    abnormalities of, 151, 215
    reference ranges, 218—219, 222
Albuminuria, 147, 205
Alcoholism, and alcoholic liver disease, see also Cirrhosis, 17, 63, 75—76, 78, 151—152, 162—163, 175, 185, 187, 201
Alcohol syndrome, 105
Aldosterone, 5—7, 17, 150, 179, 202, 208
    reference ranges, 220
Aldosteronism, primary, 202, 208, 212, 214
Alkalemia, 151—152
Alkali disease, 43
Alkaline phosphatase, 80, 96, 103—104, 206, 219, 222
    characteristics, 151—152
    pH, 199
    references ranges, 219, 222
Alkaline urine, 214
Alkalosis
    characteristics of, 7—8, 10—14, 152
    congenital, 70
    diseases and disorders associated with, 147—148, 152, 158, 201—203
    metabolic, see Metabolic alkalosis
    respiratory, see Respiratory alkalosis
    water and electrolyte metabolism studies, 7—14, 17—18
Alkaptonuria, 152—153, 189

H

Printed in the United States
by Baker & Taylor Publisher Services